Springer Japan KK

M. Ohtsu (Ed.)

Near-Field Nano/Atom Optics and Technology

With 189 Figures

Springer

Motoichi Ohtsu, Dr. Eng.
Professor
Interdisciplinary Graduate School of Science and Engineering
Tokyo Institute of Technology
4259 Nagatsuta, Midori-ku
Yokohama, 226-8502 Japan

ISBN 978-4-431-68013-0

Library of Congress Cataloging-in-Publication Data
Near-field nano/atom optics and technology / M. Ohtsu, ed.
p. cm.
Includes bibliographical references and index.
ISBN 978-4-431-68013-0 ISBN 978-4-431-67937-0 (eBook)
DOI 10.1007/978-4-431-67937-0
1. Nanostructure materials. 2. Quantum optics. 3. Near-field microscopy. 4. Photonics. 5. Semiconductors. I. Ohtsu, Motoichi.
TA418.9.N35N43 1998
620′.5—dc21

Printed on acid-free paper

Originally published by Springer -Verlag Tokyo in 1998
Softcover reprint of the hardcover 1st edition 1998

Typesetting: Camera-ready by editor
Printing & binding: Best-set Typesetter Ltd., Hong Kong
SPIN: 10676845

Preface

In the long history of optics, a dramatic event was the invention of lasers in 1960. With the precise control of the power, wavelength, pulse width, and coherence of this laser light, a wide range of applications was developed. Further, owing to the progress of quantum optics in the 1980s, even quantum fluctuations of light could be controlled. But was it possible to reduce the volume of the light to the sub-wavelength dimension? Could light confined within such a small volume be utilized? The answers derived by conventional classical and quantum optics have been "no." This book will provide information to produce the answer "yes" to these questions. That is, this book describes the science and technology of generating novel light of nanometric dimensions, and its applications.

Owing to the inability of conventional optical science and technology to surpass the diffraction limit and reach nanometric dimensions, there has always been a size incompatibility between the optical integrated circuit and its electronic counterpart. This book, dealing with near-field nano/atom optics, provides new ideas, tools, and methods to overcome this difficulty. Although the basic proposal dates from 70 years ago, real development of near-field optics has been taking place only since the mid-1980s. At the early stages of this development, most of the technical efforts were devoted to ultra-high-resolution optical microscopy. However, it has been realized recently that the intrinsic features of the optical near field (i.e., quasi-static electromagnetic interaction at a sub-wavelength distance between two nanometric-sized particles) can open a new frontier in nanometric optical science and technology.

As a professor at the Tokyo Institute of Technology and the leader of the Photon Control Project of the Kanagawa Academy of Science and Technology, I am the main author of this book. The coauthors are research staff members of the project, and they describe systematically the progress of near-field nano/atom optics and technology, with an emphasis on their own work. Therefore, the book provides advanced, instructive information for graduate students who want to study near-field nano/atom optics and technology. The book also presents new ideas and approaches not only for junior scientists and engineers working on near-field nano/atom optics and technology, but also for researchers wanting to implement the new technology in their fields. Although several books have been published in near-field optics, this one provides complete and systematic information on near-field nano/atom optics and technology.

The coauthors and I would like to express our thanks to the following individuals. Dr. N. Atoda, Prof. S. Aizawa, Dr. I. Banno, Prof. M. Fukui, Mr. T. Gozen, Dr. H. Hisamoto, Prof. S. Hisanaga, Prof. H. Hori, Dr. T. Ikeda, Prof. W. Jhe, Prof. H. Kadono, Prof. Y. Katayama, Prof. K. Kitahara, Prof. S. Koshihara, Prof. A. Kusumi, Dr. M.-B. Lee, Dr. M. Miyamoto, Dr. H. Miyazaki, Mr. Y. Narita, Mr. M. Naya, Dr. K. Nishi, Mr. M. Onishi, Dr. N. Saito, Dr. Y. Sakai, Dr. M. Sano, Dr. S. Sudo, Prof. K. Suzuki, Mr. T. Tadokoro, Prof. H. Tatsumi, Prof. K. Tsutsui, Prof. T. Ushiki, Dr. J. D. White, Prof. M. Yoshimoto, and Dr. A. Zvyagin. We also extend our thanks to our students and to the students of the individuals listed above for their collaboration, comments, and discussions in conducting the research and preparing the manuscript for the book.

February 2, 1998

Motoichi Ohtsu

Contents

List of Contributors

Makoto Ashino (Chap. 10)

"Photon Control" Project
Kanagawa Academy of Science and Technology
KSP East Rm 408, 3-2-1 Sakado, Takatsu-ku, Kawasaki
Kanagawa 213-0012, Japan

Haruhiko Ito (Chap. 11)

"Fields and Reactions", PRESTO
Japan Science and Technology Corporation
KSP East Rm 408, 3-2-1 Sakado, Takatsu-ku, Kawasaki
Kanagawa 213-0012, Japan
(also with "Photon Control" Project, Kanagawa Academy of Science and Technology, KSP East Rm 408, 3-2-1 Sakado, Takatsu-ku, Kawasaki, Kanagawa 213-0012, Japan)

Kiyoshi Kobayashi (Chap. 12)

IBM Japan Ltd.
1623-14 Shimotsuruma, Yamato, Kanagawa 242-8502, Japan

Motonobu Kourogi (Chap. 4)

Interdisciplinary Graduate School of Science and Engineering
Tokyo Institute of Technology
4259 Nagatsuta, Midori-ku, Yokohama 226-8502, Japan
(also with "Photon Control" Project, Kanagawa Academy of Science and Technology, KSP East Rm 408, 3-2-1 Sakado, Takatsu-ku, Kawasaki, Kanagawa 213-0012, Japan)

Kazuyoshi Kurihara (Chap. 5)

"Photon Control" Project
Kanagawa Academy of Science and Technology
KSP East Rm 408, 3-2-1 Sakado, Takatsu-ku, Kawasaki
Kanagawa 213-0012, Japan

Shuji Mononobe (Chap. 3)
"Photon Control" Project
Kanagawa Academy of Science and Technology
KSP East Rm 408, 3-2-1 Sakado, Takatsu-ku, Kawasaki
Kanagawa 213-0012, Japan

Motoichi Ohtsu (Preface, Chaps. 1–12)
Interdisciplinary Graduate School of Science and Engineering
Tokyo Institute of Technology
4259 Nagatsuta, Midori-ku, Yokohama 226-8502, Japan
(also with "Photon Control" Project, Kanagawa Academy of Science and Technology, KSP East Rm 408, 3-2-1 Sakado, Takatsu-ku, Kawasaki, Kanagawa 213-0012, Japan)

Toshiharu Saiki (Chaps. 2, 4, and 9)
"Photon Control" Project
Kanagawa Academy of Science and Technology
KSP East Rm 408, 3-2-1 Sakado, Takatsu-ku, Kawasaki
Kanagawa 213-0012, Japan

Rajagopalan Uma Maheswari (Chaps. 6, 7, and 8)
"Photon Control" Project
Kanagawa Academy of Science and Technology
KSP East Rm 408, 3-2-1 Sakado, Takatsu-ku, Kawasaki
Kanagawa 213-0012, Japan

Chapter 1
Introduction

1.1 Near-Field Optics and Related Technologies

This book, *Near-Field Nano/Atom Optics and Technology*, describes the progress of the authors' recent experimental and theoretical work on near-field optics, and in particular their studies of the interaction between an optical near field and atoms and materials with nanometric features. To give an idea of the scope of the book, this section offers a brief description of near-field optics and related technologies.

Conventional optical microscopes are powerful tools used in a wide range of scientific research areas. Their imaging is based on the principle of the interference of light waves. Such interference-type microscopes can use not only light waves, but also material waves, such as electron waves, to increase the microscopic resolution. The resolution is limited by diffraction, which occurs in all interference experiments.

Here, one should note that the diffraction of waves does not impose a fundamental limit. For example, one new type of device for microscopy using material waves is an interaction-type microscope called a scanning tunneling microscope. In the field of optics, this is called a "near-field optical microscope." The strength of the local electromagnetic interaction between a specimen and a scanning probe tip is mapped to produce a nanometer-scale image that lies far beyond the diffraction limit. By analogy with electron tunneling, the local electromagnetic interaction is described by photon tunneling, meaning that a photon forming a coupled mode with material excitation tunnels through a gap between specimen and probe tip. The term "optical near field" is the key to understanding the underlying physics of the novel optical science and technology for nonpropagating local electromagnetic interaction, which falls within the category of "apparatus-limited" optical processes.

For interaction-type microscopy, signal processing techniques and a theoretical background are indispensable for interpreting the images obtained. This is because topographic information about the specimen must be derived from the near-field interaction with the scanning probe tip, which involves "destructive measurement" of the optical near field associated with the illuminated specimen. The specimen's topographical features do not necessarily correspond directly to the strength of the sample–probe interaction. One has to know a considerable amount about the optical characteristics of the specimen and the probe tip as well as the means of light illumination, detection, and signal processing in order to interpret the obtained images appropriately.

Developments in near-field optics and related technologies not only include microscopy for analyzing matter, but also extend to a wide variety of more active areas of "controlling and manipulating nanometric- and/or atomic-scale objects." The title of this book, *Near-Field Nano/Atom Optics and Technology*, reflects this wide range of developments. Control and detection of a localized photochemical reaction provide an engineering technique for ultra-high-density optical data processing. The mechanical effect of an optical near field on small particles is also being studied. Control of a biological sample on the molecular level would reveal fundamental processes of excitation transfer, and result in the control of living organisms. Proposals have also been made to manipulate atomic particles and control their states with nanometric resolution. These novel applications extend into the research fields of near-field optics, photonics, and related quantum optics.

1.2 History of Near-Field Optics and Related Technologies

The rapid progress of near-field optics and related technologies in recent years has mainly been in association with advances in near-field optical microscopy, spectroscopy, fabrication, manipulation, and so on. The underlying physics and potential for applications extend into many areas of modern science and technology. However, subsequent sections of this chapter focus mainly on reviewing the history and principles of the near-field optical microscope, because it has been developed as a prototype application of near-field nano/atom optics.

The first mention of near-field optical microscopy appeared in a paper by Synge in 1928 [1], which proposed the use of a scanned microscopic aperture to construct an ultra-high-resolution microscope with a resolution far beyond the diffraction limit. The proposal was, of course, far beyond the technical limitations of those days, and experimental demonstrations in the optical region had to wait for a modern technical background before it was possible to fabricate a subwavelength-size probe tip, control it with nanometric reproducibility, and image pictures by means of computer-aided signal processing. Later in 1972, the near-field optical microscope was experimentally studied by Ash and Nicholls in the microwave region [2].

Near-field optical microscopy is considered to be a novel field of optics and photonics that is supported by modern nano-fabrication technology as well as precise control of the scanning probe tip and techniques for image processing, both of which were adapted from scanning tunneling microscopy (STM) and atomic force microscopy (AFM), which were developed in the

1980s. An important step toward realizing the near-field optical microscope was the study of a microwave version [3].

Experiments in the optical region were started by several groups in the 1980s [4, 5]. Since then, many types of probes and systems have been developed, such as ones using small apertures [6], squeezed glass pipettes [7], and sharpened optical fibers [8]. Among them, a sharpened single-mode optical fiber prepared by chemical etching demonstrated the smallest apex diameter, the highest reproducibility in fabrication, and the first nanometric resolution picture of a biological specimen [9]. An important recent advance was a demonstration of underwater operation [10], which is essential for in vivo biological measurement. A wide variety of near-field optical probes are now available for both dielectric types, metal-clad aperture types, and combined types.

Several techniques have been introduced with different names depending on their technical apparatus, such as probes and means of illuminating the specimen, and their signal detection. They include scanning near-field optical microscopy (SNOM), near-field scanning optical microscopy (NSOM), photon scanning tunneling microscopy (PSTM), scanning tunneling optical microscopy (STOM), and so on. A classification can be found in Ref. [4]. Although it might be convenient to use different names for near-field optical microscopes for the purpose of technical classification, we intend to use a unique and general name, i.e., near-field optical microscopy (or NOM for short), throughout this book in referring to all such techniques. Our objective in doing so is to describe the basic concepts underlying optical near-field problems, using a general framework of interaction-type scanning probe microscopy that includes STM, AFM, and so on. Therefore, the acronym NOM, as used in this book, should be considered as referring not to a specific microscopy technique, but to near-field optical microscopy in general.

1.3 Basic Features of an Optical Near Field

To describe the basic features of an optical near field, this section first explains the concepts of an optically "near" system and an effective field. Here, it should be noted that an evanescent field, sometimes described in textbooks on classical optics, is an example of an effective field. Next, the principle of detecting an effective field is explained. Finally, the role of a probe tip is described.

1.3.1 Optically "Near" System

Understanding the physical background of optical near-field problems is the key to the study of the principles of NOM and to further development of related technologies. The meaning and importance of optical near-field phenomena can be studied in the framework of general optical processes, such as

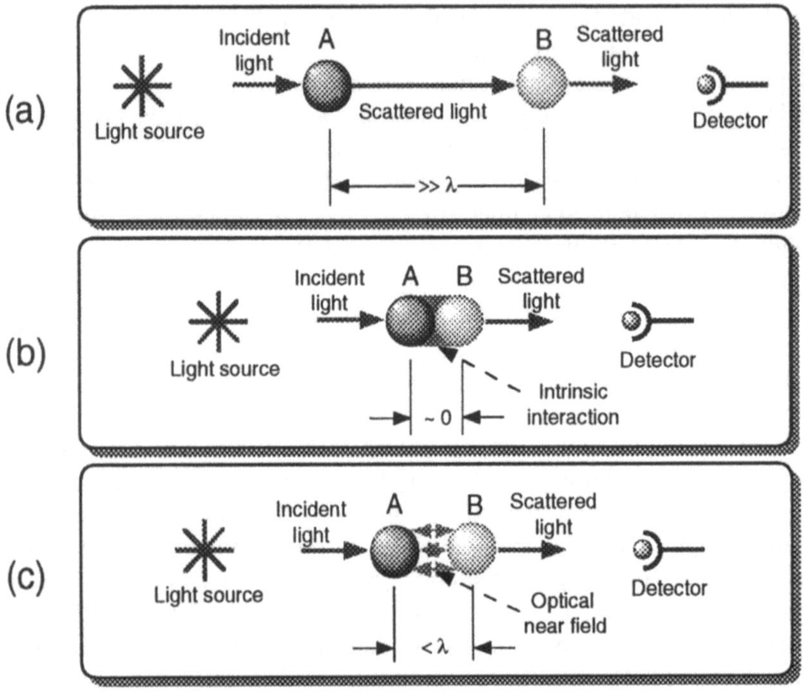

Fig. 1.1. Schematic explanation of optically **a** far, **b** close, and **c** near systems

a light-scattering process involving two atomic-scale objects, A and B. When one considers an electromagnetic interaction between two such objects, which are isolated as an electronic system, one finds that several features characterizing the nature of the interaction arise, according to the specific size of the region occupied by the electromagnetic field. That is, by noting that an optical process is a measurement of the response of a sample system placed between a light source and a photodetector, we can formulate three typical ways of characterizing the optical process according to the distance between A and B. They can be referred to as optically "far," "near," and "close" with respect to the electromagnetic interaction (see Fig. 1.1).

An optically far system is one in which A and B are placed far apart relative to the wavelength of the incident light. In this case, the interaction between A and B is unidirectional. Since the light waves or photons propagating between A and B satisfy the usual dispersion relation, the system can be separated into two successive light-scattering processes.

An optically close system is one that is electromagnetically tightly coupled owing to the existence of internal interactions between A and B even when incident light is not applied. The electromagnetic interactions in this case are due to intrinsic fluctuations of vacuum and material polarizations, which give rise to van der Waals interaction, for example. In this case, a system composed of two atomic objects, A and B, can be considered as a kind of molecular state when one studies its optical response. If one attempted to describe the interaction between A and B, one would have to use quantum electrodynamics, and would find virtual photons that do not satisfy the dispersion relation of photons in free space being exchanged back and forth between A and B.

An optically near system, which is conceptually in between optically far and close systems, is related to the optical near-field process; A and B are placed at a distance shorter than the wavelength of the incident light. However, the distance is still far with respect to the intrinsic interactions between A and B, which are also well separated with respect to their electronic states at this distance. When the system is illuminated by incident light, however, the electromagnetic interaction between A and B is not a single path. The scattered light reflected back from B to A has a considerable effect on the process, which depends on the distance between the objects: the shorter the distance, the greater the significance of the near-field effect. Therefore, the optical properties of A, such as its polarizability, depend on B. In this case, the electromagnetic interaction is quasi-static and shows a mesoscopic nature. The objects A and B are thus inseparable when they are illuminated by the incident light. If one tried to separate A from B, one would see an electromagnetic field with a curious nature that was different from that of a simple propagating light wave or photon. If one expanded the electromagnetic field existing between A and B, one would find "evanescent waves" with a complex wave number, that is to say, waves not satisfying the dispersion relation of conventional light waves or photons in a vacuum.

1.3.2 Effective Field and Evanescent Field

With respect to the internal field existing between A and B, the above-mentioned optically far, near, and close systems can also be called "macroscopic," "mesoscopic," and "microscopic," respectively. By using the characteristic parameter kl (where k and l are the wave number of the incident light and the characteristic size of the system, respectively), these systems may be roughly assigned to the ranges $kl \geq 1$, $0.01 \leq kl < 1$, and $kl < 0.01$, respectively. (Note that the numerical value 0.01 is used to represent only qualitatively that kl must be sufficiently smaller than unity. There can be some flexibility as regards the marginal value of 0.01; for example, it could vary from 0.005 to 0.05.)

In the case of a mesoscopic system, namely, an optical near-field system, one encounters an internal field of mesoscopic nature. The internal field is characterized by an effective field that represents the averaged effect of the microscopic interaction process over a volume $V \simeq l^3$. The size parameter l characterizes the optical process taking place in the subsystem, and the interaction with a shorter correlation length averaged over the volume V

behaves as an effective field. The effective field is considered to be a coupled mode of the electromagnetic field with matter in the smaller system.

To study the meaning of effective fields and scales of measurement, it is important to consider the behavior of electromagnetic waves near a dielectric surface from both the macroscopic and microscopic points of view. Let us consider a planar dielectric–air boundary where the material behavior changes abruptly relative to the optical wavelength. Here the optical wavelength corresponds to the long-range correlation length of electromagnetic interaction. There are also short-range interactions down to the atomic scale, which usually give rise to no observable effect in the macroscopic regime. However, owing to the abrupt termination of the medium, electromagnetic interactions of any correlation length are exposed to the outer half-space around the boundary.

The interactions corresponding to long-range correlations generate the common phenomenon of "total internal reflection" when the incident angle of the light is larger than the critical angle. In this case, a refracted wave showing an exponential decay in the direction normal to the boundary surface is referred to as "an evanescent wave." It exists only as a coupled mode of the electromagnetic field with material excitations on the surface, and carries energy and pseudo-momentum along the surface. The range of interaction associated with the evanescent wave is referred to as the penetration depth. It represents the decay distance of the field amplitude from the surface, and is equivalent to the inverse of the imaginary part of the complex wave number of the evanescent wave.

Evanescent waves involve two important physical meanings in the context of effective field and near field optics:

1. they provide an example of an effective field with a finite range of electromagnetic interactions;
2. they provide an example of an effective field being exposed to the outer half-space of a dielectric medium.

These two points concern the foundations of near-field nano/atom optics and technology. It should be noted that the perfect elimination of a refracted propagating wave and the appearance of an evanescent wave of a specific wave number occur only in the case of planar dielectric boundaries. In fact, these effects are due to the perfect translational symmetry of the system along its boundary surface. In the case of a nonplanar surface, there arise both a refracted propagating wave and an evanescent wave with a wave vector whose value depends on the shape of the surface.

1.3.3 Near-Field Detection of Effective Fields

Before proceeding to general cases, let us consider some observations of evanescent waves which will provide us with a simple example showing the fundamental process of NOM. First, let us consider the evanescent wave generated on a prism surface that a propagating light wave strikes at an angle of total reflection. If a second prism is held at a short distance above the surface of the first prism, a polarization wave is induced on the secondary surface as a result of electromagnetic interactions between the two closely separated facing prism surfaces, which in turn excites a propagating wave in the second prism. As a result, a part of the light wave is transmitted to the second prism, and the process of total internal reflection is said to be "frustrated." The magnitude of the intensity transmitted through the air gap depends on the gap separation, which is attributed to the exponential decay of the evanescent wave. The effect of the frustrated total internal reflection suggests that one can observe an evanescent wave by immersing a probe into the evanescent wave and measuring scattered light waves with a photodetector.

A probe is not necessarily a planar dielectric, but can be anything capable of frustrating the total internal reflection and producing a propagating field that is detectable in the far field. The idea of frustrated total internal reflection can be extended to near-field optical measurements of a confined field existing in close proximity to a material surface. In other words, one can observe even a "short-range" effective field exposed at a surface by putting a small probe tip near the surface, at a distance close to the penetration depth of the effective field. In this case, the probe tip couples directly with a local oscillating polarization induced in the illuminated surface.

If one observes the optical response of matter on a subwavelength scale, one should find that short-range electromagnetic interactions in subsystems constitute the macroscopic optical response. Then one will see that the nature of the substantial interaction depends on the scale on which one observes the subsystem. It should be noted that, in principle, one can trace the substantial interaction processes down to the atomic scale. In general, when one considers the optical response of dielectric matter, one can trace substantial interactions over a volume as large as λ^3 (where λ is the wavelength of the incident light). This gives the macroscopic dielectric characteristics. The interactions in the subwavelength range are, in this case, considered as fluctuations.

However, the situation is very different in the case of surface phenomena, because all of the substantial interactions are exposed to the outer half-space owing to the abrupt termination of the matter at the surface. It becomes possible to observe microscopic effective fields if a probe tip can interact directly with these effective fields at a small penetration depth. In this case, the probe tip scatters the local effective field into a propagating field that is observable in the far field. Such near-field detection of effective fields on a material surface is the essential characteristic of scanning probe microscopes, including STM, AFM, and NOM devices. Thus, the key to these types of near-field detection is to understand the role of local probes.

1.3.4 Role of a Probe Tip

We consider an optical process involving a physical system that can be separated into two electronically isolated subsystems, one corresponding to a specimen and the other to a probe. Here, it is assumed that an isolation scheme is implemented between a light source and a photodetector so that a signal can be transmitted from the light source to the photodetector only through interaction between the specimen and the probe tip. Such an isolation scheme is one of the fundamental requirements for near-field optical measurements.

A light wave incident on the specimen drives an effective field both inside and outside the specimen. For observation of the effective field near the specimen, the following two requirements have to be met:

1. the probe tip must have dimensions as small as the specimen (this requirement determines the penetration depth of the effective field, since the effective field corresponds to the electromagnetic correlation produced in the specimen itself);
2. the probe tip must be placed near the specimen, where an effective field with a short penetration depth maintains its significance.

If these requirements are met, the small specimen and the probe tip form an interacting subsystem. As a result, the effective field of the specimen turns out to be an internal field of a coupled system, which produces a light wave scattered from the probe tip and reaching the photodetector. This approach to detecting an optical near-field event is the essence of the near-field optical analysis of material systems.

The requirements of implementing this approach have been supported theoretically. That is, the NOM process can, to some extent, be described in terms of conventional electromagnetic theories, thus demonstrating several basic features of optical near-field interaction [11]. Among them, two results justifying these requirements are described below.

1. *Size-dependent localization.* The theory has described the short-range nature and spatial locality of the optical interaction of two closely spaced objects. An important point is that the result is scaled not by wavelength but by the sizes of the object and the probe. It should be emphasized here that a field calculation alone is not sufficient for the theoretical description of a NOM process. Instead, one should consider an interacting sample–probe system as an important subsystem separated from a global NOM process. When the sample–probe system is irradiated by an evanescent wave with a wavelength larger than that of the sample–probe system, the transferred optical intensity can be calculated on the basis of conventional Mie scattering and Kirchhof's integral. Therefore, the result is related to the far-field observation of the sample–probe scattering. A rapid decay of the transferred intensity can be clearly shown by increasing the sample–probe distance. The size-dependent localization of

optical near-field interaction is one of the most important features of a general NOM system. This feature was confirmed by accurate experimental work [12].

2. *Size-resonance behavior.* The transferred optical intensity takes a maximum when the sizes of the sample and probe are equal. This size-resonance behavior is also important in the general NOM process. That is, the near-field optical interaction between sample and probe exhibits a kind of resonance character with respect to spatial frequency.

To extract optical near-field phenomena from the whole light-scattering process, one needs to use a probe tip that is sensitive only to high-spatial-frequency components relevant to the local effective field around the specimen, as well as a signal-processing scheme that eliminates the signal background. This is why the technique of fabricating fine probes is of primary importance. It should be noted that the probe tip as a spatial-frequency filter must have a specific bandwidth that corresponds to the locality of the effective field confined around the specimen.

1.4 Building Blocks of Near-Field Optical Systems

Since near-field detection of a near-field event is the fundamental process that allows microscopy and related systems to go beyond the diffraction limit, the NOM system can be divided into several fundamental building blocks, as listed below. Each of these has a different characteristic size with respect to the electromagnetic interaction. Interpretations of any NOM image and the output signals from near-field optical systems require an appropriate evaluation of all of the following characteristics of the building blocks of the specific system. Although only the building blocks for a near-field optical microscope are shown here, it should be noted that the following list is also effective for general near-field nano/atom optical systems such as spectroscopy, fabrication, and manipulation.

1. *Local electromagnetic interaction.* The primary part is the local electromagnetic interaction between the nanometer-sized specimen and the probe tip, which exhibits a short-range nature. That is, the interaction is effective only when they are in very close proximity to each other. Such a short-range electromagnetic interaction corresponds to a very local perturbation of the electromagnetic background due to the presence of subwavelength objects, namely, the specimen and the probe tip.
2. *Coupling from near to far field.* To observe the local electromagnetic interaction between the specimen and the probe tip, one has to realize an adequate coupling of the local event to some propagating field that extends to the light source and photodetector. The typical coupling scheme employs irradiation of the specimen with an evanescent wave and the

connection of the probe tip to a single-mode optical fiber. The combination of the probe tip and the coupling scheme should have the character of a "spatial-frequency filter" that picks up the spatial Fourier component lying within an adequate window fitting the geometrical shape of the specimen.

3. *Signal collection and transfer.* To send a message to the local specimen and observe its response, one uses a light source and a photodetector placed at each end of the optical system. The important elements of the NOM system are a scheme for isolating the source field from the detector field and an extremely sensitive photodetection technique. In particular, the former reduces the signal background containing information about the averaged optical properties of the system.
4. *Production of a NOM image.* A NOM image is provided by means of a computer-aided mapping of the signal for the positions of the probe tip scanned two-dimensionally.
5. *Interpretation of the image.* To interpret the NOM signal, one needs to seek clues from theoretical analyses, such as the nature of the interaction between the specimen and the probe tip, the polarization dependence on the illumination and signal collection scheme, and the signal transfer function of the probe tip as a function of the spatial modulation of the local field described on a spatial Fourier frequency basis.

There is current interest in combined NOM operations using several different types of scanning probe microscopes, such as STM and AFM. One of the most popular techniques is a shear force-controlled positioning of the NOM probe tip, which utilizes the atomic force between the specimen's surface and a vibrating probe. The amplitude modulation of the resonant vibration of a NOM probe due to the atomic force exerted by the specimen's surface provides a measure of the sample–probe separation. Although the combined operation might be attractive, one must pay attention to interference between fundamental processes, that is, one must be aware that interaction-type microscopes employ a process of destructive measurement. When the types of interaction combined are very different in their natures and in their scales of relevant spatial sizes, time constants, and material characters, and so on, the combined apparatus provides much useful information. However, if their ranges are in competition with each other, the results obtained are very different from the expected ones. In fact, in some situations, a seemingly different measurement of a specimen may result in a similar image owing to interference. This difficulty is often referred to as an "artifact" problem.

NOM and related techniques involve a wide variety of interesting physics and a novel field of applications. This is because a specific internal field exhibits a significant effect on the optical property of matter when the effective field exhibits a resonance behavior. Here, the resonance behavior represents the state in which a specific space–time correlation of internal interaction processes or a cumulative motion of the internal degree of freedom becomes significant. In these cases, one can describe the corresponding effective field as a well-established mode, such as plasma oscillation in a metallic medium. The remaining parts of internal interactions with shorter correlation ranges are considered as fluctuations in the optical property of the system. For example, a surface plasmon has been used for NOM techniques, which involves a resonance interaction with a metallic surface. Since the character of the resonance strongly modifies the near-field interaction between the observed surface and probe tip, one can use it to obtain specific information about the sample by means of the probe tip. This type of probe tip is generally referred to as a "super-tip." With super-tips, one can make intensive use of spectroscopic measurement, taking advantage of a NOM with illumination from a tunable laser light. It should be noted, however, that a resonance-type NOM rejects all signals coming from interactions beyond its resonance range.

1.5 Comments on the Theory of Near-Field Optics

As is generally required for any kind of interaction-type scanning probe microscopy, one has to deconvolute the specimen–probe tip interaction into the characteristics of the unperturbed state of a specimen. Thus, a theoretical background is important for image interpretation. The NOM system is a complicated version of a light-scattering problem that involves electromagnetic interactions of matter on several different characteristic scales. However, in principle, if NOM does work at all, it should allow one to arrive at a simple understanding and theoretical description of the local electromagnetic interaction, since it would otherwise be impossible to discuss any NOM image in relation to the topographic nature of an observed specimen.

It is important to note that a NOM system involves several subsystems with different characteristic scales as regards electromagnetic interaction. In this case, we can describe the most important subsystem of the specimen and probe tip in terms of a local theoretical treatment, or nonglobal theory, which in turn allows a simple interpretation of the NOM image. Two important schemes with different scales make it possible to connect the nanometer to micrometer ranges, namely, the schemes of near-to-far coupling and light-wave transmission. Thus, the process of the near-field detection of near-field events has an essential meaning in the light-scattering process of the NOM system. It should also be noted that such a nonglobal theoretical treatment is general for mesoscopic systems, and that an appropriate model is indispensable for understanding the nature of the near-field phenomena as well as their application.

To support experimental work using NOM, a theoretical evaluation of the intensity scattered from a sample–probe system is not sufficient. One of the most useful theoretical descriptions is the angular spectrum representation of the scattered field. The angular spectrum provides a way of evaluating the electromagnetic interaction of a near-field regime in terms of waves with very high spatial frequency and correspondingly short decay length. The scattered wave from an object is represented as the sum of the propagating (homogeneous) plane waves in all spatial directions and evanescent (inhomogeneous) waves with an entire set of values of penetration depth. The dominant spatial frequency can be found from the peak of the angular spectrum, and the lateral locality of the interaction can be found from the spectral width around the peak.

An important requirement for further development of NOM theory is to establish an empirical or intuitive model describing a local optical process that is quasi-static and short-range as regards the nanometer-scale specimen–probe tip interaction. The model description would be more fruitful if it could provide some novel evidence of the physical importance of the local interacting subsystem, or more generally of optical near-field problems, in some way such as an excitation transfer or tunneling of quasi-particles. As an example, a quasi-particle model of a local specimen–probe tip interaction can be described in terms of Yukawa-type screened potential [13]. Such a model provides a convenient way of evaluating the signal transfer function in terms of the spatial Fourier frequency and understanding the NOM process as a spatial frequency filter. The model description has the potential to extract some physical importance from the problems of localized interaction and optical near fields.

It should also be noted that there is a relation between the tunneling phenomenon and optical near-field problems. The basic idea in Bardeen's picture describing the tunneling process in STM is the separation of the sample wave function and the probe wave function, which extend to the source and detector, respectively. From these wave functions, Bardeen extracted the most fundamental cross term of the tunneling current, and provided a comprehensive description of the STM process [14]. Bardeen's idea can be expected to be extended or developed by studying the conditions under which the separated evanescent fields on the sample and probe tip can be related to the idea of a tunneling current. Once an analogy between NOM and STM has been successfully established, one can give such tunneling photons or excitations a special importance that would lead to further developments in optical near-field physics. The overlap integral of evanescent waves describes a coupling of a decaying wave from the source surface with an exponentially increasing wave to the detector surface. This behavior in surface-to-surface transmission of evanescent light waves shows a similarity to the tunneling nature of the Shrödinger wave [15, 16]. It has been shown that the angular spectrum representation of scattered waves and the introduction of evanescent waves are not mere mathematical tricks, but also have physical effects on atoms [17]. It has also been shown that any scattered light waves from an object of arbitrary shape can be expressed in terms of plane waves when they include both homogeneous and inhomogeneous waves [18]. Such an expansion of the arbitrary form of scattered waves into plane waves with a complex wave number corresponds to the angular spectrum representation reviewed above. The amplitude distribution in the angular spectrum provides a measure of the range of penetration and locality of the optical near field.

1.6 Composition of This Book

It was explained in Sect. 1.3 that the optical near-field process involves an optically "near" system, in other words, an optically "mesoscopic" system in which two nanometer-sized particles are coupled through a short-range electromagnetic interaction by exchanging near-field photons. The basic idea of this book is that, in view of these features, *the optical near-field process has intrinsic and promising applications not only in optical microscopy and imaging, but also in spectroscopy, fabrication, and manipulation, and so on. They can be realized by utilizing the interaction between the specimen and probe, which may sometimes be resonant.*

Accordingly, it is argued that near-field optics should be extended to the study of the interaction between nanometric matter and atoms, and to possible applications. Therefore, the title of this book, *Near-Field Nano/Atom Optics and Technology*, represents a field extending from microscopy to spectroscopy, fabrication, and manipulation. Thus, the main part of the book is devoted to spectroscopy, fabrication, and manipulation of nanometric and/or atomic objects. For applications in these areas, the book starts by reviewing the principle and fabrication of a probe tip, which is the key device for generating and detecting an optical near field (Chaps. 2–5). For readers who are interested in general near-field optical microscopes, the second part of the book is devoted to a description of instrumentation and a review of imaging experimental results demonstrating high spatial resolution capability (Chaps. 6–8). The next chapters, which form the main part of the book, review recent progress in spectroscopy, fabrication, and atom manipulation, including several related plasmon technologies (Chaps. 9–11). Finally, existing theories are reviewed and future theoretical problems are outlined (Chap. 12).

References

1. E. A. Synge, Phil. Mag. **6**, 356 (1928)
2. E. Ash, G. Nicholls, Nature **237**, 510 (1972)
3. M. Fee, S. Chu, T. W. Hänsch, Opt. Commun. **69**, 219 (1989)
4. D. W. Pohl, D. Courjon (eds.), *Near Field Optics* (Kluwer, Dordrecht, 1993)
5. M. Ohtsu, J. Lightwave Technol. **13**, 1200 (1995)
6. U. Dürig, D. W. Pohl, F. Rohner, J. Appl. Phys. **59**, 3318 (1986)

7. E. Betzig, M. Isaacson, A. Lewis, Appl. Phys. Lett. **51**, 2088 (1987)
8. T. Pangaribuan, K. Yamada, S. Jiang, H. Ohsawa, M. Ohtsu, Jpn. J. Appl. Phys. **31**, L1302 (1992)
9. S. Jiang, H. Ohsawa, K. Yamada, T. Pangaribuan, M. Ohtsu, K. Imai, A. Ikai, Jpn. J. Appl. Phys. **31**, 2282 (1992)
10. M. Naya, R. Micheletto, S. Mononobe, R. Uma Maheswari, M. Ohtsu, Appl. Opt. **36**, 1681 (1997)
11. K. Jang, W. Jhe, Opt. Lett. **21**, 236 (1996)
12. T. Saiki, M. Ohtsu, K. Jang, W. Jhe, Opt. Lett. **21**, 674 (1996)
13. H. Hori, in *Near Field Optics*, 105 (Kluwer, Dordrecht, 1993)
14. J. Bardeen, Phys. Rev. Lett. **6**, 57 (1961)
15. T. Martin, R. Landauer, Phys. Rev. A **45**, 2611 (1992)
16. A. M. Steinberg and R. Chiao, Phys. Rev. A **49**, 3283 (1994)
17. T. Matsudo, H. Hori, T. Inoue, H. Iwata, Y. Inoue, T. Sakurai, Phys. Rev. A **55**, 2406 (1997)
18. E. Wolf, M. Nieto-Vesparinas, J. Opt. Soc. Am. A **2**, 886 (1985)

Chapter 2
Principles of the Probe

2.1 Basic Probe

2.1.1 Optical Fiber Probe for the Near-Field Optical Microscope

As mentioned in Chap. 1, the principle of the near-field optical microscope (NOM) is short-range electromagnetic interaction between two antennas, a probe antenna and a sample antenna, which are much smaller than the wavelength of driving field. It is apparent that the fabrication and manipulation of the small antenna are the most important factors for the successful development of NOM. One of the most realistic and commonly used method for the preparation of the small antenna is sharpening an optical fiber to a very small apex. By employing the scanning technique already established in scanning tunneling microscope (STM) and atomic force microscope (AFM), the antenna at the apex of sharpened fiber works as a probe on a sample surface under precise distance control. The following sections contain the principal roles of the basic probe relating to the functions of each part of the probe.

Figure 2.1 shows schematics of typical fiber probes, introducing the terminology for each part. A conventional optical fiber is sharpened by chemical etching in buffered hydrofluoric acid (see Chap. 3). An apex (diameter $d=2a$), which is the antenna of the fiber probe, can easily be made small, and a minimum apex size of only a few nanometers has been already achieved. In NOM, only the apex works as an antenna, which interacts electromagnetically with the sample. The tapered part has a simple conical shape and the cone angle θ is clearly defined. The apex radius and the cone angle can be varied by controlling the etching condition. The exterior surface of the probe is coated with opaque metal, such as aluminum or gold, to avoid the illumination of excitation light or the detection of the background signal through the side wall of the tapered part. The very end of the metal coating is removed to allow the sharpened glass part, including the apex, to protrude. The small aperture with diameter of d_f (also called the foot diameter) makes an important contribution, as will be discussed later. We call the protruded glass part and the metal-coated part the tapered core and the metallized tapered core, respectively. The main objective of this chapter is to describe the functions of the apex, tapered core, and aperture in the imaging process.

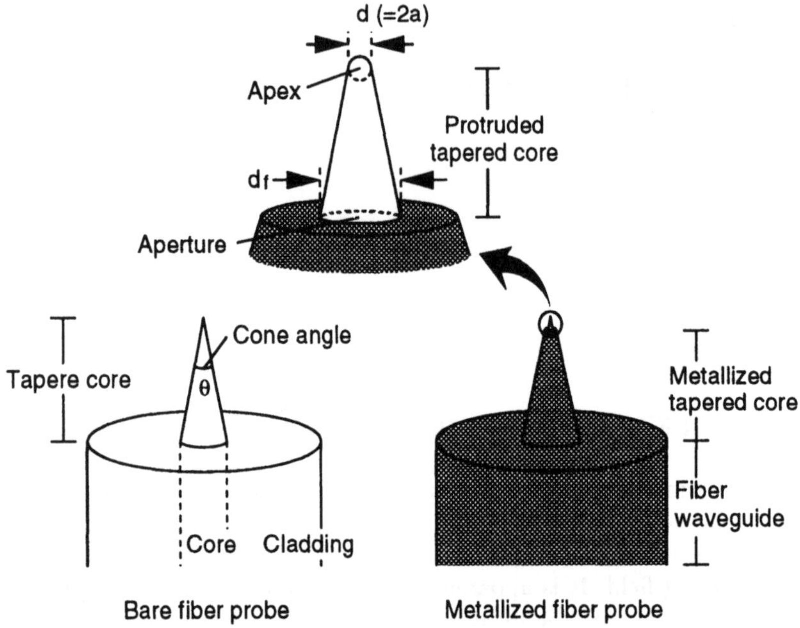

Fig. 2.1. Schematics of typical fiber probes

The metallized tapered core is regarded as a metal-cladding (metallic) waveguide where excitation or signal light passes through. Since the metallic waveguide has a complex loss mechanism, e.g., the existence of a cutoff diameter and absorption by the metal cladding, the optimization of the structure of the waveguide is essential. We discuss the transmission characteristics of the metallic waveguide in Chap. 4. The constitutive material and structure of the fiber waveguide also determine several other important factors, including polarization preservation and the propagation feature of ultraviolet light. An optimized design of a fiber probe for ultraviolet spectroscopy appears in Chap. 3.

2.1.2 Principle of the Imaging Mechanism: Dipole–Dipole Interaction

The spatial resolution and contrast in NOM imaging strongly depend on the structure of the fiber probe, e.g., the apex diameter, cone angle, foot diameter, and so on. In this section, we design an optimized structure for a fiber probe for the simultaneous achievement of high spatial resolution, high contrast, and high sensitivity. Since the apex size and the sample size are much smaller than the wavelength of the irradiation light (Rayleigh particles), the framework of the design is based on the simple theory of two interacting electric dipoles [1–3].

As shown in Fig. 2.2, the apex and the sample are assumed to be small spheres with finite radii of a_a and a_s respectively. Two dipoles with polarizabilities of α_a and α_s are driven by an external electric field E. The modulation of polarizability due to the dipole–dipole coupling is described as

$$\Delta\alpha = 2a_a a_s/R^3 \tag{2.1}$$

for a broadside illumination (Fig. 2.3a) and

$$\Delta\alpha = -a_a a_s/R^3 \tag{2.2}$$

for an end-on illumination (Fig. 2.3b), where R is the distance between the centers of two spheres. The polarizabilities can be written in terms of the sizes of the spheres and their refractive indices, n_a and n_s,

$$\begin{aligned} \alpha_i &= g_i a_i^3 \\ g_i &= (n_i^2 - 1)/(n_i^2 + 1) \qquad (i = a, s) \end{aligned} \tag{2.3}$$

In the derivation of Eq. 2.3, the influence of the substrate on which the sample exists is not taken into account. In the standard operation of NOM, we detect the electric field intensity as

$$\begin{aligned} |E_s + \Delta E_s|^2 &\propto [(\alpha_a + \Delta\alpha) + (\alpha_s + \Delta\alpha)]^2|E|^2 \\ &\approx (\alpha_a + \alpha_s)^2|E|^2 + 4(\alpha_a + \alpha_s)\Delta\alpha|E|^2 \end{aligned} \tag{2.4}$$

where E_s and ΔE_a are the unmodulated and modulated electric fields, respectively, scattered by the coupled dipoles. In Eq. 2.4, the first term is background signal, which does not have any dependence on the relative position of the apex and the sample. The second term is strongly dependent on the separation of the two dipoles, which contribute to the near-field signal with spatial resolving power.

Several important features of the interacting dipoles in the near-field region can be derived immediately from these equations. In the following sections, on the basis of the coupled dipoles model, we investigate the desirable shape of a fiber probe for the achievement of high resolution, high contrast, and high sensitivity.

2.1.3 Resolution

From the simple coupled dipoles model, it can be inferred that the ultimate spatial resolution of NOM is determined only by the apex size. The other parts of the probe, such as the aperture and the tapered core, contribute to the enhancement of contrast rather than to the achievement of high resolution. This section is devoted to a discussion on the optimization of the apex size to obtain the sufficient resolution.

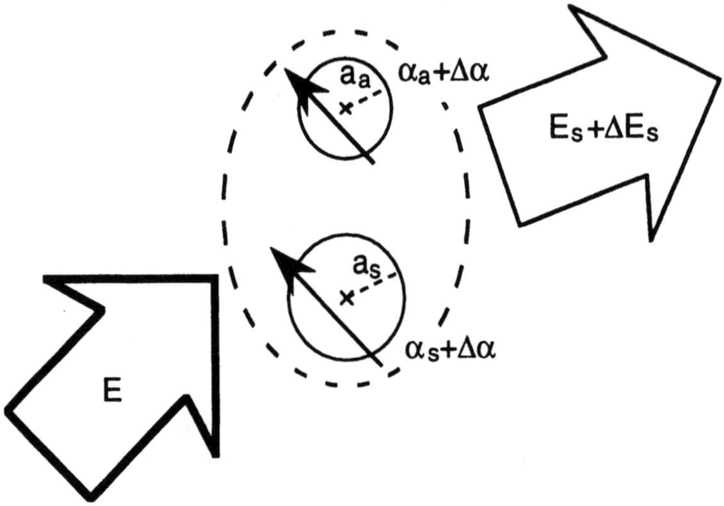

Fig. 2.2. Simplified configuration of NOM; illumination light, two interacting spheres, and scattered light

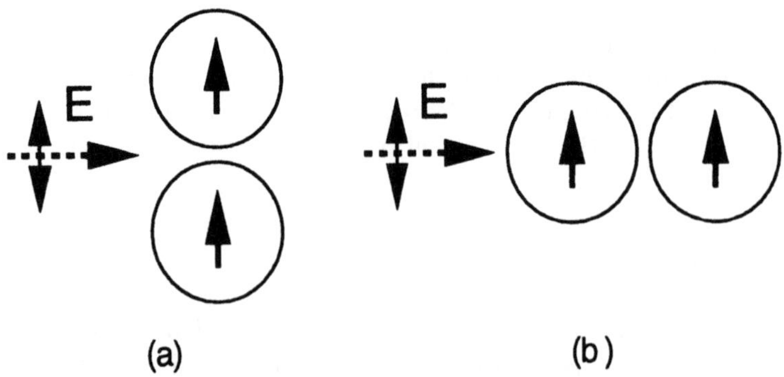

Fig. 2.3. a Broadside and b end-on illumination of two particles

Using Eq. 2.1, where the broadside illumination is assumed, we consider the apex-size-dependent feature of spatial resolution relating to the sample size. The size of the sample is fixed as $a_s = a$ ($\ll \lambda$). For an apex with radii $2a$, a, and $0.5a$, the change in the polarizability modulation $\Delta\alpha$ in a line scan is depicted as a function of the horizontal separation, x, from the sample (Fig. 2.4). The strength of the modulation $\Delta\alpha$ is normalized by the polarizability of the sample α_s. In this calculation, the NOM is operated in constant-height mode, where the distance between the apex and the substrate is constant. The closest spacing between the apex and the sample is zero in the three cases. With decreasing apex radius, the image size becomes smaller while the signal intensity also decreases. The width of the signal profile and the peak intensity of the signal are plotted as a function of the apex radius in Fig. 2.5. As far as the resolution is concerned, this result suggests that the smaller apex is more suitable. To determine the optimized apex size, however, the image contrast should be taken into account. The image contrast, the ratio of modulated to unmodulated scattering field intensity, is derived from Eq. 2.4 as

$$4(\alpha_a + \alpha_s)\Delta\alpha|E|^2/(\alpha_a + \alpha_s)^2|E|^2 = 4\Delta\alpha/(\alpha_a + \alpha_s) \qquad (2.5)$$

Figure 2.6 shows the calculated result of the contrast as a function of apex radius. The material of the sample is assumed to be glass (n_s=1.5, g_s=0.3) and the apex is chosen to be g_s=0.3 (glass) and g_s=3 (metal). In both cases, the contrast becomes highest when the apex size is similar to the sample size. This behavior is easy to explain qualitatively. The modulated intensity saturates with increasing apex size because only part of the large apex, which is near the sample, contributes to the polarizability modulation. Since the unmodulated intensity increases in proportion to the apex volume, in contrast to the saturation of the modulated signal, the contrast decreases monotonously. On the other hand, the modulated intensity decreases with a reduction of the apex size to zero. The contrast also deteriorates because the unmodulated scattering signal from the sample remains constant, and is much larger than the modulated signal.

One general conclusion can be derived from these simple considerations, which is that the optimized apex size is the same as the characteristic size of the sample investigated. Taking into account the signal intensity and the contrast, an apex size much smaller than the sample size is unsuitable even though it realizes better resolution.

2.1.4 Contrast

In the previous section, on the basis of the coupled dipoles model, we concluded that the apex size should be the same as the sample size. As long as the apex works as a scattering antenna, the contrast given by Eq. 2.5 is almost independent of the refractive index of the apex (a change by a factor

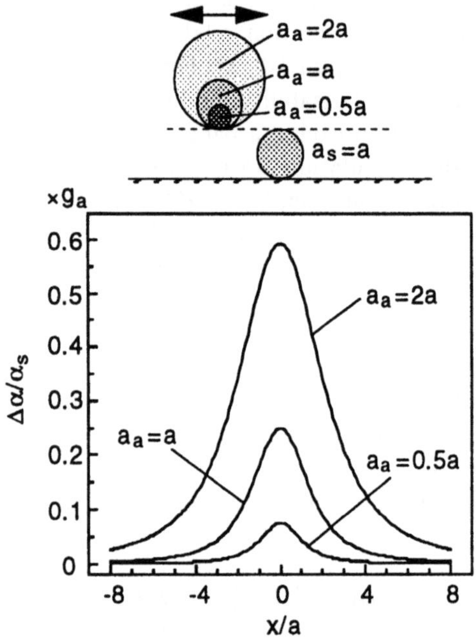

Fig. 2.4. Dependence of modulation signal profile on the apex size

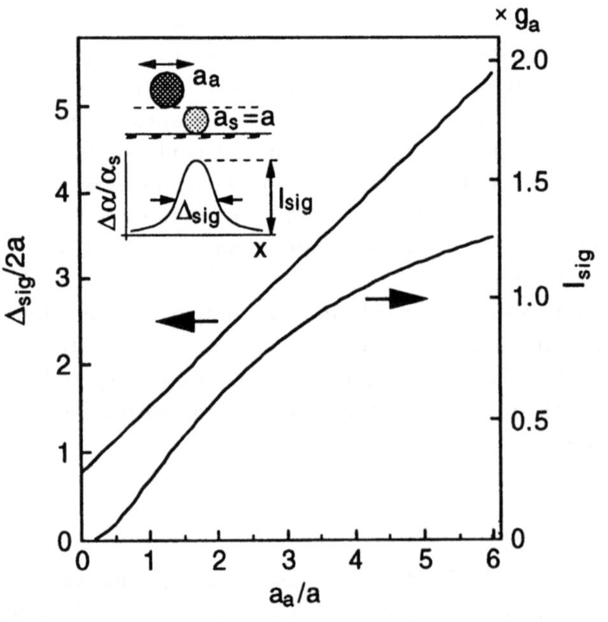

Fig. 2.5. Resolution and signal intensity as a function of apex radius

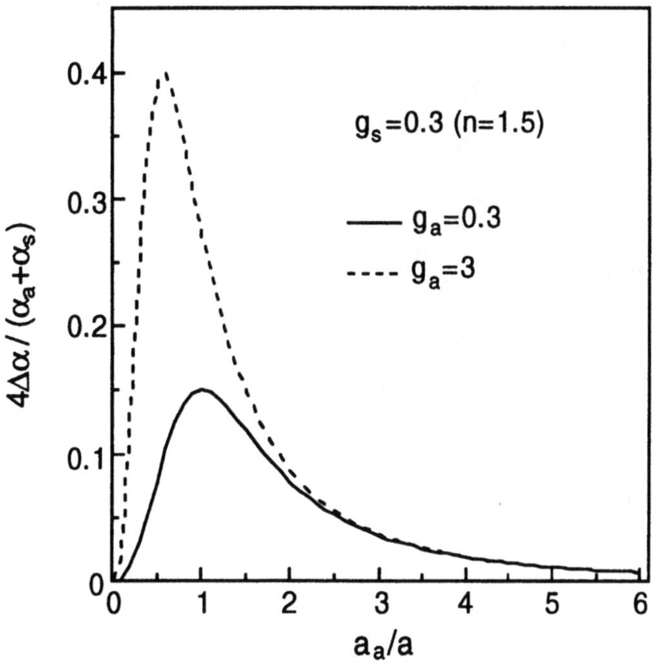

Fig. 2.6. Contrast as a function of apex radius

of three at the most, as shown in Fig. 2.6). To enhance the contrast more effectively, it is necessary to introduce a novel probe, such as a fluorescent probe with wavelength conversion, as described in Sect. 2.2. Prior to such consideration, in this section, we discuss how the contrast can be improved for a fiber probe with a tapered structure.

In the coupled dipoles model, the apex and the sample are treated as small spheres. In actual operation, however, the small apex is fabricated at the very top part of the tapered structure. This tapered part, in general, spoils the contrast of the image since it also couples to the sample dipole and generates an intense background signal. To estimate this influence qualitatively, we use a simplified model for the tapered probe, as shown in Fig. 2.7, where the tapered part is replaced by a composite of spheres. The radii of the apex and the sample are equal to a. A comparison of the signal profiles between probes with different cone angles is made in Fig. 2.8. Here the coupling between the apex and the tapered part is neglected. While the resolution and intensity of the required signal are the same in both cases, the intensity of the background signal due to the coupling between the sample and the tapered part is generally different. When the cone angle is large, the interaction between the sample and the large sphere generates an intense background signal and as a result the contrast deteriorates. This estimation suggests that the cone angle should be small to suppress the background signal.

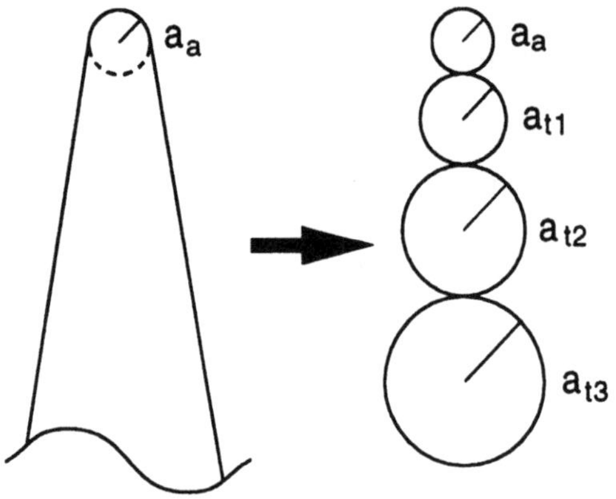

Fig. 2.7. Simplified model for a tapered probe

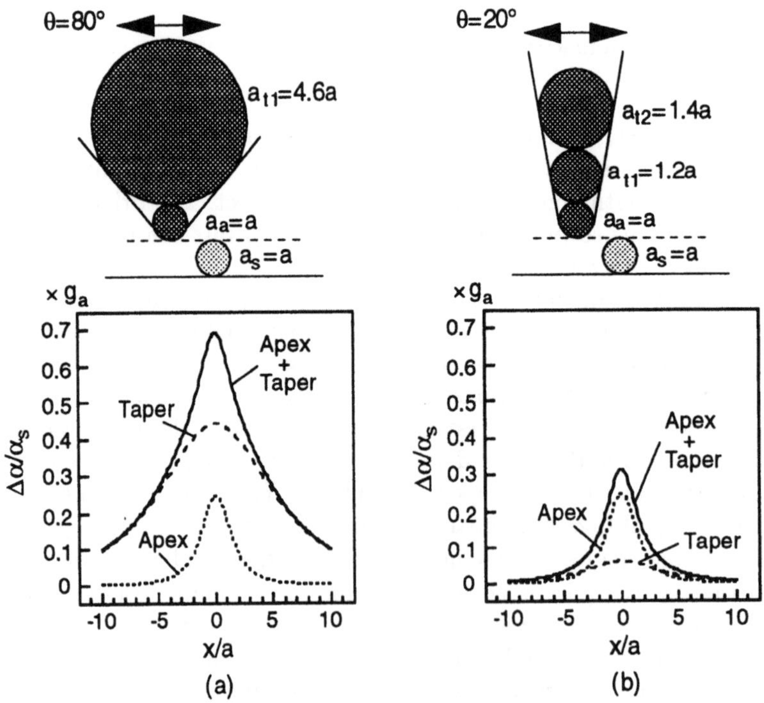

Fig. 2.8. Signal profiles obtained by tapered probes with cone angles of **a** 80° and **b** 20°

As a more effective method to eliminate the undesirable signal, especially in the case of a collection-mode NOM operation (see Sect. 6.1.1), the contribution of a small metal aperture is considered in this section. An aperture with a diameter smaller than wavelength of light can be employed as a spatial frequency band-pass filter, whose feature is determined principally by the aperture size (foot diameter d_f). Recently, based on numerical calculations, Jang [4] found the resonance (band-pass) enhancement in the detection efficiency of the signal when the sample size is similar to the aperture size. In Sect. 1.3.4, this resonance has been called as size-dependent behavior. If the aperture is incorporated within the tapered probe, the contrast of the image can generally be improved. Figure 2.9 shows the expected signal profile when comparing probes with and without an aperture. Due to the band-pass behavior of the aperture, the coupling between the sample and the tapered part is effectively suppressed, and the resultant background signal is much decreased. (In this calculation, −5 dB suppression of the coupling efficiency between the sample and the tapered part is assumed.) It should be added that in actual operation, the aperture also rejects the scattered stray light from other small structures existing near the objective sample. We can conclude that an apertured probe with a small cone angle is the best structure for obtaining high-contrast images.

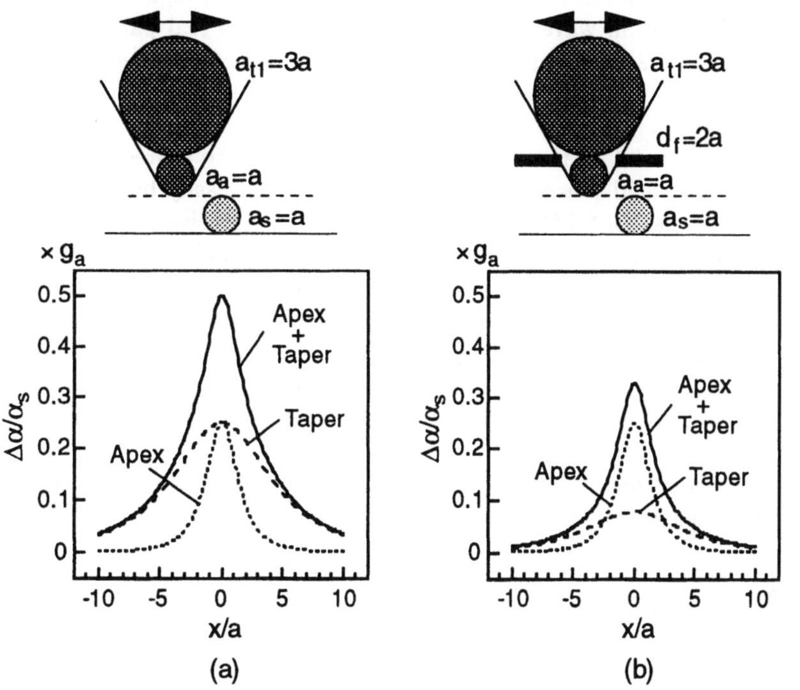

Fig. 2.9. Signal profiles obtained by tapered probes **a** without aperture and **b** with aperture

2.1.5 Sensitivity

From Eq. 2.3, we have two ways to acquire the maximum signal (modulation) intensity. The first method, as already mentioned in Sect. 2.1.3, is to make the apex size as large as the characteristic size of the sample investigated. By optimizing the apex size, we obtain large polarizability from Eq. 2.3 and the resultant modulation is also enhanced without serious deterioration in the spatial resolution. The second method is to use a material with large refractive index n_a as the apex. We have some candidate materials such as semiconductors, metals, and so on. For example, silicon has a refractive index of $n_a \sim 4$ in visible region and the polarizability modulation is enhanced by a factor of three compared with the glass apex ($n_a \sim 1.5$). A few groups have applied a silicon probe [5, 6] in NOM operations although the apex size was not sufficiently small to obtain nanometer-size resolution owing to the difficulty of sharpening silicon. A metal sphere (plasmon probe; see Chap. 10) is also efficient owing to its negative dielectric constant and its resonant behavior at the irradiation wavelength λ where $n^2(\lambda)$ equals -2. Figure 2.10 shows the plots of calculated values of $|g|^2$ for silver as a function of irradiation wavelength using numerical data of the complex refractive index of silver [7]. In order to estimate the enhancement factor compared with a glass apex, $|g|^2$ for silver is normalized against that for glass. An enhancement factor of more than ten will be expected to obtain all over the visible region. Moreover, around an irradiation wavelength of 350 nm, where the real part of $|n|^2$ is almost equal to -2, $|g|^2$ increases resonantly. The Q factor of the resonance is determined by the imaginary part of $|n|^2$, which is strongly dependent on the kind of metal.

The improvement of the sensitivity by a metal probe is frequently employed as a near-field plasmon microscope since it can easily be realized by using metal probes for STM or metal-coated glass fiber probes. In the case of collection-mode NOM, however, when the protrusion part is completely metallized, the transmission intensity of the signal light into the fiber is generally decreased. Therefore the thickness of the metal film, which also influences the resonant behavior of the plasmon probe, is an important design factor. The strong optical confinement effect of a metal apex is promising technique, which can be applied as a small and intense light source for nanostructure fabrication.

In conclusion, the important components for the optimization of the probe are summarized as follows:

1. the apex size should be the same as the characteristic size of the sample to be investigated;
2. the smaller the cone angle, the smaller the undesirable background signal due to the dipole coupling between the sample and the tapered part;
3. the coupling can also be eliminated more efficiently by the combination of a metal aperture with a glass tapered part;

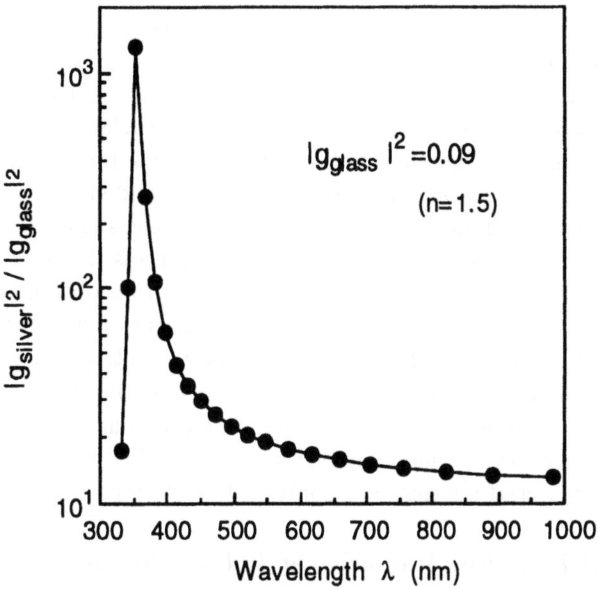

Fig. 2.10. Enhancement factor of polarizability of silver particle as a function of wavelength of irradiation light

4. by proper metallization of the apex, the signal intensity is much enhanced compared with that from a bare glass apex.

2.2 Functional Probe: New Contrast Mechanisms

2.2.1 Signal Conversion by Functional Probes

In Sect. 2.1 we considered the most standard operation of NOM, where the apex works as a simple scattering antenna to interact with the sample and contribute to the slight modulation of the mutual polarizabilities. If the apex has other functions, such as being absorptive, fluorescent, or chemically sensitive, the possible applications of NOM become considerably increased. For this purpose, several groups have been trying to fabricate sharpened micropipettes filled with dye molecules [8] or molecular microcrystals [9], and cleaved semiconductor crystals [10]. One advantageous way to use new functions is signal conversion to achieve a high-contrast image. For example, fluorescent material absorbs the near-field light and emits the fluorescence at a different wavelength, which enables us to extract the modulated signal from the intense unmodulated background signal. To give the probe these functions in a very small region, however, several new techniques will need to be introduced and developed. One promising fabrication technique for functional probes and their applications is described in Chap. 5. In this section, we describe the importance of signal conversion techniques using functional probes from the viewpoint of new contrast mechanisms.

The difficulty of generating high-contrast imaging is the most serious and general problem encountered in the study of small features of surface structures. From Eq. 2.5, it is inferred that the smaller the sample, the more difficult it is to extract the required signal obscured in the background signal. In fact, since the tapered part also generates a background scattered signal, the contrast is even lower than that given by Eq. 2.5. If we can suppress the background signal in some way, a highly-resolved, clear image will be obtained more easily. The techniques of signal conversion, such as absorption and the resultant fluorescence, or a nonlinear change in polarization, contributes to the background rejection as well as the filtering method, such as the employment of a band-pass filter or a polarizer. Some of these techniques are related to near-field spectroscopy, which is widely applied in various fields including device characterization, material physics, and analytical chemistry.

2.2.2 Absorption and Emission: Radiative and Nonradiative Energy Transfer

In this section, by introducing the principle of radiative energy transfer between two single molecules [11], we describe how the image contrast can be improved through the wavelength conversion process. We consider the energy transfer between a donor and an acceptor which have absorption and emission spectra as shown in Fig. 2.11. Here, the donor and acceptor molecules are given the role of the sample and apex, respectively. If the wavelength of light for the donor excitation is set at λ_{exc}, an excitation of the acceptor molecule can be avoided. Due to the wide overlap of donor emission and acceptor absorption spectra, however, the excitation of the donor transfers to the acceptor when the apex (acceptor) approaches the sample (donor) within the characteristic Förster dipole–dipole resonance energy transfer radius R_0. If we detect the fluorescence from the acceptor through a band-pass filter, a background-free signal is obtained, where the emission from the donor (background) directly excited by the excitation light is effectively rejected. This corresponds to the situation in which only the modulated component of scattered light is extracted in standard near-field signal detection. A high-resolution image will be also obtained since the interaction distance R_0 lies typically within a few nanometer range. The successful rejection of background and resultant high contrast is attributed to the wavelength conversion technique; the direct excitation of the acceptor is avoided by selecting the excitation wavelength, and the fluorescence from the donor is rejected by the filtering method. Although single molecule imaging with an energy transfer mechanism is the ideal method, and the principal model of near-field dipole–dipole interaction, in practical experiments, serious technical difficulties are encountered: it is almost impossible to handle a single acceptor molecule arbitrarily at the present time. It is expected that progress in the fabrication technique of a functional probe will result in a new technology for single molecule manipulation.

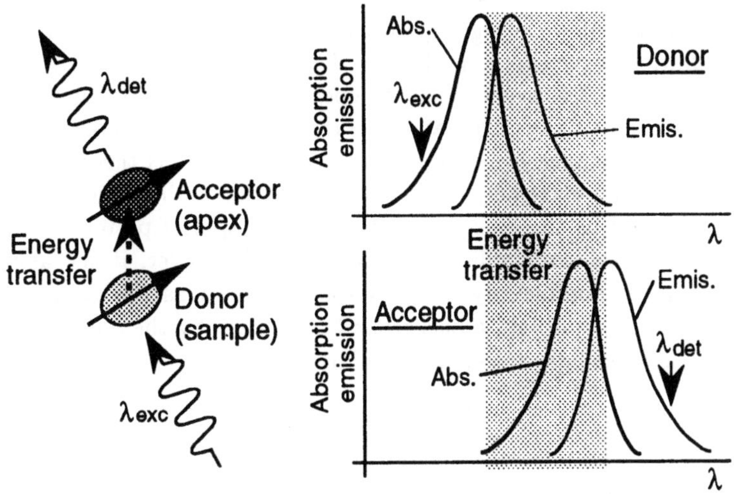

Fig. 2.11. Principle of energy transfer between two single molecules

Related to the energy transfer between two single molecules, a nonradiative energy transfer mechanism is also expected to bring about a similar effect of enhancement of the contrast. In general, many kinds of metal absorb in the visible region. If a metal particle is employed in place of the single acceptor molecule, nonradiative energy transfer occurs from the donor molecule to the acceptor metal within an interaction distance of around a few nanometers. Since the energy transfer probability is large enough, emission from the donor molecule is easily quenched by the approach of the metal within the interaction region. By scanning the metal apex, therefore, the donor molecule can be imaged as a hole with a spatial resolution of a few nanometers in the image of the donor fluorescence. Compared with energy transfer between two single molecules, this method can easily be realized: the apex of the metal-coated apex works as the absorbing acceptor.

2.2.3 Resonance, Nonlinearity, and Other Mechanisms

Using the resonant and nonlinear features of the apex material is another promising way to enhance the image contrast. If the apex has a very sharp resonance structure of its refractive index in the spectral region, the apex–sample interaction in the near-field region induces effective modulation of signal properties, since it cooperates with the nonlinearity of the material in various ways (Fig. 2.12). When the apex approaches the sample to within the distance of the sample size, the resonance wavelength of the apex changes due to the modification of the effective refractive index of the environment around the apex (in the case of a plasmon probe, see Sect. 10.2). The configurational resonant effect can be of practical use. The wavelength of the irradiation light is adjusted to slightly off resonance for the apex. With the approach of the sample, the resonant wavelength gradually changes, and finally coincides with the irradiation wavelength in the near-field region of the sample. In the framework of Sect. 2.1, a new position-dependent feature is added to the modulated polarizability: the polarizability of the apex has a sharp resonance at an apex–sample distance of less than the sample size. It is theoretically predicted that when an extremely sharp resonance, such as the exciton state in CuCl, is employed as the apex and the sample, a spatial pattern much smaller than their real sizes will be obtained [12]. Moreover, if the apex material has a large nonlinearity at the resonance, such as the nonlinear polarization rotation, we obtain a background-free image by the rejection of unmodulated scattered light through the polarizer.

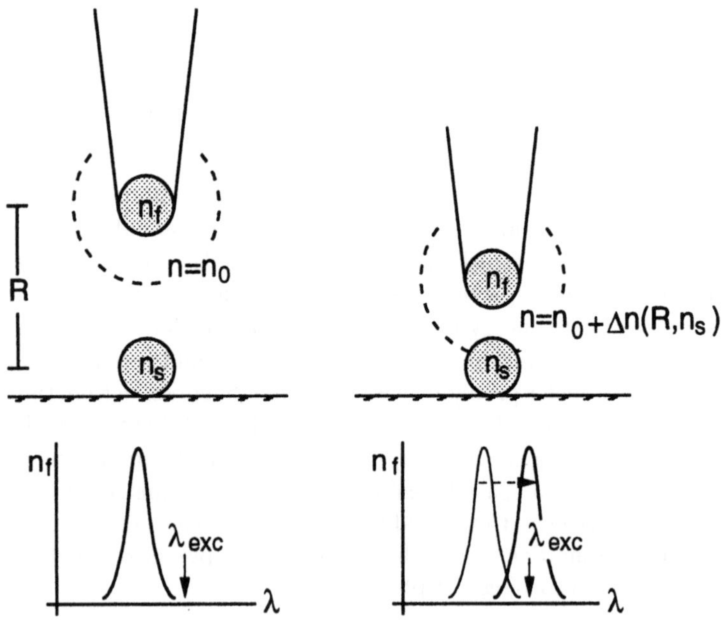

Fig. 2.12. Near-field interaction between sample and functional probe with sharp optical resonance

The use of chemically sensitive material as the apex will open up a new type of scanning microscope. When light is used for the probe of the chemical reaction with the sample, this technique also belongs to one of the family of NOM. An apex which has such functions is called a nano-chemical sensing probe. Since, in general, the reaction occurs in the stage of the interaction between single molecules within the contact region, the molecular size resolution is achievable in principle. Furthermore, the image contrast will be improved when adequate detection of the response signal succeeds, such as a change in the emission spectrum due to a reaction with the sample. Details of the fabrication and application of a chemical sensing probe will be described in Chap. 5.

References

1. J. D. Jackson, *Classical Electrodynamics* (Wiley, New York, 1975)
2. S. Levine, G. O. Olaofe, J. Colloid and Interface Sci. **27**, 442 (1968)
3. F. Zenhausern, Y. Martin, H. K. Wickramasinghe, Science **269**, 1083 (1995)
4. K. Jang, W. Jhe, Opt. Lett. **21**, 236 (1996)
5. F. Zenhausern, M. P. O'Boyle, H. K. Wickramasinghe, Appl. Phys. Lett. **65**, 1623 (1994)
6. H. U. Danzebrink, G. Wilkening, O. Ohlsson, Appl. Phys. Lett. **67**, 1981 (1995)
7. P. B. Johnson, R. W. Christy, Phys. Rev. B **6**, 4370 (1972)
8. W. Tan, Z. -Y. Shi, S. Smith, D. Birnbaum, R. Kopelman, Science **258**, 778 (1992)
9. K. Lieberman, S. Harush, A. Lewis, R. Kopelman, Science **247**, 59 (1990)
10. A. Lewis, U. Ben-Ami, N. Kuck, G. Fish, D. Diamant, L. Lubovsky, K. Lieberman, S. Katz, A. Saar, M. Roth, Scanning **17**, 3 (1995)
11. T. Förster, in *Modern Quantum Chemistry*, O. Sinanoglu (ed.) (Academic, New York, 1965)
12. Y. Ohfuti, K. Cho, Phys. Rev. B **51**, 14379 (1995)

Chapter 3

Probe Fabrication

3.1 Introduction

To perform a near-field optical microscopy (NOM) application such as optical imaging of nanometric biological specimens and spatially resolved spectroscopy of optical devices, one has to fabricate a nanometric probe which works a sensitive scatterer and/or a selective generator of high spatial frequency components of a localized optical near field. NOM employing a scatterer-type probe and a generator-type probe are defined as collection-mode (c-mode) NOM and illumination-mode (i-mode) NOM,[1] respectively. Instruments of i-mode NOM and c-mode NOM will be described in Chap. 6. For illumination-collection hybrid mode (i-c mode) NOM,[2] one requires a probe functioning as both a generator and a scatterer of the optical near field.

In order to realize such a scatterer-type probe and generator-type probe, forming a dielectric taper with a nanometric apex and also metallizing the taper has been used as an effective method. Popular methods for fabricating dielectric tapers have been etching of a quartz crystal rod [1], pulling of a pipette [2], meniscus-etching of an silica fiber [3–6], selective etching of a fiber [7–11], pulling of a fiber [12–14], pulling and etching of a fiber [5, 15, 16], and meniscus-etching/selective etching of a fiber [17–19]. For the metallization of such a dielectric taper, a vacuum evaporation method [1, 2, 12] has been used in which the dielectric taper is coated with an aluminum or gold film by a vacuum evaporation unit. Among the fabricated probes, an optical fiber probe has high throughput due to its waveguide structure and has been widely employed for various applications in visible and near infrared regions.

---

[1] In the c-mode NOM, the light is incident to the total internal reflection. The three dimensional optical near field generated and localized on the sample surface is scattered by a probe and part of the scattered field is detected and collected through the sample and the probe The principle of operation of i-mode NOM is similar except that the probe acts as a generator of the optical near field which illuminates the sample surface. The field scattered by the sample is collected by conventional optics.

[2] In the i-c mode NOM, a sample is excited by an optical near field on the probe. The light generated on the sample is scattered and collected by the probe.

Three basic techniques, i.e., pulling, meniscus-etching, and selective etching have been used for tapering an optical fiber.

By a pulling technique [12], in which an optical fiber is heated and pulled by a micropipette puller combined with a CO_2 laser, one can fabricate a tapered fiber with an apex diameter of 50 nm and a cone angle of 20–40° using a commercial micropipette puller. However, it is difficult to control the cone angle while maintaining an apex diameter as small as 50 nm. With regard to the pulled fiber, the thermal properties of the metallized probe have been studied [20–22]. In meniscus-etching, originally developed to fabricate a fiber-optic microlens [3], a single-mode fiber is immersed in HF acid with a surface layer of an organic solution. It is tapered with a cone angle since the height of meniscus formed around the fiber is reduced depending on the fiber diameter. Although the cone angle can be increased up to 35–40°, the obtained tapered fiber has a geometrically eccentric apex with an elliptical cross section. The longer and shorter principal diameters of this elliptical apex take values of 200 nm and 10–20 nm, respectively. To fabricate a probe with an apex diameter less than 10 nm, the selective etching technique was developed [7, 8]. The core is tapered by immersing a high-GeO_2-doped fiber in a buffered hydrogen fluoride solution. The cone angle is controlled from 20° to 180° for an apex diameter less than 10 nm. Further, selective etching is the most highly reproducible technique among the three. The characteristics of the basic techniques are summarized in Table 3.1.

Table 3.1. Characteristics of the basic techniques for tapering an optical fiber

Technique	Cone angle, θ	Apex diameter, d	Reproducibility
Meniscus-etching	9–40°	60 nm or more	80% or less
Selective etching	14–180°	10 nm	Almost 100%
Pulling	20–40°	50 nm	Around 80%

In imaging applications requiring high spatial resolution, a tapered probe or a metallized probe with a nanometric apex diameter and a small cone angle has to be fabricated because the resolution capability of the probe is determined by its parameters such as the cone angle and the apex diameter (see Chap. 2.). However, the throughput [3] of a metallized probe is decreased by decreasing the cone angle. In spectroscopic applications where one must cope with extremely low detected power, the probe should have high throughput in i-mode and be highly sensitive in c-mode. In the i-mode NOM, to avoid thermal damage to the sample and the probe, the probe should be used with an input power as low as possible. Therefore, the resolution capability and throughput of the tapered probe have to be optimized depending on the ap-

[3] When a metallized tapered fiber is used as a generator-type probe for an i-mode NOM employing an objective lens, the throughput is defined as the relative output power to the input power (see Chap. 4).

plication of NOM to imaging or spectroscopy. This optimization should be done through tailoring the probe. Furthermore to obtain a highly resolved image, a metallized probe must be tailored so as to have a protruding tip emerging from a metallic film. However, it is difficult to fabricate such a protrusion-type probe with a nanometric apex diameter by means of the vacuum evaporation method because the apex is covered with a thin metallic film due to the throwing of the evaporated metal vapor.

In order to fabricate application-oriented probes, we recently proposed some methods [16, 23–27] which were based on hybrid selective etching of specially designed multistep index fibers [25, 26] and preferential etching of the metal covering the apex [17, 23, 27]. We succeeded in fabricating application-oriented probes such as a protrusion-type probe with high resolution capability [17, 23, 27], a double-tapered with high throughput [24], a triple-tapered probe with high resolution capability and high throughput [25, 26], and an ultraviolet probe with a pure silica core [16, 25]. In this chapter, these methods, which are based on selective etching and its basic techniques, are described. For reference, Table 3.2 summarizes the advantages of the probes described in this chapter.

Table 3.2. Advantages of the probes described

Advantage	Probe	Figure	Section
High resolution	Protrusion-type shoulder-shaped probe	3.18	3.4.1
	[SP with a controlled cladding diameter]	3.5	3.3.1.1
	[SP with a nanometric flattened apex]	3.7	3.3.1.2
	Protrusion-type pencil-shaped probe	3.23	3.4.2
	[PP with a nanometric apex diameter]	3.17 3.26a	3.3.2.2 3.5.3
	[PP with a ultra small cone angle]	3.14	3.3.2.1
High throughput	Double-tapered probe	3.11	3.3.1
	PP with a large cone angle	3.26e	3.5.3
High resolution & high throughput	Triple-tapered probe	3.26d	3.5.1
UV	UV single-tapered probe	3.29	3.6.1
	UV triple-tapered probe	3.32	3.6.2

SP, shoulder-shaped probe; PP, pencil-shaped probe.

3.2 Selective Etching of a Silica Fiber Composed of a Core and Cladding

Prior to describing methods for fabricating application-oriented probes, we present a geometrical model of selective etching which is a basic technique for tapering a fiber composed of a core and a cladding. Further, we discuss several fibers which were originally produced for an optical transmission system by vapor-phase axial deposition (VAD) [28].

3.2.1 Geometrical Model of Selective Etching

Figure 3.1a shows a cross-sectional profile of the refractive index of a silica fiber with a germanosilicate (GeO_2)-doped core and a pure silica cladding. Here, n_1 and n_2 are the refractive indices of the core and cladding, respectively; r_1 and r_2 are the radii of the core and cladding, respectively. When the fiber is immersed in a buffered hydrogen fluoride solution (BHF) with a volume ratio of [NH_4F aqueous solution (40wt.%)]: [HF acid (50wt.%)]: [H_2O]= X:1:1 at 25°C, the core is hollowed at X=0 and tapered at X=10. Figure 3.1b and c show schematic explanation of the geometrical models for hollowing and tapering, respectively. Bright shading and dark shading in the upper parts represent the cross-sectional profiles of the fiber before and after the etching, with **b** an etching time T and **c** an etching time τ. ϕ is the convex angle of the hollow, τ is the etching time required for making the apex diameter zero, d is the apex diameter, and θ is the cone angle of the tapered core. The lower parts show the dissolution rates R_1 and R_2 of the core and

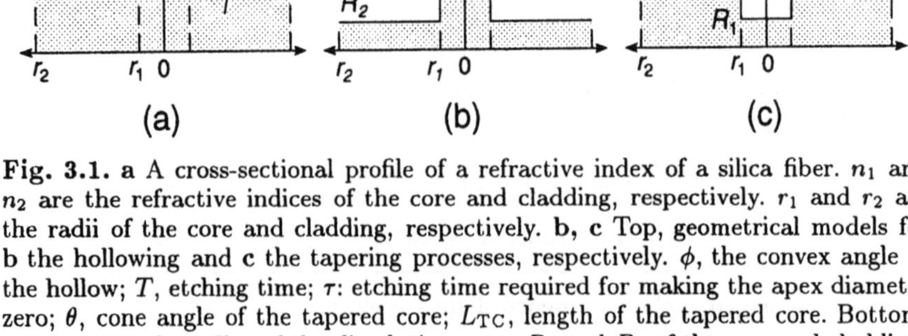

Fig. 3.1. **a** A cross-sectional profile of a refractive index of a silica fiber. n_1 and n_2 are the refractive indices of the core and cladding, respectively. r_1 and r_2 are the radii of the core and cladding, respectively. **b**, **c** Top, geometrical models for **b** the hollowing and **c** the tapering processes, respectively. ϕ, the convex angle of the hollow; T, etching time; τ: etching time required for making the apex diameter zero; θ, cone angle of the tapered core; L_{TC}, length of the tapered core. Bottom, cross-sectional profiles of the dissolution rates R_1 and R_2 of the core and cladding, respectively

cladding, respectively, where $R_1 > R_2$ in b and $R_1 < R_2$ in c. Assuming that the dissolution rates R_1 and R_2 are constant, the convex angle ϕ is expressed as

$$\sin(\phi/2) = R_2/R_1 \tag{3.1}$$

Relations between the cone angle θ, the length L_{TC} of the tapered core, and the apex diameter d are represented by

$$\sin(\theta/2) = R_1/R_2 \tag{3.2}$$

$$L_{\mathrm{TC}} = (r_1 - d/2)/\tan(\theta/2) \tag{3.3}$$

and

$$d(T) = \begin{cases} 2r_1(1 - T/\tau) & (T < \tau) \\ 0 & (T \geq \tau) \end{cases} \tag{3.4}$$

Here the etching time τ required for making the apex diameter zero is represented by

$$\tau = (r_1/R_1)\left[(R_1 + R_2)/(R_2 - R_1)\right]^{1/2} \tag{3.5}$$

3.2.2 Pure Silica Fiber with a Fluorine Doped Cladding

By etching a pure silica core fiber with a fluorine-doped cladding with BHF, a tapered fiber with a constant cone angle is realized. Here, the ratio $[R_1/R_2]$ of the dissolution rates is independent of the concentration of BHF. For the following discussion, we define the relative refractive index difference Δn of doped glass to pure silica, which is expressed as $(n_2^2 - n_1^2)/2n_2^2$ and

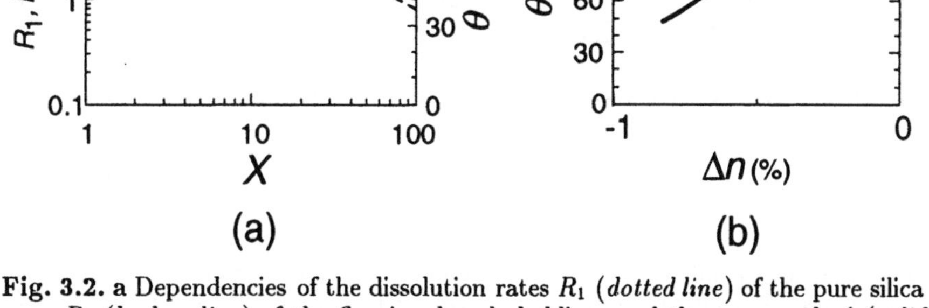

Fig. 3.2. a Dependencies of the dissolution rates R_1 (*dotted line*) of the pure silica core, R_2 (*broken line*) of the fluorine-doped cladding, and the cone angle θ (*solid line*) on the volume ratio X of NH_4F aqueous solution. The value of Δn defined as $(n_2^2 - n_1^2)/2n_2^2$ is -0.7%. b Dependency of θ on Δn at X=10

$(n_1^2 - n_2^2)/2n_1^2$ for a pure silica core fiber and a pure silica cladding fiber, respectively. For the case of selective etching of a fiber with a relative refractive index difference Δn=−0.7% in BHF with a volume ratio X:1:1, the dissolution rates R_1 and R_2 of the pure silica core and fluorine-doped cladding and the cone angle θ, depend on the volume ratio X of NH_4F aqueous solution, as shown in Fig. 3.2a.

Figure 3.2b shows the dependence of θ on Δn at X=10, which reveals that the cone angle θ is controlled by varying Δn or the fluorine doping ratio. However, it is difficult to produce a VAD fiber with a fluorine-doped cladding and a large index difference of $|\Delta n| > 0.7\%$.

3.2.3 GeO_2 Doped Fiber

When a fiber with a GeO_2-doped core and a pure silica cladding is immersed in BHF with volume ratios of X:1:1, the fiber is hollowed in $X < 1.7$ and is tapered in $X > 1.7$. Figure 3.3a shows the dependencies of the dissolution rates R_1, R_2, and the cone angle θ on X for a fiber with Δn=2.5%. The ratio R_1/R_2 decreases with increasing X and approaches a constant value at X=10–30. The cone angle θ, determined by the ratio R_1/R_2, takes a minimum at X=10.

If X is fixed, the cone angle of the tapered core is determined by the relative refractive index difference Δn. Since the value of Δn (> 0) is increased by increasing the GeO_2 doping ratio, VAD is an effective method to produce a GeO_2-doped fiber with Δn as large as 3.0%. Figure 3.3b shows the dependence of θ on Δn at X=10. Furthermore, by etching a fiber with a

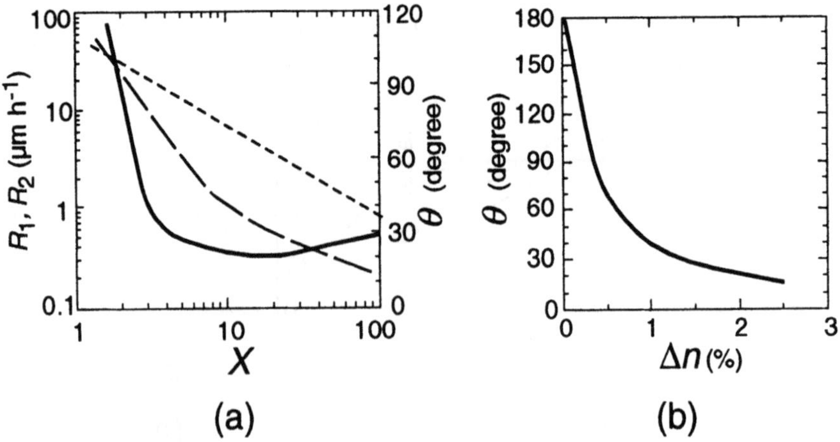

Fig. 3.3. a Dependencies of the dissolution rate R_1 (*broken curve*) of the GeO_2-doped core, R_2 (*dotted line*) of the pure silica cladding, and the cone angle θ (*solid curve*) on X. The value of Δn defined as $(n_1^2 - n_2^2)/2n_1^2$ is 2.5%. b Dependency of θ on Δn at X=10

high-GeO_2-doped core and a fluorine-doped cladding, a tapered fiber with a very small cone angle of 14° [8, 9] is fabricated.

3.2.4 Tapered Fibers for Optical Transmission Systems

When the following five types of fibers, fabricated originally for optical transmission systems, are selectively etched in BHF with a volume ratio of [NH_4F aqueous solution (40wt.%)]: [HF acid (50wt.%)]: [deionized water]= 10:1:1, one can fabricate tapered fibers with parameters as summarized in Table 3.3. These fibers are: (1) a 1.3-μm single-mode fiber (1.3 μm-PSF) with a pure silica core and a fluorine-doped cladding; (2) a 1.3-μm single-mode fiber (1.3 μm-SF) with a GeO_2-doped core and a pure silica cladding; (3) a 0.6-μm single-mode fiber (0.6 μm-SF) with a GeO_2-doped core; (4) a dispersion-shifted fiber (DSF) with a double GeO_2-doped core; (5) a dispersion-compensating fiber (DCF) with a high-GeO_2-doped core. While the cone angle is determined by the relative refractive index difference Δn, the apex diameter is affected by the core radius r_1. To fabricate a tapered fiber with a small cone angle and a nanometric apex diameter, one has to use a GeO_2-doped fiber such as DCF with Δn as high as 2.5% and a small core radius of r_1=1 μm.

Table 3.3. Structural parameters of tapered fibers fabricated originally for optical transmission systems

Fiber type	Δn (%)	r_1 (μm)	θ (at X=10)	Apex diameter d (nm)
1.3 μm-PSF	−0.35	5	120°	<200
1.3 μm-SF	0.3	5	105°	<180
0.6 μm-SF	0.3	2	105°	<60
DSF	0.9	2	60°	<20
DCF	2.5	1	20°	<10

PSF, single mode fiber with a pure silica core; SF, single-mode fiber with a GeO_2-doped silica core; DSF, dispersion-shifted fiber; DCF, dispersion-compensating fiber.

With a DCF with Δn=2.5% and r_1=1 μm, one can obtain a tapered fiber with a small cone angle of 20° and an apex diameter less than 10 nm with almost 100% reproducibility. Such a tapered fiber can be employed as a scatterer-type probe with high resolution capability for a collection-mode NOM [7]. Further, the cone angle can be controlled as $20° \leq \theta < 180°$ for $d <$ 10 nm by varying the concentration of BHF, as shown in Fig. 3.3a. Such high controllability of the cone angle is indispensable for tailoring a high-throughput probe and a high-resolution probe. Such a probe with a tapered core and flat cladding-end will be called a shoulder-shaped probe.

3.3 Selective Etching of a Dispersion Compensating Fiber

Using a dispersion-compensating fiber (DCF), we have developed three types of probe. These are (1) a shoulder-shaped probe with a small cladding diameter, (2) a probe with a nanometric flattened apex, and (3) a probe with a double-tapered core. Further, to fabricate a pencil-shaped probe which has the tapered core and cladding with a controllable cone angle and a nanometric apex, we developed two methods based on meniscus-etching and selective etching. Methods for fabricating such shoulder-shaped and pencil-shaped probes are reviewed in Sects. 3.3.1 and 3.3.2, respectively.

3.3.1 Shoulder-Shaped Probe

3.3.1.1 Shoulder-Shaped Probe with a Controlled Cladding Diameter

When a shear force technique is employed for NOM to regulate the sample–probe separation, the cladding diameter is one of the main parameters governing the resonance frequency of a dithering probe. Although the cladding diameter of a shoulder-shaped probe can be reduced by increasing the etching time T, there is a more effective method for controlling it. Figure 3.4 shows this method schematically, where r_2 is the cladding radius before etching, D is the cladding diameter of the probe, θ is the cone angle of the tapered core, d is the apex diameter, and L_{TC} is the length of the tapered core. This method involves two steps: (A) reducing the cladding thickness and (B) tapering the core.

In step A, the fiber is immersed in BHF with volume ratio of [NH_4F aqueous solution (40wt.%)]: [HF acid (50wt.%)]: [H_2O]= 1.7:1:1 for an etching time T_{A}. The cladding diameter is reduced to $[2r_2 - 2R_{2\mathrm{A}}T_{\mathrm{A}}]$. The core end is kept flat since the dissolution rates $R_{1\mathrm{A}}$ and $R_{2\mathrm{A}}$ of the core and cladding are equal.

In step B, the fiber is selectively etched in X_{B}:1:1 (where X_{B} >1.7) for etching time T_{B}. The core is tapered, with the cone angle θ expressed as

$$\sin(\theta/2) = R_{1\mathrm{B}}/R_{2\mathrm{B}} \tag{3.6}$$

To make the apex diameter zero, the etching time T_{B} must be longer than the time τ which is given by

$$\tau = (r_1/R_{1\mathrm{B}})\,[(R_{1\mathrm{B}} + R_{2\mathrm{B}})/(R_{2\mathrm{B}} - R_{1\mathrm{B}})]^{1/2} \tag{3.7}$$

The cladding diameter D of the tapered probe is expressed as

$$D = 2r_2 - 2(R_{2\mathrm{A}}T_{\mathrm{A}} + R_{2\mathrm{B}}T_{\mathrm{B}}) \tag{3.8}$$

which is proportional to the etching time T_{B}.

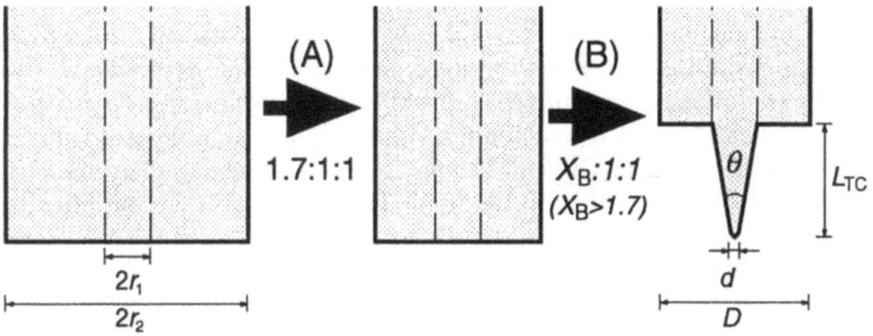

Fig. 3.4. Etching method to fabricate a shoulder-shaped probe. r_1, r_2, radii of the core and cladding, respectively. D, reduced cladding diameter. θ, cone angle of the tapered core. L_{TC}, length of the tapered core

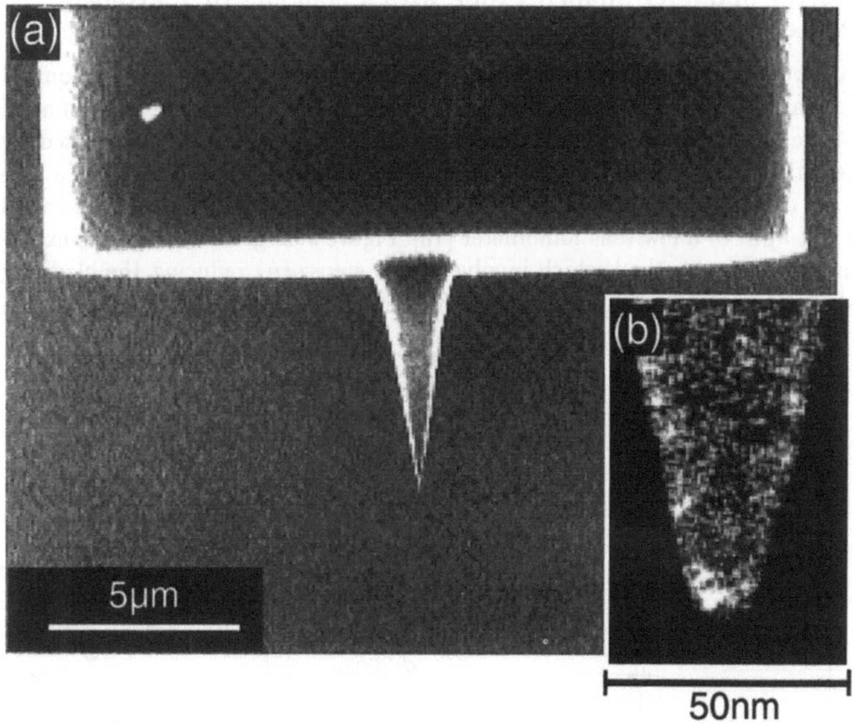

Fig. 3.5. SEM micrographs of **a** a shoulder-shaped probe and **b** the magnified apex region. D=20 μm; θ=20°; $d < 10$ nm

Using a DCF with r_1=1 μm, r_2=62.5 μm, and Δn=2.5%, we obtained a shoulder-shaped probe with a cone angle of θ=20° and a cladding diameter of D=20 μm, and an apex diameter of $d < 10$ nm. Figure 3.5a and b show SEM micrographs of the probe and the magnified apex region, respectively. The etching conditions are summarized in Table 3.4. By substituting r_1=1 μm, R_{1B}=1.1 μm h^{-1}, and R_{2B}=6.5 μm h^{-1} into Eq. 3.7, an estimated value of τ=64 min is obtained. For T_B=75 min ($> \tau$), the cladding diameter D is represented by $D = -60T_A + 106$ [μm]. The cone angle θ is controlled by varying the volume ratio X_B in step B. For the dependence of the cone angle θ on the volume ratio X_B, see Fig. 3.3a.

Table 3.4. Conditions for fabricating a probe as in Fig. 3.5

Step	Volume ratio of BHF	Etching time (min)	Dissolution rates (μm h^{-1})
A	1.7:1:1	86	$R_{1A}=R_{2A}=30$
B	10:1:1	75	$R_{1B}=1.1$, $R_{2B}=6.5$

3.3.1.2 Shoulder-Shaped Probe with a Nanometric Flattened Apex

A tapered probe which has a nanometric flat apex with diameter d is required for fabricating an aperture-like probe with a flat end. A shoulder-shaped probe with a flat apex with diameter d defined by Eq. 3.4 is fabricated by selective etching of a fiber for etching time T shorter than τ in Eq. 3.5. We have developed an etching method to fabricate a shoulder-shaped probe with a flat apex of a few tens nanometer [10]. Figure 3.6a gives a schematic explanation of the method which involves three steps: (A) reducing the cladding thickness; (B) tapering the core; (C) flattening the apex. Here, steps A and B are the same as in Fig. 3.4a.

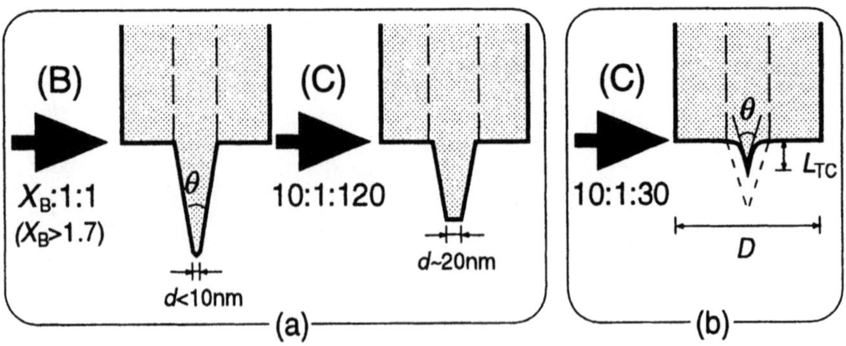

Fig. 3.6. Etching methods for fabricating a shoulder-shaped probe with **a** a nanometric flat apex and **b** a probe with a reduced length of tapered core

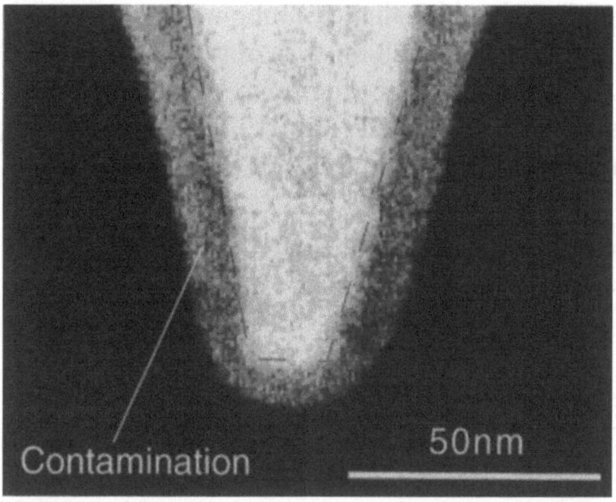

Fig. 3.7. SEM micrograph of a shoulder-shaped probe with a flattened apex. d=20 nm; θ=20°

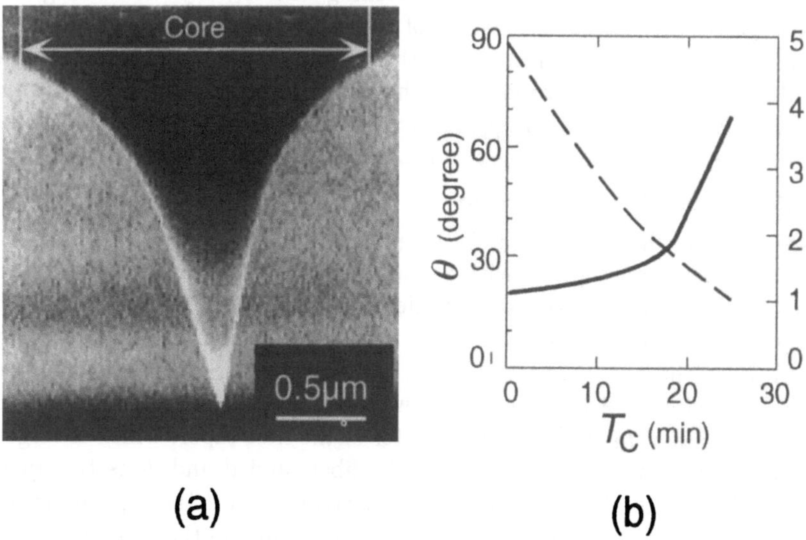

Fig. 3.8. a SEM micrograph of the tapered core region of a shoulder-shaped probe with a reduced length L_{TC}=2.2 µm of the tapered core. **b** Dependencies of the cone angle θ (*solid curve*) and L_{TC} (*broken curve*) on the etching time T_C in step C

In step C, the shoulder-shaped probe with $d < 10$ nm is immersed in an extremely low concentration of BHF with X_C:1:Y_C=10:1:120. In actual conditions, as shown in Table 3.5, a shoulder-shaped probe with a cone angle of $\theta=20°$ and a flat apex with a diameter of 20 nm was obtained. Figure 3.7 shows a SEM micrograph of the apex region of the probe. While such mesoscopic flattening is often observed at $R_{2C} < R_{1C}$ ($\ll 1$) for a short etching time, the etching mechanism behind the flattening is not yet well known.

Table 3.5. Conditions for fabricating a probe as in Fig. 3.7

Step	Volume ratio of BHF	Etching time (min)
A	1.7:1:1	60
B	10:1:1	75
C	10:1:120	2

Further, by etching in a BHF with a volume ratio of X_C:1:Y_C=10:1:30 (where $R_{1C}=R_{2C}$), one can fabricate a shoulder-shaped probe with a reduced length of tapered core as schematically shown in Fig. 3.6b. Here, L_{TC} is the length of the tapered core, and θ is the cone angle. Figure 3.8a shows a SEM micrograph of the fabricated probe at T_C=15 min. Here, the probe has a length L_{TC}=2.2 μm, a cone angle of 28°, and $d < 20$ nm. L_{TC} is about 0.6 times that of the regularly conical core, with θ=28°. Figure 3.8b shows the dependencies of θ and L_{TC} on the etching time T_C. By metallizing the tapered probe with a reduced length of the tapered core, one can fabricate a generator-type probe with increased throughput. However, it is sometimes difficult to employ such a probe with the reduced aspect ratio [L_{TC}/D] for scanning a sample with a bumpy surface structure because the edge of the cladding can scratch the sample surface.

3.3.1.3 Double-Tapered Probe

For spectroscopic applications and near-field imaging of a bumpy sample surface, a shoulder-shaped probe with a double-tapered core was developed [24]. Figure 3.9 shows a cross-sectional profile of a double-tapered probe with the apex region protruding from a metallic film. Here, θ_1 and θ_2 ($< \theta_1$) are the cone angles of the first and second taper, respectively. L_{TC} is the length of the double-tapered core, λ_{air} is the optical wavelength in air, n_1 is the refractive index of the fiber. d_B is the base diameter of the first taper, which is larger than the optical wavelength size λ in the fiber, and d and d_f is the apex and foot diameters of the protrusion, respectively. Since the light entering the tapered core is strongly attenuated by metallic film in the subwavelength cross-sectional portion, one has to decrease the cone angle θ_1 in order to increase the throughput of the probe. We now describe a tapering technique for fabricating such a double-tapered probe. For the metallizing technique, evaluating the throughput, and spectroscopic applications employing the probe, see Sect. 3.4.1, Chap. 4, and Chap. 9, respectively.

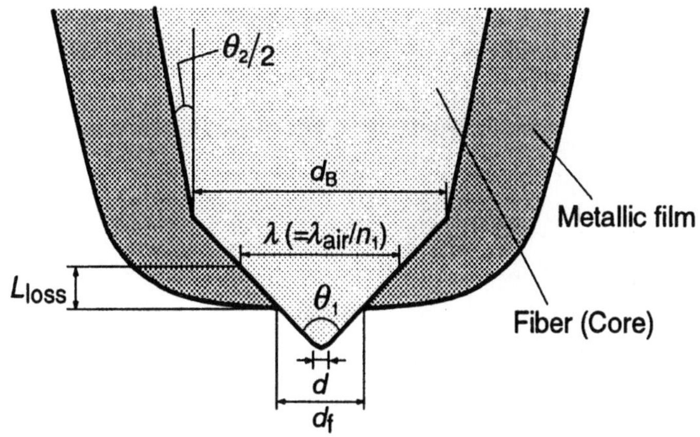

Fig. 3.9. Cross sectional profile of a double-tapered probe with the apex region protruded from a metallic film. θ_1, θ_2, cone angle of the first and second tapers, respectively; d, apex diameter; d_f, foot diameter; d_B, base diameter of the first taper; λ, optical wavelength in the core; L_{loss}, length of the lossy part between the cross-sectional diameters of λ and d_f

The method for fabricating the probe involves two steps: (A) controlling the cladding diameter and forming the first taper; (B) forming the second taper. Figure 3.10 shows the method schematically. In step A, by immersing the fiber in BHF with X_A:1:1 (where $X_A > 1.7$), the core is tapered, with the cone angle θ_1 represented by

$$\sin(\theta_1/2) = R_{1A}/R_{2A} \tag{3.9}$$

In step B, immersing the fiber in BHF with 10:1:1 the core is tapered with cone angles of θ_1 and θ_2, where

$$\sin(\theta_2/2) = R_{1B}/R_{2B} \tag{3.10}$$

The base diameter d_B is expressed as

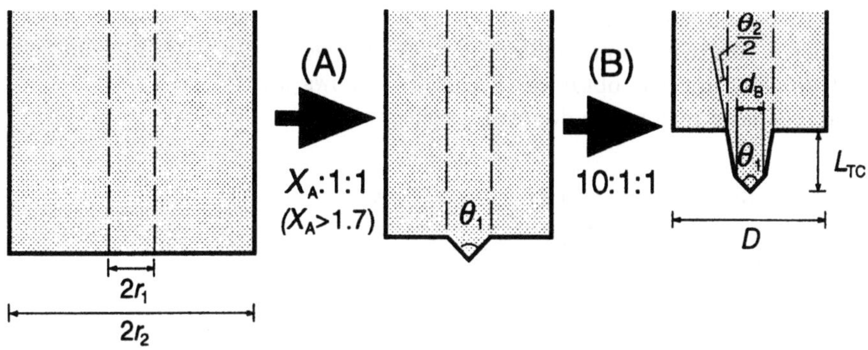

Fig. 3.10. Etching method for fabricating a double-tapered probe

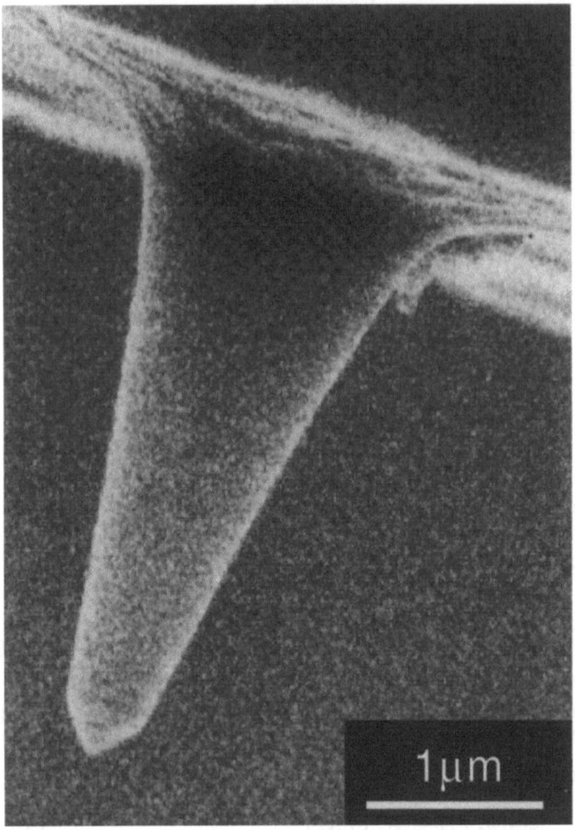

Fig. 3.11. SEM micrograph of a double-tapered core. θ_1=90°; θ_2=20°; d_B=500 nm

$$d_B = 2r_1(1 - T_B/\tau) \qquad (T_B < \tau) \tag{3.11}$$

which is controlled by varying the etching time T_C. Here, τ is defined by Eq. 3.7. The cladding diameter D is given by Eq. 3.8.

In actual conditions as shown in Table 3.6, a probe with a double-tapered core with cone angles of θ_1=90° and θ_2=20°, and a base diameter of d_B=500 nm was obtained. Figure 3.11 shows a SEM micrograph of the tapered core region of a double-tapered probe.

Table 3.6. Conditions for fabricating a probe as in Fig. 3.11

Step	Volume ratio of BHF	Etching time (min)
A	1.8:1:1	70
B	10:1:1	40

3.3.2 Pencil-Shaped Probe

For scanning on a bumpy surface, a pencil-shaped probe with a tapered cladding and a tapered core should be fabricated rather than a shoulder-shaped probe with a flat cladding end. However, it is impossible to fabricate a pencil-shaped probe simply by selective etching of a DCF. In order to overcome this difficulty, we have developed the following method.

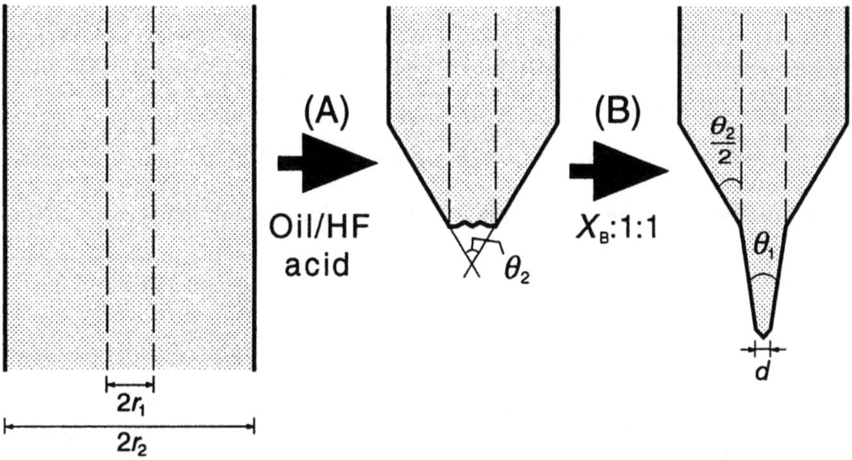

Fig. 3.12. Etching method for fabricating a pencil-shaped probe with an ultra-small cone angle. θ_1, θ_2, cone angles of the core and cladding, respectively

3.3.2.1 Pencil-Shaped Probe with an Ultra-Small Cone Angle

A method based on meniscus-etching and selective etching was developed for fabricating a pencil-shaped probe with an ultra-small cone angle of 3–20°. The method which is shown in Fig. 3.12 schematically consists of two steps: (A) tapering the cladding, and (B) tapering the core. In this figure, oil/HF acid represents an etching solution of HF acid with a surface layer of organic liquid such as silicone oil. θ_1 and θ_2 are the cone angles of the core and the cladding, respectively, and the radii of the core and cladding of the original fiber are r_1 and r_2, respectively. The dissolution rates of the core and cladding in step A are defined as R_{1A} and R_{2A}, respectively.

In step A, the cladding of the GeO_2-doped fiber is tapered by immersing the fiber in oil/HF, and the time T_A of withdrawing [4] the fiber from the oil/HF acid is represented by

[4] In a previous meniscus-etching technique developed by Turner [3], the fiber is not withdrawn at T_A defined by Eq. 3.12 but after the meniscus height has become zero. However, if a high-GeO_2-doped fiber such as a DCF has been immersed for a time longer than T_A, the fiber is hollowed due to the large core-dissolution rate $R_{1A} \gg R_{2A}$. Therefore, a DCF cannot be tapered by Turner's technique.

$$T_A = [r_2 - r_1]/R_{2A} + r_1/R_{1A} \tag{3.12}$$

In tapering the cladding, the height of meniscus formed at the interface around the fiber is decreased by decreasing the cladding diameter. The oil on the surface slows down the evaporation of the HF acid and determines the cone angle θ_2. If dimethylsilicone oil with a density of 0.935 g cm^{-3} is used, the cone angle θ_2 is 20°. Since the dissolved silica contaminates the interface region of oil/HF acid, one cannot repeat the process using the same etching solution.

In step B, the core is tapered in BHF with a volume ratio of [NH_4F aqueous solution (40wt.%)]: [HF acid (50wt.%)]: [H_2O]=X_B:1:1. The cone angle θ_1 is represented by

$$\sin(\theta_1/2) = [R_{1B}/R_{2B}]\sin(\theta_2/2) \tag{3.13}$$

and is smaller than the cone angle θ in Eq. 3.2 for the volume ratio $X = X_B$.

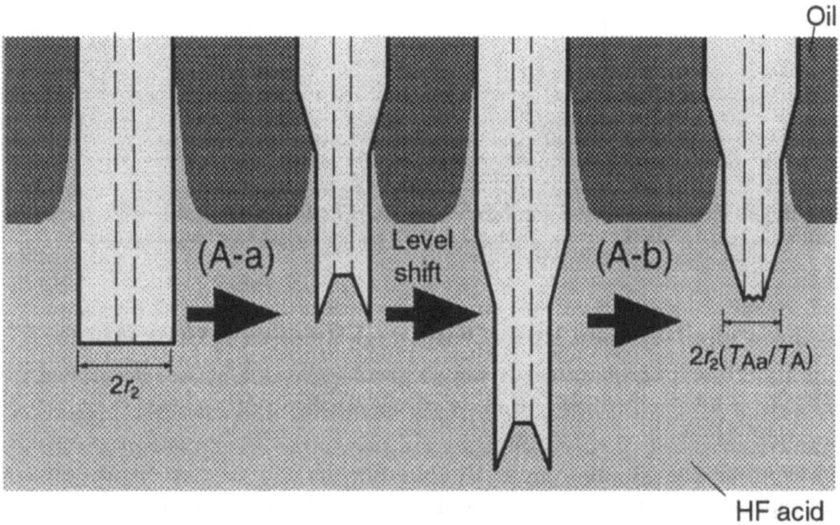

Fig. 3.13. A level shift method for controlling the cladding diameter of a pencil-shaped probe. T_{Aa}, etching time of the substep A-a; T_A, total etching time of substeps A-a and A-b

Further, to control the cladding diameter of the pencil-shaped probe, we developed a level shift method based on meniscus-etching in oil/HF acid. This consists of substeps A-a and A-b as shown in Fig. 3.13, which replace step A of Fig. 3.12. Here, r_2 is the radius of the cladding. The etching time for substeps A-a and A-b are represented by T_{Aa} and T_{Ab} $(= T_A - T_{Aa})$, respectively. In this technique, the fiber is only lowered into the oil/HF acid for the etching time T_{Aa} $(0 < T_{Aa} < T_A)$. The cladding diameter D of the

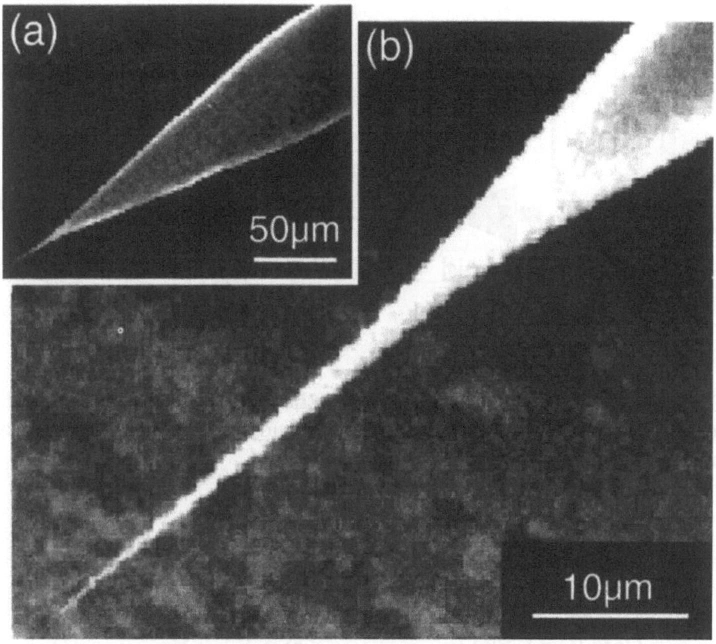

Fig. 3.14. SEM micrographs of a pencil-shaped probe with **a** an ultra-small cone angle. **b** Its magnified apex region: θ_1=4°; θ_2=20°

pencil-shaped probe is proportional to the etching time T_{Aa} and expressed as $D = 2r_2(T_{Aa}/T_A) - 2R_{2B}T_B$.

This method used a DCF fiber with a refractive index difference of Δn=2.5% and a core radius of r_1=1 μm. Conditions for this application were summarized in Table 3.7. A pencil-shaped probe with cone angles of θ_1=3.5° and θ_2=20°, and an apex diameter $d < 30$ nm was fabricated. Figure 3.14a and b show SEM micrographs of the probe and the magnified tapered core region. The cone angle can be increased by decreasing the volume ratio X_B as shown in Figure 3.14.

Table 3.7. Conditions for fabricating the probe in Fig. 3.14

Step	Volume ratio of BHF	Etching time (min)	Dissolution rates (μm h^{-1})
A	Silicone oil/HF acid	22	R_{1A}=1950, R_{2A}=170
B	10:1:1	120	R_{1B}=1.1, R_{2B}=6.5

3.3.2.2 Pencil-Shaped Probe with a Nanometric Apex Diameter

The method of fabricating this probe, consists of three steps, shown schematically in Fig. 3.15, in which oil/HF acid represents HF acid with a surface

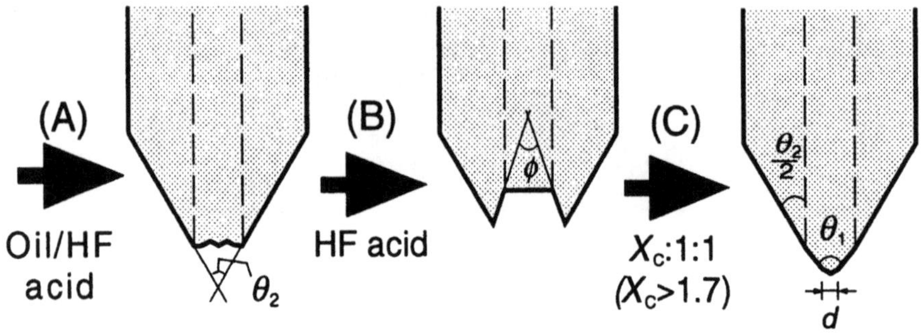

Fig. 3.15. Etching method for fabricating a pencil-shaped probe with a nanometric apex diameter. θ_1, θ_2, cone angles of the tapered core and the cladding, respectively; ϕ, convex angle of the hollowed core

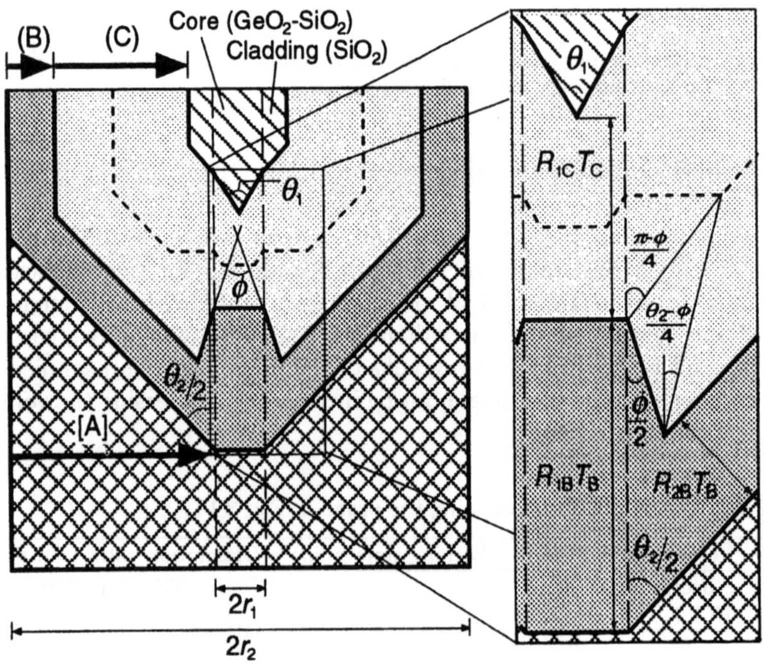

Fig. 3.16. Left. Geometrical model of the etching method in Fig. 3.15. Right. Magnified figure of the central part of the left-hand figure. R_{1i}, R_{2i}, dissolution rates of the core and the cladding in step i

layer of organic liquid such as silicone oil. In step A, the cladding of a high-GeO_2-doped fiber is tapered by etching at the interface between oil and HF acid, as in step A of Fig. 3.12. In step B, the tapered fiber end is hollowed by immersing the fiber in HF acid. Here, oil/HF acid can also be used. However, the etching has to be done not at the interface but in the HF acid. In the third step C, the peaks on either side of the taper end are flattened, and the core is then tapered with a cone angle θ_2 given by

$$\sin(\theta_2/2) = R_{1C}/R_{2C} \tag{3.14}$$

The total etching time for fabricating a probe with a zero apex diameter can be determined based on the geometrical model in Fig. 3.16. In this model, we introduce a minimum etching time T_{Bmin} which is defined as the time required to reach a zero apex diameter. The etching time T_{A} is given by Eq. 3.12. T_{B} and T_{C} are given by

$$T_{\mathrm{B}} = [\mathcal{B}/\mathcal{A}(\mathcal{C}+1)][r_1/R_{2\mathrm{B}}] \tag{3.15}$$

and

$$T_{\mathrm{C}} = \left[\frac{\cos(\theta_1/2)}{\sin(\theta_1/2)\{1-\sin(\theta_1/2)\}}\right]\left(\frac{r_1}{R_{2\mathrm{D}}}\right) \tag{3.16}$$

where

$$\begin{aligned}\mathcal{A} &= \left[\frac{\sin(\theta_2/2)-\sin(\phi/2)}{\sin(\theta_2/2)\sin(\phi/2)}\right]\left[\frac{\tan(\theta_2/2)}{\tan(\theta_2/2)+\tan(\phi/2)}\right] \\ &\quad\times\left[\frac{\tan(\phi/2)+\tan\big((\theta_2-\phi)/4\big)}{\tan\big((\pi-\phi)/4\big)-\tan\big((\theta_2-\phi)/4\big)}\right]\end{aligned} \tag{3.17}$$

$$\mathcal{B} = \left[\tan\left(\frac{\pi-\phi}{4}\right)\right]^{-1}\left[\frac{\{\cos(\theta_2/2)\}^{-1}-\tan(\theta_2/2)}{\tan(\theta_1/2)\{1-\sin(\theta_1/2)\}}\right] \tag{3.18}$$

and

$$\mathcal{C} = \left[\tan\left(\frac{\pi-\phi}{4}\right)\right]^{-1}\left[\left\{\cos\left(\frac{\theta_2}{2}\right)\right\}^{-1}-\tan\left(\frac{\theta_2}{2}\right)\right] \tag{3.19}$$

This method requires a high-GeO_2-doped fiber, which makes it possible for the core to be hollowed, and a large difference between the dissolution rates $R_{1\mathrm{B}}$ and $[R_{2\mathrm{B}}/\sin(\theta_2/2)]$.

Table 3.8. Conditions for fabricating a probe of Fig. 3.17

Step	Volume ratio of BHF	Etching time (min)	Dissolution rates (μm h^{-1})
A	Silicone oil/HF acid	22	$R_{1\mathrm{A}}$=1950, $R_{2\mathrm{A}}$=170
B	HF acid	2	$R_{1\mathrm{B}}$=1950, $R_{2\mathrm{B}}$=170
C	10:1:1	90	$R_{1\mathrm{C}}$=1.1, $R_{2\mathrm{C}}$=6.5

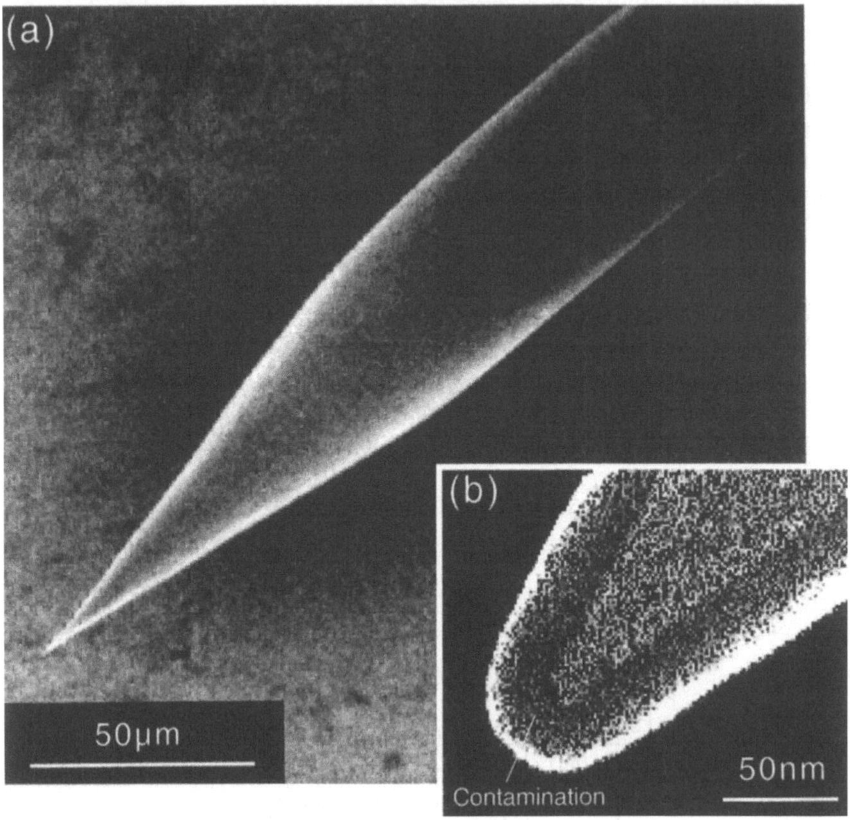

Fig. 3.17. SEM micrographs of a pencil-shaped probe with **a** a nanometric apex diameter. **b** Its magnified apex region: θ_1=20°; $d < 10$ nm

Using a dispersion-compensating fiber with a relative index difference of 2.5%, a pencil-shaped probe with an apex diameter less than 10 nm and cone angles of θ_1=20° and θ_2=20° was fabricated in the conditions summarized in Table 3.8. Figure 3.17a and b show SEM micrographs of the probe obtained and the magnified apex region. The cladding diameter of the probe can be controlled by means of the level shift method in Fig. 3.13.

The cone angle θ_1 is controlled by varying the NH_4F volume ratio X_C in the region $\theta_1 \geq 20°$, as for a shoulder-shaped probe. The dependence of θ_1 on X_C corresponds to that of θ on X, which was shown in Fig. 3.3a.

More advanced methods and multistep index fibers have been developed based on the geometrical models described above. They employ hybrid selective etching of multistep index fibers, which are described in Sects. 3.4.2 and 3.5.2.

3.4 Protrusion-Type Probe

As described in Sect. 3.3, a tapered fiber with a nanometric apex can be formed by etching a dispersion-compensating fiber. When such a tapered fiber is used as a scatterer-type probe for a c-mode NOM, the principal factor governing the resolution of the imaging is the apex size of the probe. However, the resolution is affected by propagating components and the scattering of the low-spatial-frequency components of the optical near field. To increase the resolution of the NOM, a metallized probe with an apex region protruding from metallic film is required.

In order to improve the resolution of a c-mode NOM, we have proposed and fabricated a new type of probe with a conical tip protruding from a metallic film [17, 23]. Figure 3.18 shows a cross section of the protrusion-type probe, where θ, d, d_f, and t_M represent the cone angle, the apex diameter, the foot diameter, and the thickness of the metallic film, respectively. This protrusion-type probe selectively scatters frequency components between $(1/d_f)$ and $(1/d)$. To realize such a probe, a selective resin coating (SRC) method and a chemical polishing method were developed for fabricating a protrusion-type shoulder-shape probe and a protrusion-type pencil-shaped probe, respectively.

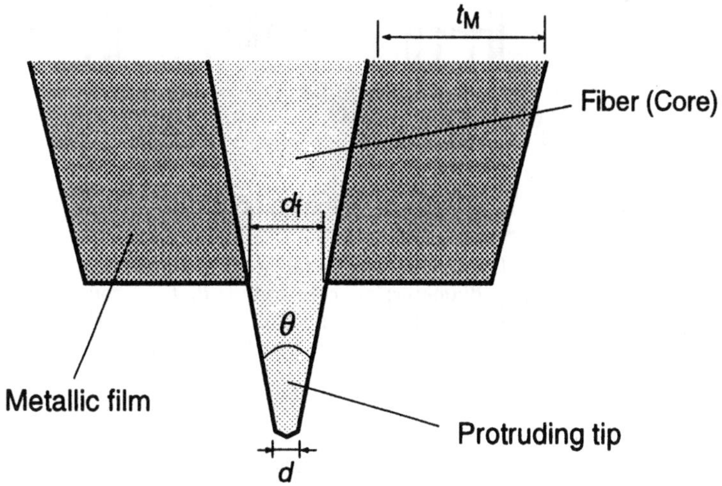

Fig. 3.18. Cross-sectional profile of a protrusion-type probe. θ, cone angle of the taper; d, d_f, apex and foot diameters of the protruding tip, respectively; t_M, thickness of the metallic film

3.4.1 Selective Resin Coating Method

Figure 3.19 shows a schematic diagram of the SRC method involving four steps: (A) metal coating; (B) selective resin coating; (C) preferential etching of metal covering the apex region; (D) removal of resin. Figure 3.20 shows a SEM micrograph of the apex region of the fabricated protrusion-type probe. The probe has a conical protrusion from a gold film and an apex diameter d of less than 10 nm and a foot diameter d_f of less than 30 nm.

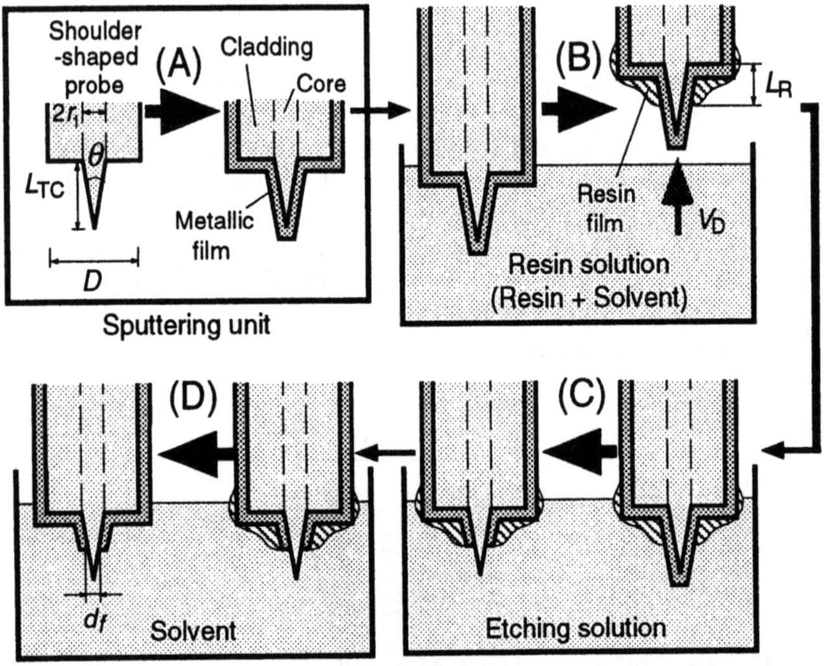

Fig. 3.19. Selective resin coating (SRC) method for fabricating a protrusion-type shoulder-shaped probe. D, cladding diameter; r_1, core radius; θ, L_{TC}, cone angle and length of the tapered core, respectively; V_D, speed of withdrawing the fiber from the resin solution; L_R, length of the tapered core on which the resin is coated; d_f, foot diameter of the protrusion

Prior to applying the SRC method, a shoulder-shaped probe was fabricated with a cladding diameter D=45 μm and a cone angle θ=20°. In step A the probe is coated with 120 nm-thick gold by a magnetron sputtering unit. In step B, the probe is dipped in an acrylic resin solution and was removed with a withdrawal speed of V_D=5 cm s^{-1}. The resin solution has a viscosity coefficient of 11 cP and a density of 0.85 g cm^{-3} at 25°C. In step C the probe is etched for 2 min in a solution KI-I_2-H_2O, mixed with a weight ratio of 20:1:400 and diluted 50 times with water. In step D the acrylic film is removed by dipping the probe in acetone.

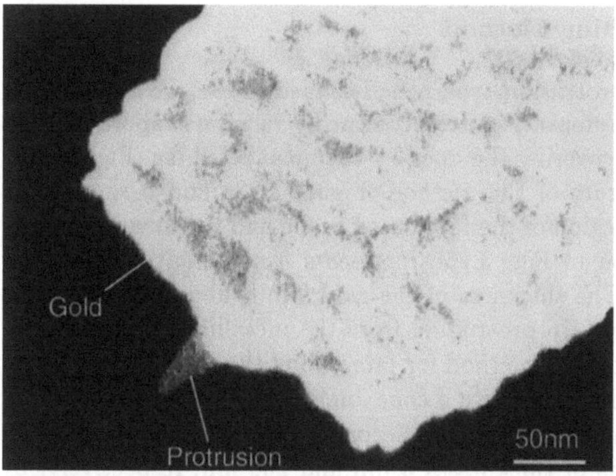

Fig. 3.20. SEM micrograph of a protrusion-type probe fabricated by the SRC method. θ=20°; $d < 10$ nm; d_f=30 nm

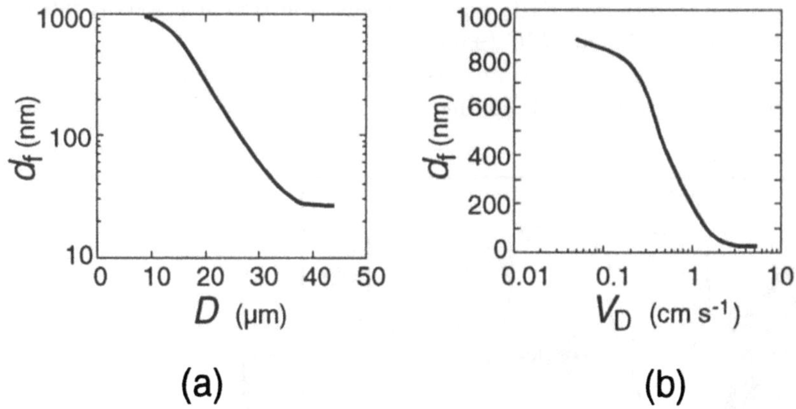

Fig. 3.21. Dependencies of the foot diameter d_f on **a** the cladding diameter D and **b** the withdrawal speed V_D

The foot diameter can be controlled by varying the viscosity and density of the resin solution, the withdrawing speed, the cladding diameter and so on. Figure 3.21a and b show the dependencies of the foot diameter d_f on the cladding diameter D and the withdrawal speed V_D, respectively. This SRC method can be applied to other shoulder-shaped probes such as the double-tapered core described in Sect. 3.3.1.

Coating the resin film can also be realized by another technique based on dip-coating of photoresist and near-field photolithography. For details of the technique, see Matsumoto and Ohtsu [29].

3.4.2 Chemical Polishing Method

In order to fabricate a protrusion-type pencil-shaped probe, we developed a method consisting of two steps: (A) metal coating by vacuum evaporation and (B) removing the metal covering the apex by chemical polishing. Figure 3.22 shows a schematic diagram of the method. Figure 3.23a and b show SEM micrographs of the top region of the fabricated probe and the magnified apex region. The dark portion in Fig. 3.23b represents a nanometric protrusion from the metallic film. The thickness of the gold film is about 150 nm. The foot diameter of the silica tip protruding from the metallic film is less than 20 nm. We describe below the method for fabricating the probe in Fig. 3.23.

Firstly, a pencil-shaped fiber with a cone angle θ_1=20° was fabricated by the etching technique shown in Fig. 3.15. Secondly, the fiber was coated with chromium and gold films of thicknesses 3 nm and 200 nm, respectively, by a vacuum evaporation unit. Here, the pressure was maintained at 10^{-6} to 10^{-7} Torr. We used electron-beam evaporation for gold coating. The fiber was tilted with an angle Ψ=50° and rotated. Finally, the metallized fiber was immersed for 15 min in a KI-I_2 aqueous solution mixed at a weight ratio of KI:I_2:H_2O=20:1:80000 at 25°C (±0.5°C).

We performed step A with various angles Ψ in the region of 55°–90° and observed the probe by SEM. In the case of a thin coating of 120 nm and

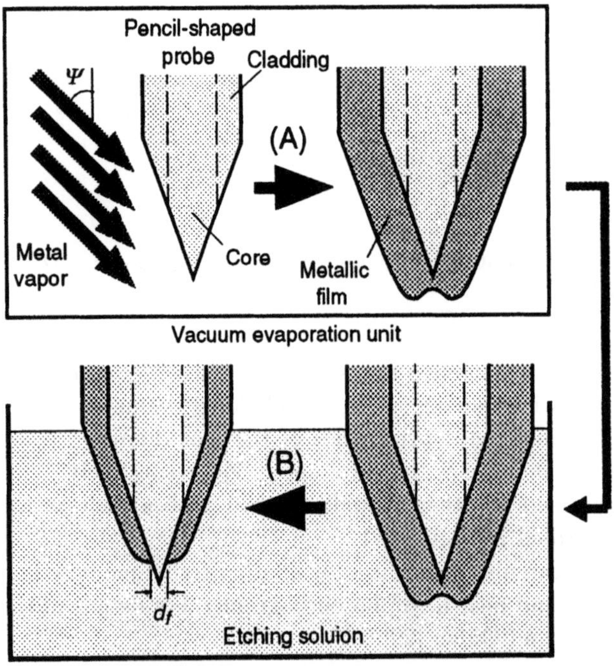

Fig. 3.22. A method for fabricating a protrusion-type pencil-shaped probe. Ψ tilt angle of the fiber axis for evaporating a metal vapor

a tilt angle of Ψ=50° we evaluated a metal thickness of 30 nm covering the fiber apex from the SEM micrographs before and after chemical polishing. At Ψ=50° we obtained a protrusion-type probe with a foot diameter of 20 nm (±15 nm) with reproducibility higher than 75% using 12 fiber samples with apex diameters less than 10 nm.

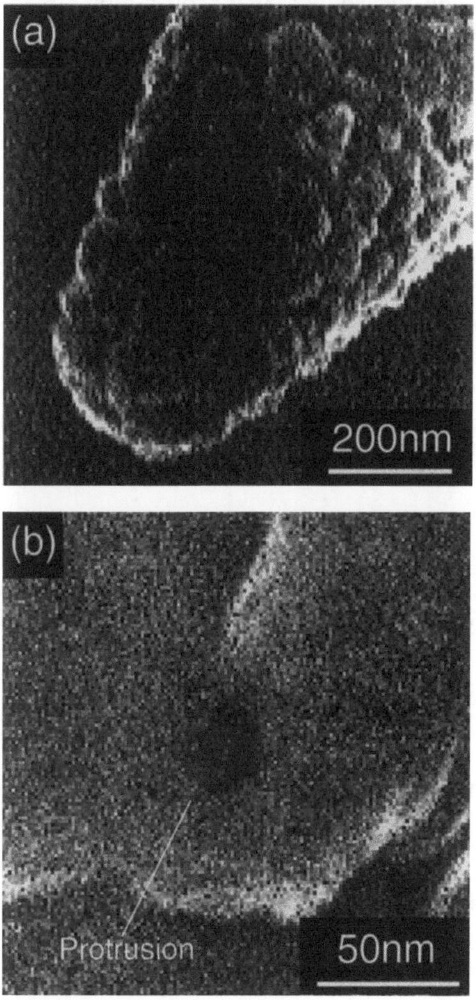

Fig. 3.23. SEM micrographs of **a** the top region of the fabricated protrusion-type pencil-shaped probe and **b** its magnified top region. The thickness of the gold film is 150 nm. $d_f < 20$ nm

3.5 Hybrid Selective Etching of a Double-Cladding Fiber

As described in previous sections, by using a dispersion-compensating fiber with a high-GeO_2-doped core, we succeeded in fabricating some application-oriented probes such as protrusion-type probe with high resolution capability and a double-tapered probe with high throughput. Although a high resolution high throughput probe is required for some spectroscopic applications, one cannot fabricate a high resolution high throughput probe using DCF. Further, even though a pencil-shaped probe can be fabricated by the method in Fig. 3.15, it is difficult to increase the production efficiency of this method because the etching solution of oil/HF is not re-usable.

To overcome these difficulties, we recently developed a multistep index fiber [26] and fabricated application-oriented probes such as pencil-shaped probes and a triple-tapered probe. By advanced methods based on hybrid selective etching of the fiber with a double cladding [26], we fabricated the application-oriented probes with almost 100% reproducibility and high production efficiency. In this section, we describe the structure of a triple-tapered probe and design/fabrication of the fiber. The multistep index fiber is called a double-cladding fiber from now on. We describe another triple-tapered probe with a pure silica core in Sect. 3.6.2.

3.5.1 Triple-Tapered Probe

Figure 3.24a and b show cross-sectional profiles of two metallized probes with the (a) single- and (b) triple-tapered structure. In (a), θ is the cone angle and

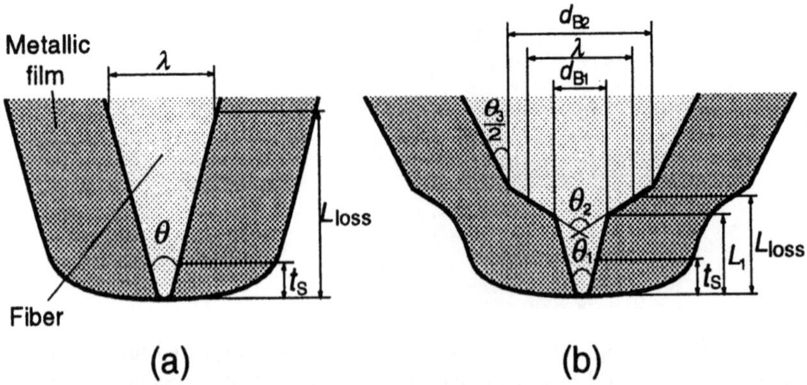

Fig. 3.24. a Cross-sectional profiles of a probe with a single-tapered structure. θ, cone angle of the tapered core; λ, optical wavelength; L_{loss}, length of the tapered core with the foot diameter equal to λ; t_S, skin depth of the metal, b Cross-sectional profiles of a probe with a triple-tapered structure. θ_1, θ_2, θ_3, cone angles of the first, second , and third tapers, respectively; d_{B1}, d_{B2}, base diameters of the first and second tapers, respectively; L_1, length of the first taper

λ is the optical wavelength in the fiber. In (b), L_{loss} is defined as the length of the portion with a cross-sectional diameter of λ to the apex. t_S is the skin depth of metal, L_1 is the length of the first taper, θ_1, θ_2, and θ_3 are the cone angles of the first, second, and third tapers, respectively, and d_{B1} and d_{B2} are the base diameters of the first and second tapers, respectively. The light entering in the single-tapered probe is strongly attenuated by the coating metal from the portion with diameter λ to the apex. To reduce the attenuation, one has to decrease the length L_{loss}. In using a high-throughput probe with an enlarged cone angle, one must accept the limited resolution affected by optical leaking out of the metal around the apex region of the probe. In the triple-tapered probe, we can decrease the length L_{loss} by increasing θ_2 in order to increase the throughput. Thus, the resolution capability is increased by simultaneously decreasing the cone angle and decreasing the first taper length L_1 to a few hundreds nanometers, which corresponds to several times the skin depth t_S.

3.5.2 Geometrical Model of Selective Etching of a Double-Cladding Fiber

Figure 3.25a shows the cross-sectional profile of the relative refractive index difference Δn of the proposed fiber. Here, the fiber is composed of three sections: (1) a GeO_2-doped silica core, (2) a pure silica cladding, and (3) a fluorine-doped silica support. The values of Δn of sections 1 and 3 with

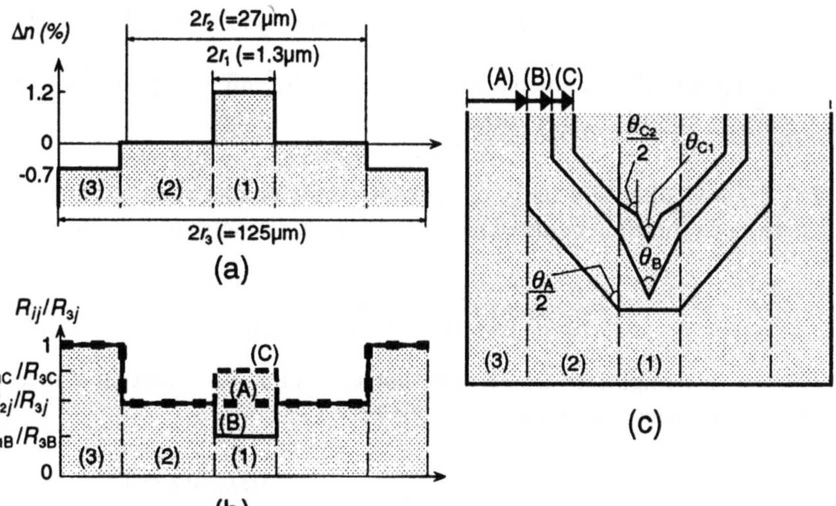

Fig. 3.25. **a** Cross-sectional profile of the relative refractive index difference, Δn. r_1, radius of the GeO_2-doped silica core; r_2, radius of the pure silica cladding; r_3, radius of a fluorine-doped silica support. **b** Etching method for fabricating application-oriented probes. **c** Cross sectional profile of the relative dissolution rate

respect to section 2 are 1.2% and −0.7%, respectively. The radii of sections 1–3 are r_1=0.65 μm, r_2=13.5 μm, and r_3=62.5 μm, respectively. Figure 3.25b shows a schematic diagram of the etching process. The process consists of three steps A–C. By denoting the dissolution rate of section i in step j as R_{ij} (where i=1, 2, 3; j=A, B, C), the cross-sectional profile of the relative dissolution rates R_{ij}/R_{3j} can be seen (Fig. 3.25c). As described in Sect. 3.2.3, the relative dissolution rate of GeO_2-doped glass to pure silica glass depends on the concentration of buffered hydrogen fluoride solution (BHF) mixed with a volume ratio of NH_4F solution (40wt.%): HF acid (50wt.%): H_2O being X:1:Y. We can, for example, use 1.7:1:1, 10:1:1, and 1.7:1:5, in which the dissolution rates of sections 1 and 2 are experimentally found to satisfy the relation

$$\frac{R_{1\mathrm{B}}}{R_{2\mathrm{B}}} = 0.29 < \frac{R_{1\mathrm{A}}}{R_{2\mathrm{A}}} = 1.0 < \frac{R_{1\mathrm{C}}}{R_{2\mathrm{C}}} = 1.48 \tag{3.20}$$

On the other hand, the relative dissolution rates of fluorine-doped silica glass to pure silica glass are independent of the concentration of BHF as described in Sect. 3.2.2. Therefore, the dissolution rates of section 3 relative to pure silica take a constant value of

$$R_{2j}/R_{3j} = 0.51 \qquad (j = \mathrm{A,\ B,\ C}) \tag{3.21}$$

Based on the geometrical model described in Sect. 3.3.2, we now discuss the etching process using 1.7:1:1, 10:1:1, and 1.7:1:5 in steps A–C, respectively.

In step A, the fiber is tapered to an angle of θ_A given by

$$\sin(\theta_\mathrm{A}/2) = R_{2j}/R_{3j} \qquad (\text{where} \quad j = \mathrm{A,\ B,\ C}) \tag{3.22}$$

Assuming the fiber diameter to be equal to $[2r_2]$ after step A, the etching time T_A is given by

$$T_\mathrm{A} = (r_3 - r_2)/R_{3\mathrm{A}} \tag{3.23}$$

The tapered fiber will have an apex diameter smaller than $2r_1$ if $T_\mathrm{A} \geq \tau_\mathrm{A}$, where τ_A is the time required to make an apex diameter of $2r_1$ and is

$$\tau_\mathrm{A} = \frac{r_2 - r_1}{R_{3\mathrm{A}}}\sqrt{\frac{R_{2\mathrm{A}} + R_{3\mathrm{A}}}{R_{3\mathrm{A}} - R_{2\mathrm{A}}}} \tag{3.24}$$

Thus, it is straightforward to find that the radius r_2 of section 2 must be smaller than the critical radius $r_{2\mathrm{P}}$ expressed as

$$r_{2\mathrm{P}} = \frac{r_3 + \xi \cdot r_1}{1 + \xi}, \qquad \text{where} \quad \xi = \sqrt{\frac{R_{2\mathrm{A}}/R_{3\mathrm{A}} + 1}{1 - R_{2\mathrm{A}}/R_{3\mathrm{A}}}} \tag{3.25}$$

In step B, section 1 is sharpened with a different angle θ_B, given by

$$\sin(\theta_\mathrm{B}/2) = R_{1\mathrm{B}}/R_{3\mathrm{B}} \tag{3.26}$$

We obtain pencil-shaped probes with zero apex diameter and cone angles θ_A and θ_B when the etching time T_B is larger than τ_B, expressed as

$$\tau_B = (r_1/R_{1B})[(R_{1B} + R_{3B})/(R_{3B} - R_{1B})]^{1/2} \tag{3.27}$$

Further, to obtain a triple-tapered probe, we perform step C, when the largest cone angle θ_{C2} is given by

$$\sin(\theta_{C2}/2) = R_{1C}/R_{3C} \qquad (\text{where} \quad R_{1C} > R_{2C}) \tag{3.28}$$

The cone angle θ_{C1} is increased from θ_B by increasing the etching time T_C, and is equal to θ_{C2} at $T_C > \tau_C$, where τ_C is given by

$$\tau_C = (r_1/R_{1C})[(R_{1C} + R_{3C})/(R_{3C} - R_{1C})]^{1/2} \tag{3.29}$$

Therefore, we can obtain a triple-tapered probe and a pencil-shaped probe with a cone angle $\theta_B = \theta_{C2}$ at $0 < T_C < \tau_C$ and $T_C > \tau_C$, respectively.

To decrease the cone angles θ_A and θ_B given by Eqs. 3.22 and 3.26, we have to increase $|\Delta n|$ of sections 3 and 1 (or the doping ratios of fluorine and GeO_2), respectively. To obtain a cone angle θ_A as small as 62°, the value of $|\Delta n|$ in section 3 is estimated from Eqs. 3.22 and 3.26 to be as high as 0.7%. To obtain a cutoff wavelength of around 400 nm, section 1 is tailored with a relative refractive index difference of 1.2% and a core radius of r_1=0.65 μm. Then we obtain an estimated value of θ_A=17° from Eqs. 3.20 and 3.26. Further, when the outer radius r_3 is a standard value of 62.5 μm, we obtain a critical radius of r_{2P}=23 μm from Eqs. 3.21 and 3.25. We make the radius r_2 have a value of 13.5 μm, which is smaller than the critical radius.

To realize the designed fiber in Fig. 3.25a, we produced a preform glass rod by vapor-phase axial deposition [28] and drew the fiber using the preform. To suppress the diffusion of GeO_2 and fluorine, the drawing tension should be as high as possible. However, in the case of drawing the fiber with a high tension of 60 g, we could not reproducibly cleave the fiber to obtain a flat facet with a commercial fiber cleaver. We consider that the low reproducibility can be attributed to the remaining stress between sections 2 and 3. To suppress this remained stress, we kept a low tension of less than 30 g during the drawing.

3.5.3 Application-Oriented Probes: Pencil-Shaped Probe and Triple-Tapered Probe

To demonstrate the tailoring capability of the different types of probes, we actually performed the etching process using the fabricated fiber. We prepared 30 samples with flat ends. The fibers were etched consecutively for T_A=40 min in a 1.7:1:1 solution and for T_B=20 min in a 10:1:1 solution. We obtained a pencil-shaped probe with a small cone angle for high resolution. Figure 3.26a–c show SEM micrographs of the probe, the magnified tapered core, and the magnified apex region, respectively. The cone angles are θ_A=62°

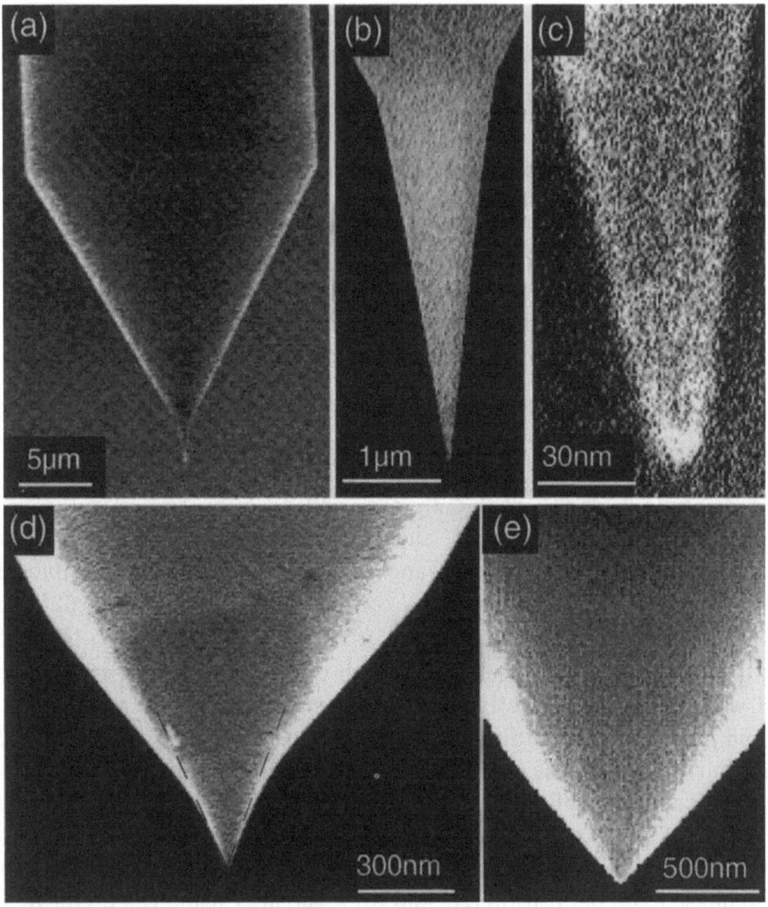

Fig. 3.26. a–c SEM micrographs of a pencil-shaped probe with high resolution capability, its magnified tapered core, and its magnified apex region. θ_A=62°; θ_B=17°; $d < 10$ **nm; d SEM micrograph of a triple-tapered probe with high resolution capability and high throughput. θ_A=62°; θ_{C1}=50°; θ_{C2}=85°. e SEM micrograph of a high throughput pencil-shaped probe. θ_A=62°; θ_B=85°**

and θ_B=17°. These agree with the estimated values from Eqs. 3.22 and 3.26. The apex diameter is less than 10 nm. For the 30 samples, the values of θ_A and θ_B coincided with those of Fig. 3.26a and c within measurement errors of (±0.5°) and (±1°), respectively. This reveals that this kind of probe can be fabricated with almost 100% reproducibility.

Further, by etching the pencil-shaped probe in a 1.7:1:5 solution for T_C=2 min, we obtained a triple-tapered probe with high resolution capability and high throughput. Figure 3.26d shows a SEM micrograph of the triple-tapered probe. The probe has three cone angles of θ_{C1}=50°, θ_{C2}=85°, and θ_A=62° and an apex diameter of less than 10 nm. The two base diame-

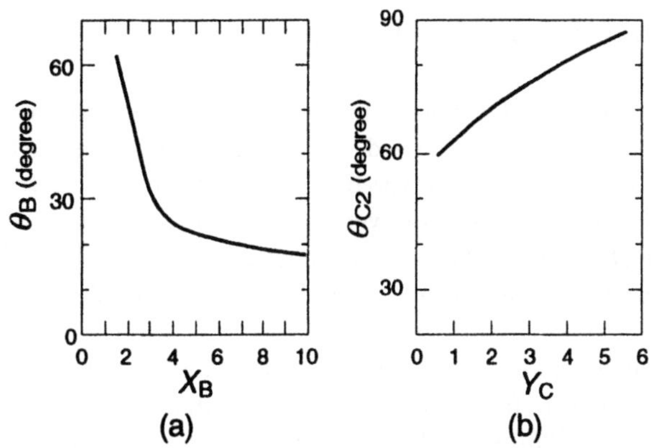

Fig. 3.27. a Dependencies of the taper angle θ_B on the volume ratio X_B of NH_4F aqueous solution in the etching solution. b Dependency of the taper angle θ_{C2} on the volume ratio of H_2O in the etching solution

ters of tapers with θ_{C1}=50° and θ_{C2}=85° are d_{B1}=250 nm and d_{B2}=1.28 μm, respectively.

At T_C=2.75 min, we obtained another type of pencil-shaped probe with a large obtuse angle near the apex region. This is suitable for fabricating a metallized probe with high throughput. The result obtained is shown in Fig. 3.26e, in which θ_A=62° (±0.5°) and θ_B=85° (±1°). The same probe can also be obtained by etching a fiber with a flat end for T_A=40 min in a 1.7:1:1 solution and in a 1.7:1:5 solution for T_B=2.75 min.

Figure 3.27a and b show the dependencies of θ_B and θ_{C2}, respectively, on the concentrations of BHF. Here, 1.7:1:1, X_B:1:1, and 1.7:1:Y_C are used in steps A–C. The cone angles θ_B and θ_{C2} are controlled by varying X_B and Y_C, respectively. However, at around X=1.7 (where $R_{1B}/R_{2B}\approx 1$), the apex diameter increased up to several tens of nanometers. We consider that the increase can be attributed to the geometrical eccentricity between sections 1 and 2. Actually, the fabricated fiber has an eccentric radius of about 200 nm.

To produce the probe having θ_B=62° and apex diameter less than 10 nm, we apply steps B and C in Fig. 3.15 to this fiber. We obtained a probe having θ_B=62° apex diameter less than 10 nm with almost 100% reproducibility by the technique consisting of hollowing in 1.2:1:1 and sharpening in 3:1:1. The cone angle θ_B is given by

$$\sin(\theta_B/2) = R_{1B}/R_{2B} \tag{3.30}$$

and is independent of the geometric structure of section 3. So, the apex diameter is not affected by the eccentric radius of the fiber.

3.6 Probe for Ultraviolet NOM Applications

Using GeO_2-doped fibers such as DCF and a double-cladding fiber, we fabricated application-oriented probes and carried out NOM applications such as imaging a biological sample, and a spectroscopic study of semiconductor devices in visible and infrared regions. However, UV-emitting devices and materials have been little studied by these probes because the GeO_2-doped fiber has optical absorption and luminescence at around 363 nm and 394 nm, respectively.

To overcome this difficulty and to realize highly spatially resolved imaging at near-UV region, we succeeded in fabricating a UV single-tapered probe [16] and a UV triple-tapered probe [25] which have a pure silica core and a double core, respectively. Fabrication methods of these probes are describes in Sects. 3.6.1 and 3.6.2, respectively. Employing these probes, UV near-field spectroscopy was performed (see Chap. 9.).

3.6.1 UV Single-Tapered Probe

Among the fibers described in Table 3.3, only 1.3 μm-PSF with a pure silica core can be used at near-UV light. We developed a tapering method based on pulling/etching of a 1.3 μm-PSF. The method involves two steps: (A) heating and pulling the fiber by a micropipette puller [12–14]; (B) etching the fiber in buffered hydrogen fluoride solution as shown schematically in Fig. 3.28a. Figure 3.28b shows the magnified top region of the tapered shape formed by step B. Here, θ and $2r_{1E}$ are the cone angle in the apex region of the fiber and the reduced core diameter at the end of the tapered core, respectively. Figure 3.28c shows the cross-sectional profile obtained by increasing the etching time in step B. Details of this profile are discussed later. Figure 3.29a–c show SEM micrographs of the tapered fiber, the magnified top region, and the apex region, respectively. The cone angle and the apex diameter are θ=65° and less than 10 nm, respectively. By investigating 20 fiber samples, we obtained 80% reproducibility for an cone angle θ=60° (±5°) and an apex diameter less than 10 nm. We now describe the method for producing this pure silica core fiber.

Table 3.9. Parameters of the puller for a pure silica core fiber

Loop	Heat $\mathcal{H}$	Filament $\mathcal{F}$	Velocity $\mathcal{V}$	Delay $\mathcal{D}$	Pull $\mathcal{P}$
1	300	0	20	124	0
2	400	0	20	140	125

In step A, a 125-μm-diameter PSF was heated and pulled by a micropipette puller (Sutter Instrument, P-2000) combined with a CO_2 laser.

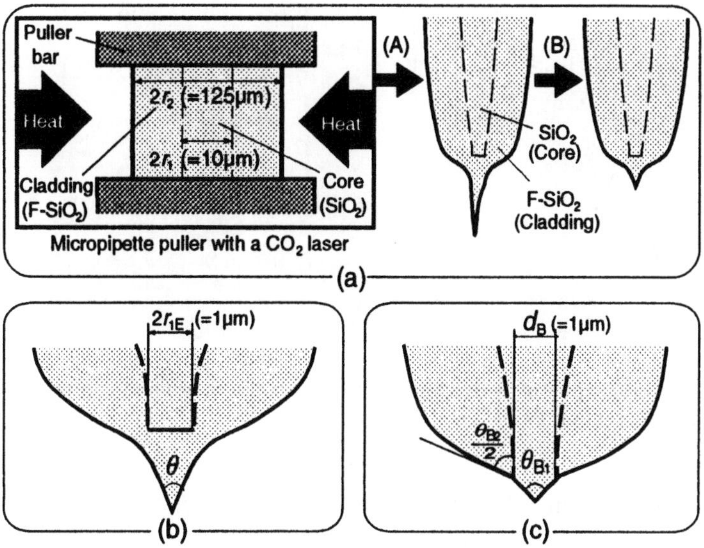

Fig. 3.28. **a** A method for fabricating a tapered fiber with the pure silica core. **b** Cross-sectional profile of the top of the fiber formed by step A in **a**. r_{1E}, reduced core diameter at the end of the tapered core; θ_{B1}, θ_{B2}, cone angles of the core and cladding, respectively. **c** Cross-sectional profile obtained by increasing the etching time in step B

During the heating and pulling, the puller is adjusted so that its parameters [5] are as shown in Table 3.9. Here, the heat parameter $\mathcal{H}$ ($0 \leq \mathcal{H} \leq 999$) decides the CO_2 laser power. The filament parameter $\mathcal{F}$ is the length of the fiber which is scanned with the CO_2 laser beam. The velocity parameter $\mathcal{V}$ ($1 \leq \mathcal{V} \leq 255$) shows the velocity of the puller bar at the end of the heating time. The delay parameter $\mathcal{D}$ ($0 \leq \mathcal{D} \leq 255$) represents the delay time between the end of the heating and the beginning of the pulling (in millisecond units). The puller is mechanically adjusted to make the delay zero at $\mathcal{D}=125$. The pull parameter $\mathcal{P}$ decides the strength of pull and is controlled in a region of $0 \leq \mathcal{P} \leq 255$. The process has two cycled loops.

In step B, the fiber was etched by immersing it for 30 min in a buffered hydrogen fluoride solution (BHF) with a volume ratio of [40wt.%-NH_4F aqueous solution]: [50wt.%-HF acid]: [deionized water]= 10:1:1. The temperature of the BHF was 25°C (±0.1°C). The fluorine-doped cladding and the pure

[5] Sutter has mechanically adjusted commercial pullers (P-2000) to fabricate a micropipette with a diameter of 1 mm and prepares an optional puller bars on which a 125 μm bare fiber can be attached. In our case, to fabricate a probe with a bare-portion length as small as 1cm, all fiber samples were carefully attached to a plastic coated portion on one bar and a bare portion on another. However, we could not realize a reproducibility of step A of more than 80%. This is attributed to mechanical misalignment. If step A is repeated using the same puller and the same fiber, among the parameters shown in Table 3.9, some, such as $\mathcal{H}$ and $\mathcal{D}$, have to be changed in order to fabricate a pulled fiber as seen in Fig. 3.28a.

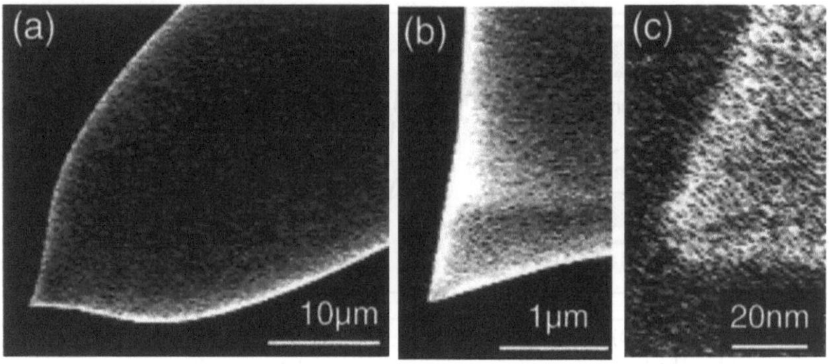

Fig. 3.29. SEM micrographs of **a** the tapered probe, **b** the magnified top region, and **c** magnified apex region. θ=65°; $d < 10$ nm

silica core have dissolution rates of R_1=6.6 μm h^{-1} and R_2=7.6 μm h^{-1}, respectively.

Next, by comparing Figs. 3.28b and 3.28c, we discuss the size of the reduced core at the taper end. In step B, etching is performed until the core is exposed from the cladding of fluoride-doped glass. Then, the cone angle θ is increased by increasing the etching time T. In our experiments, the cone angle were 35° and 65° for etching times of 15 and 30 min, respectively. On the other hand, once the core is exposed from fluorine-doped silica, the core is selectively etched due to the difference of the dissolution rate R_1 of the core and the dissolution rate R_2 of the cladding ($> R_1$). The fiber then has two cone angles of θ_{B1} and θ_{B2}, as shown in Fig. 3.28c, where d_B is the base diameter of the conical core, [6] with the angle θ_{B1} given by

$$\sin(\theta_{B1}/2) = (R_1/R_2)\sin(\theta_{B2}/2) \tag{3.31}$$

When the conical core is formed, the value of d_B is equal to the reduced core diameter $2r_{1E}$ of the fiber, as shown in Fig. 3.28b. For the tapered probe in Fig. 3.28a, we evaluated a reduced core diameter of 1 μm from a SEM image of selectively etched fibers which were immersed in BHF of 10:1:1 for times longer than 50 min. It is straightforward to obtain a minimum ratio of r_{1E}/r_1=10. To increase the confined optical power density at the core end, the diameter r_{1E} should be reduced to the wavelength size λ in the fiber. To fabricate a probe with a small core diameter of less than UV wavelength size (200 nm) in the fiber, one can use a new fiber with a core diameter smaller than 2 μm.

[6] In the actual case of using a PSF, the cross section of the apex region was elliptical, with a longer principal diameter of around 200 nm. Therefore, we could not employ the selectively etched PSF as a probe.

3.6.2 UV Triple-Tapered Probe

To fabricate a UV triple-tapered probe, we developed a multistep index fiber with the double core involving a subwavelength core and a pure silica core. In the following text, this fiber is called a double core fiber.

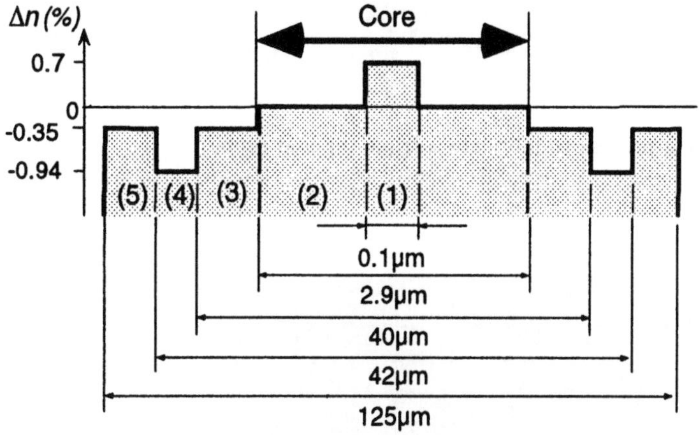

Fig. 3.30. Cross-sectional profile of the relative refractive index difference Δn of a multistep index fiber produced to fabricate the probe

3.6.2.1 Advanced Method Based on Hybrid Selective Etching of a Double Core Fiber

Figure 3.30 shows the cross-sectional profile of the relative refractive index difference Δn of a developed double-core fiber. The diameters of sections 1–5 are 100 nm, 2.9 μm, 40 μm, 42 μm, and 125 μm, respectively. The core consists of sections 1 and 2 made of GeO_2-doped silica and pure silica, respectively. The value of Δn of section 1 is 0.7%. Since the diameter of section 1 is only 0.1 μm, 99.9% cross-sectional area of the core is occupied by the pure silica of section 2. The cladding is made of fluorine-doped silica. Sections 3 and 5 are made of low-fluorine-doped silica with an index difference of −0.35%. Section 4 is made of the high fluorine-doped silica with an index difference of −0.94%. We fabricated a preform glass rod with a diameter of 25.5 mm by a combination of vapor-phase axial deposition [28] and plasma-activated chemical vapor deposition, and then drew the 125-μm-diameter fiber at a speed of 150 m min^{-1} by heating the preform to a temperature of 2106°C while maintaining the drawing tension of the fiber as high as 75 g.

Figure 3.31 shows a schematic explanation of the etching process for the fabrication of a triple-tapered fiber. Here, we denote a buffered hydrogen fluoride solution (BHF) with a volume ratio of NH_4F solution (40wt.%): HF acid (50wt.%): H_2O as X:1:1. The values of X in the three steps are defined as $X_A < 1.7$, $X_B = 1.7$, and $X_C > 1.7$. The process involves three steps: (A)

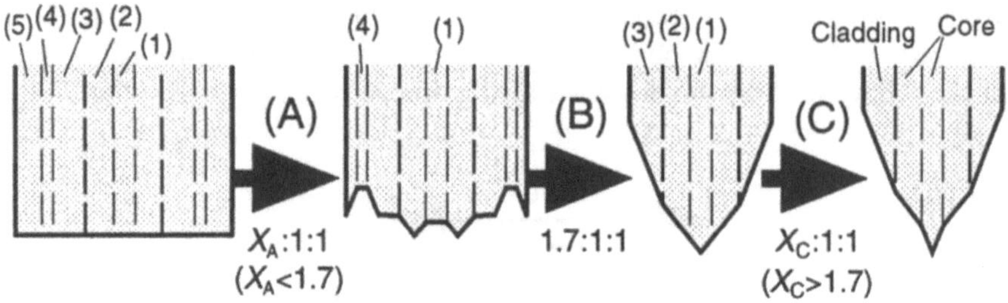

Fig. 3.31. Etching process to fabricate a triple-tapered probe

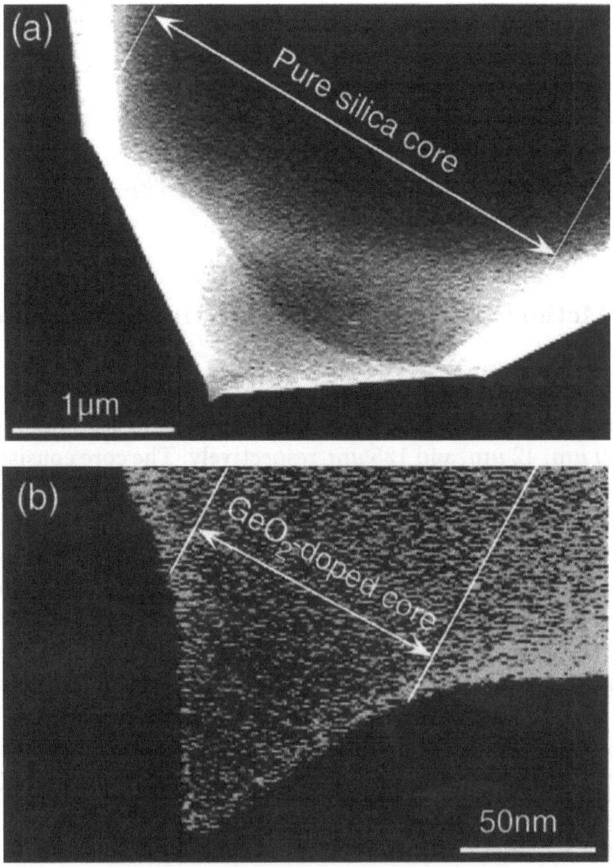

Fig. 3.32. SEM micrographs of **a** a triple-tapered probe and **b** its magnified apex region. $\theta_1=60°$; $\theta_2=120°$; $\theta_3=60°$; $d < 10$ nm

hollowing of sections 1 and 4 in X_A:1:1; (B) tapering of sections 2 and 3 in X_B:1:1; (C) sharpening of section 1 in X_C:1:1. The etching time in step A and X_A strongly affect both the total etching time and the diameter of the fabricated triple-tapered fiber. In step C, the cone angle θ_1 is controlled by varying X_C.

To obtain a triple-tapered probe with a cladding diameter of about 35 μm, the fiber in Fig. 3.30 was consecutively etched by X_A=0.6, X_B=1.7, and X_C=10 for 40 min, 20 min, and 10 min, respectively. All steps were performed at 25°C. Figure 3.32a and b show scanning electron micrographs of the triple-tapered probe and the apex region, respectively. The probe has an apex diameter of less than 10 nm (Fig. 3.32b). The values of the base diameters of d_{B1} and d_{B2} are 100 nm and 2 μm, respectively. The first cone angle θ_1, the second cone angle θ_2, and the third cone angle θ_3 are 60° (±1°), 120° (±1°), and 60° (±1°), respectively. After completing the etching process, we coated the triple-tapered probe with a 200-nm-thick aluminum film by vacuum evaporation.

3.6.2.2 Geometrical Model

We discuss the dissolution rates of a double-core fiber to tailor a triple-tapered probe. From now on, the dissolution rates of sections 1–5 in step i (where i=A, B, C) is defined as R_{1i}, R_{2i}, R_{3i}, R_{4i}, and R_{5i} ($= R_{3i}$), respectively. The dissolution rate of the fluorine-doped silica is a smaller than that of pure silica, and decreases with increasing doping ratio. Therefore, the dissolution rates satisfy the relation $R_{3i}/R_{2i} = R_{5i}/R_{2i} < R_{4i}/R_{2i}$. On the other hand, the ratio of GeO_2-doped silica to pure silica decreases with increasing values X of BHF, and converges to a constant value at X=10–30. At X=1.7, the ratio is equal to unity. Thus, for $X_A < 1.7$, $X_B = 1.7$, and $X_C > 1.7$, the dissolution rates satisfy the relations $R_{1A} > R_{2A} < R_{3A} = R_{5A} < R_{4A}$, $R_{1B} = R_{2B} < R_{3B} = R_{5B} < R_{4B}$, and $R_{1C} < R_{2C} < R_{3C} = R_{5C} < R_{4C}$, respectively. Based on our geometrical model as described in Sect. 3.3.2, the cone angles of θ_1, θ_2, and θ_3 can be expressed as

$$\sin(\theta_1/2) = R_{1C}/R_{3C} \qquad (3.32)$$

$$\sin(\theta_2/2) = R_{2i}/R_{3i} \qquad (i = \text{A, B, C}) \qquad (3.33)$$

and

$$\sin(\theta_3/2) = R_{3i}/R_{4i} \qquad (i = \text{A, B, C}) \qquad (3.34)$$

respectively. For the fiber in Fig. 3.30, the right-hand side of Eq. 3.32 is 0.5 at X=10, which decreases with increasing index differences. The right-hand sides of Eqs. 3.33 and 3.34 take constant values of 0.87 and 0.50 in all steps using different concentrations of BHF. To obtain a probe with the sharp angle of θ_1=20°, we fabricated a preform in which sections 1 and 2 had a value of Δn as large as 2.0%. However, the value of Δn of the drawn fiber was smaller, i.e., 0.7%. This decrease in Δn can be attributed to diffusion of GeO_2 occurred at the drawing of the fiber.

References

1. D. W. Pohl, W. Denk, M. Lanz, Appl. Phys. Lett. **44**, 651 (1984)
2. K. Lieberman, S. Harush, A. Lewis, R. Kopelman Science **247**, 59 (1990)
3. D. R. Turner, US Patent, 4,469,554 (1983); K. M. Takahashi, J. Colloid Interface Sci. **134**, 181 (1990)
4. T. Hartmann, R. Gatz, W. Wiegräbe, A. Kramer, A. Hillebrand, K. Lieberman, W. Baumeister, R. Guckenberger, in *Near field optics*, Vol. 242, NATO ASI series E, pp. 35–44 (Kluwer, Dordrecht,1993)
5. J. Radojewski, N. Stonik, H. Paginia, International J. Electron. **76**, 973 (1994)
6. P. Hoffmann, B. Dutoit, R.-P. Salathé, Ultramicroscopy **61**, 165 (1995)
7. S. Jiang, H. Ohsawa, K. Yamada, T. Pangaribuan, M. Ohtsu, K. Imai, A. Ikai, Jpn. J. Appl. Phys. **31**, 2282 (1992)
8. T. Pangaribuan, K. Yamada, S. Jiang, H. Ohsawa, M. Ohtsu, Jpn. J. Appl. Phys. **31**, L1302 (1992)
9. T. Pangaribuan, S. Jiang, M. Ohtsu, Electron. Lett. **29**, 1978 (1993); T. Pangaribuan, S. Jiang, M. Ohtsu, Scanning **16**, 362 (1994)
10. R. Umà Maheswari, S. Mononobe, M. Ohtsu, J. Lightwave Technol. **13**, 2308 (1995)
11. P. Tomanek, in *Near field optics*, Vol. 242, NATO ASI series E, pp. 87–96 (Kluwer, Dordrecht, 1993)
12. E. Betzig, in *Near field optics*, Vol 242, NATO ASI series E, pp. 7–15 (Kluwer, Dordrecht, 1993); E. Betzig, J. K. Trautman, Science **257**, 189 (1992)
13. G. A. Valaskovic, M. Holton, G. H. Morrison, Appl. Opt. **34**, 1215 (1995)
14. M. Garcia-Parajo, T. Tate, Y. Chen, Ultramicroscopy **61**, 155 (1995)
15. M. Spajor, A. Jalocha, in *Near field optics*, Vol. 242, NATO ASI series E, pp. 87–96 (Kluwer, Dordrecht, 1993)
16. S. Mononobe, T. Saiki, R. Uma Maheswari, T. Suzuki, S. Koshihara, H. Miyazaki, M. Ohtsu, Opt. Express
17. S. Mononobe, M. Naya, R. Uma Maheswari, T. Saiki, M. Ohtsu, in *Near field optics and related techniques*, Vol. 8, EOS Topical Meeting Digest series, pp. 105–106 (European Optical Society, Orsay, 1995)
18. S. Mononobe, M. Ohtsu, J. Lightwave Technol. **14**, 2231 (1996); Erratum J. Lightwave Technol. **15**, 162 (1997)
19. S. Mononobe, M. Ohtsu, J. Lightwave Technol. **15** 1051 (1997)
20. B. I. Yakobson, A. LaRosa, H. D. Hallen, M. A. Paesler, Ultramicroscopy **61**, 179 (1995); M. Paesler, P. Moyer, in *Near-field optics: theory, instrumentation, and applications*, pp. 46–53 (Wiley-Interscience, New York, 1996)
21. V. Kurpas, M. Libenson, G. Martsinovsky, Ultramicroscopy **61**, 187 (1995)
22. M. Stähelin, M. A. Bopp, G. Tarrach, A. J. Meixner, I. Zschokke-Gränacher, Appl. Phys. Lett. **68**, 2603 (1996)

23. S. Mononobe, M. Naya, T. Saiki, M. Ohtsu, Appl. Opt. **36**, 1496 (1997)
24. T. Saiki, S. Mononobe, M. Ohtsu, N. Saito, J. Kusano, Appl. Phys. Lett. **68**, 2612 (1996)
25. S. Mononobe, T. Saiki, T. Suzuki, S. Koshihara, M. Ohtsu, Opt. Commun. **126**, 45 (1998)
26. S. Mononobe, M. Ohtsu, IEEE Photon. Technol. Lett. **10**, 99 (1998)
27. S. Mononobe, R. Uma Maheswari, M. Ohtsu, Opt. Express **1**, 229 (1997), http://epubs.osa.org/opticsexpress/
28. T. Izawa, S. Sudo, in *Optical fibers: materials and fabrication* (KTK Scientific, Tokyo, 1987)
29. T. Matsumoto, M. Ohtsu, J. Lightwave Technol. **14**, 2224 (1996)

Chapter 4
High-Throughput Probes

4.1 Introduction

In general, a serious problem of the fiber probe is its low throughput (in the case of illumination-mode operation, the throughput is defined as the ratio of the output light power at the apex of the fiber probe to the light power coupled into the other end of the probe). The essential cause of the low throughput is the guiding loss along (or inside) the metallized tapered core. In order to study this loss mechanism and to realize high-throughput probes, we focus our discussion on the characteristics of the tapered core. Since the nanometric protruded part at the top of the fiber probe does not contribute to this loss, this chapter treats only the fiber probe without a protruded part, i.e., a flat-apertured probe. Its cross-sectional profile is shown in Fig. 4.1a, for which the foot diameter d_f of the protruded probe in Fig. 4.1b can be called the aperture diameter.

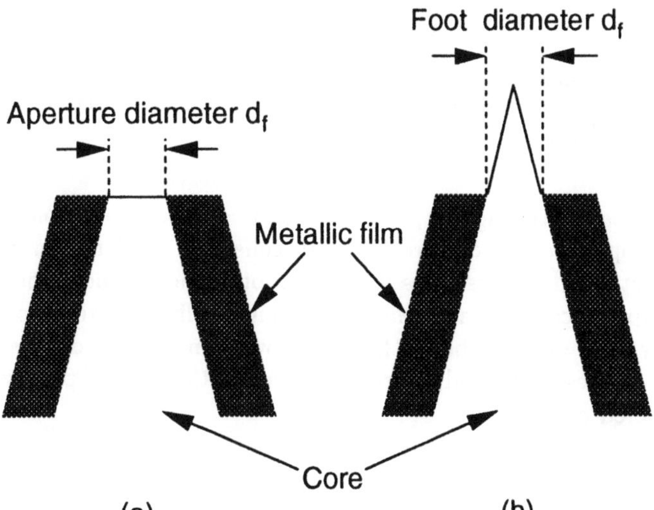

Fig. 4.1. Schematic comparison between **a** an apertured probe and **b** a protruded probe

As an example, Fig. 4.2 shows the measured relation between the aperture diameter d_f and the throughput of the probe in Fig. 4.1a. To measure the throughput, a diode laser light (830 nm wavelength) of 2 mW power is coupled into the fiber with a coupling efficiency of 2.5%. The output power is measured with a photodiode of 1 cm×1 cm active area which is located 1 mm away from the probe (see inset in Fig. 4.2). This figure shows that the throughput decreases rapidly with a decrease in the size of the probe [1, 2].

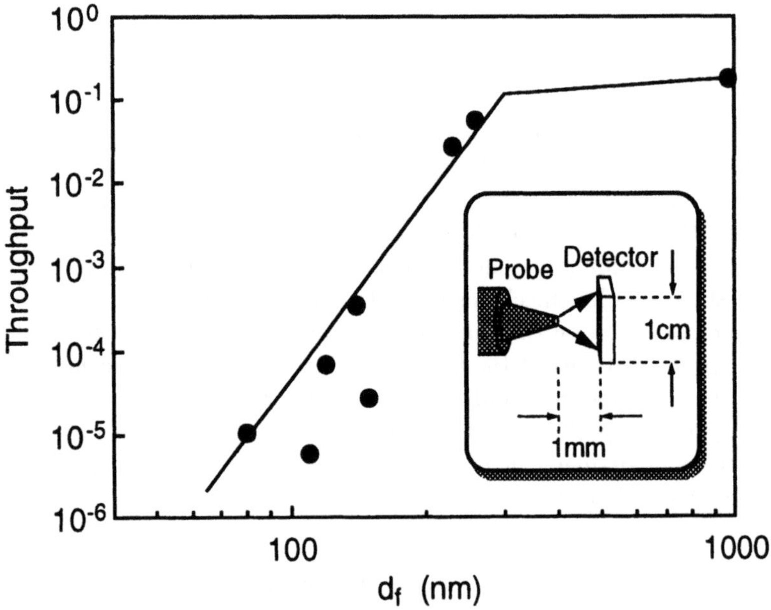

Fig. 4.2. Measured relation between the aperture diameter d_f and the throughput of a probe with a flat aperture at the top (see Fig. 4.1a). The inset shows schematically the method of measurement

In order to increase the throughput, i.e., to generate a strong optical near field under i-mode operation, and to realize high collection efficiency in c-mode operation, Sect. 4.2 describes a possible excitation of the plasmon mode in the metallized tapered core. We also demonstrate highly efficient excitation of the optical near field on a probe. In Sect. 4.3, we demonstrate double- and triple-tapered probes fabricated in order to increase the throughput by shortening the tapered core.

4.2 Excitation of the HE–Plasmon Mode

4.2.1 Mode Analysis

Mode analysis has been carried out by approximating a tapered core as a cylindrical core with a metal cladding. The result shows that this core can guide the light even if its diameter is smaller than half the wavelength [3]. Figure 4.3 shows the equivalent refractive indices of relevant modes (at 830 nm wavelength) derived by mode analysis for an infinitely thick gold coated core. Refractive indices of the glass and gold used for this derivation take real and complex values, which are 1.53 and $0.17+\imath 5.2$, respectively. Definitions of the HE_{11} and EH_{11} modes in this figure are based on those in Ref. [3].

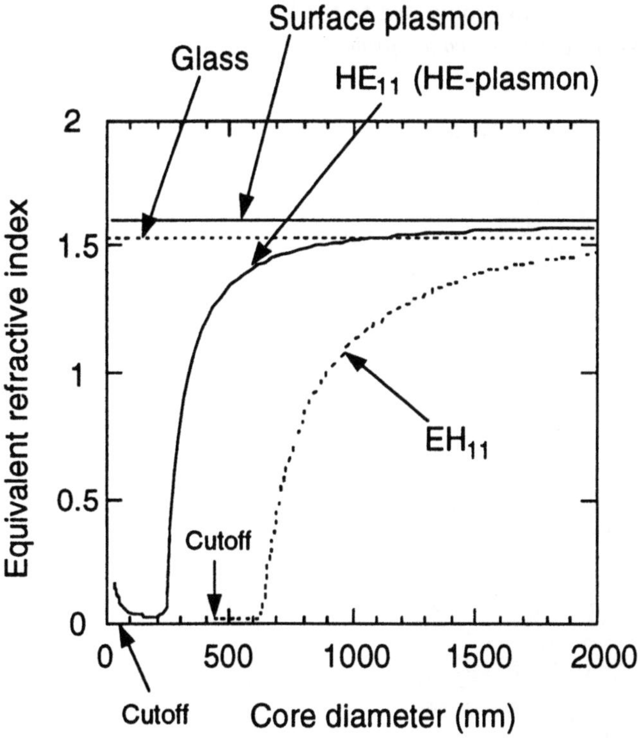

Fig. 4.3. Equivalent refractive indices of the relevant modes guided through a glass core with a gold metallic film

This figure shows that the cutoff core diameter of the HE_{11} mode is as small as 30 nm, while that of the EH_{11} mode is 450 nm. This means that only the HE_{11} mode can excite the optical near field efficiently when the apex size of the probe is below 100 nm. This figure also shows that the equivalent refractive index of the HE_{11} mode approaches that of a surface plasmon, and is between those of glass and gold. This means that the origin of the HE_{11} mode in a metallized core is the surface plasmon. Thus, we call the HE_{11} mode the *HE–plasmon mode* from now on. However, the HE–plasmon mode is not easily excited in the conventional core because its coupling efficiency with the lowest order optical fiber guided mode is very low due to mode mismatching between the HE_{11} and HE–plasmon modes at the foot of the tapered core.

4.2.2 Edged Probes for Exciting the HE–Plasmon Mode

An effective way of exciting the HE–plasmon mode is to utilize the coupling of the plasmon by scattering at the edge of the metal [4]. If the tapered core has a sharp edge at its foot, part of the guided light inside the single-mode fiber can be scattered at this edge and converted to the HE–plasmon mode [5]. We call the probe with such a core an *edged probe*.

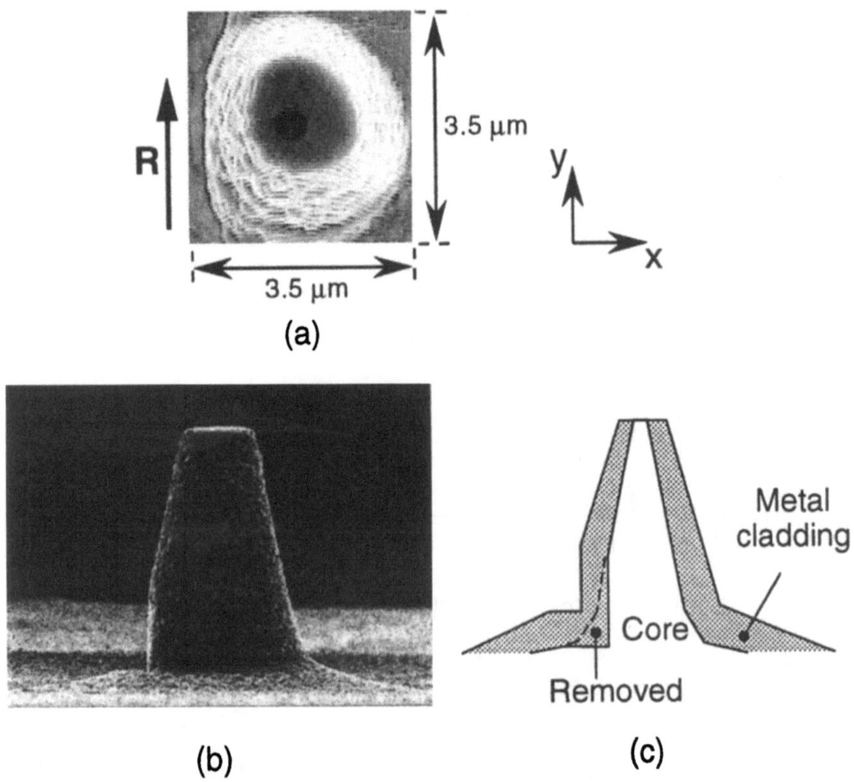

Fig. 4.4. Scanning electron micrographs of an edged probe with a flat aperture at the top. **a** Top view. **b** Side view. The arrow R in **a** indicates the direction parallel to the surface from which part of the core was removed. The x and y axes are the directions normal and parallel to the arrow R, respectively

Figure 4.4 shows scanning electron micrographs of a fabricated edged probe with a flat aperture at the top (the aperture diameter d_f is 500 nm). The arrow R in Fig. 4.4a indicates the direction parallel to the surface from which a part of the core was removed. The x and y axes are the directions normal and parallel to the arrow R, respectively. The white lines in Fig. 4.4c represents the profile of the tapered core buried in a gold metallic film. A part of the foot of the core was removed to form a sharp edge, where the height of the part removed was 1.5 μm. This probe was fabricated in the following steps.

1. A GeO_2-doped fiber core was tapered by selective chemical etching [6] to realize a cone angle of 20°.
2. The foot of the tapered core was removed by using a focused ion beam (FIB) to form a sharp edge.
3. The core was coated with a 500-nm-thick gold film.
4. The gold film was removed from the top of the core by using a FIB in order to form a planar aperture.

Probes with an aperture diameter, d_f, as small as 30 nm have been realized by these methods. Figure 4.5 shows the throughput of these probes, which were measured by the same method as in Fig. 4.2. Comparison with Fig. 4.2 confirms the realization of a high throughput by the edged probe. The spatial distribution of the output light power was also measured by using the experimental set-up in Fig. 4.6. A linearly polarized light from a diode laser of 830 nm wavelength was coupled into a single-mode fiber probe. To examine the polarization dependency of the distribution, the direction of the incident light polarization was varied by a half-wave plate. The light power on the edged probe (probe A in this figure) was measured by scanning with another fiber probe (probe B in this figure, which is a conventional axially symmetric fiber probe). Separation between the two probes was regulated to several nanometers by using a shear-force technique (see Chap. 6 for details).

We compared the output light power of the edged and conventional fiber probes. A protruded fiber probe [7] with an apex diameter and foot diameter of 10 nm and 60 nm, respectively, was used as probe B. Figure 4.7a and b show the calculated spatial distributions of the EH_{11} and HE–plasmon mode powers, respectively, for an edged probe with d_f=500 nm. The vectors $\boldsymbol{E}$ in these figures represent the direction of the incident light polarization. Figure 4.7c and d show the measured results, for which the directions of the incident light polarization are orthogonal to each other. Agreements of Fig. 4.7a and b with Fig. 4.7c and d, respectively, can be clearly seen. As a reference, the spatial distribution for an axially symmetric fiber probe with d_f=500 nm (fabricated without step 2) was also measured. It had a double-peaked profile, corresponding to the EH_{11} mode, and the positions of the peaks moved when the direction of the incident light polarization was varied. These results indicate that the edge at the foot of the tapered core successfully

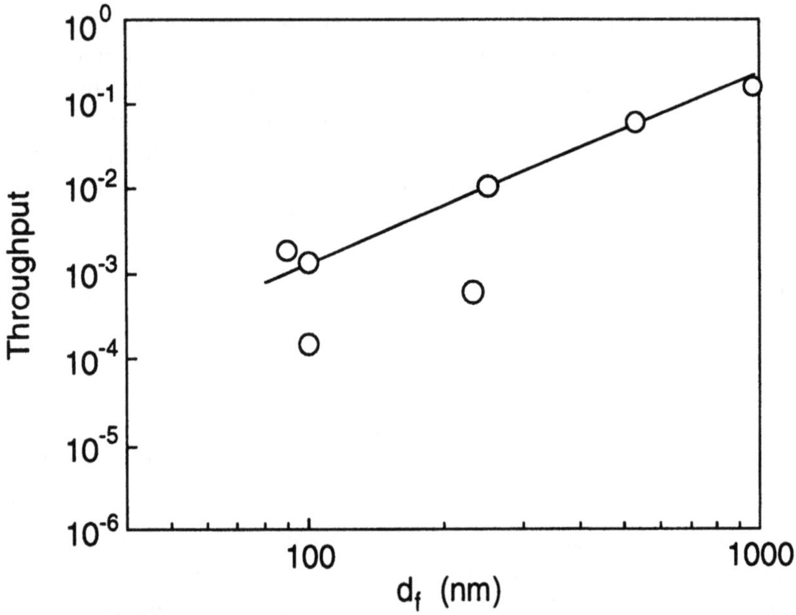

Fig. 4.5. Measured relation between the aperture diameter d_f and the throughput of an edged probe

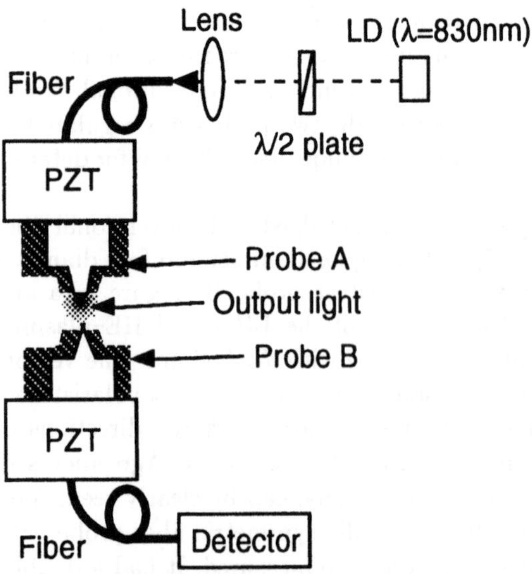

Fig. 4.6. Experimental set-up to measure the spatial distribution of the output light power on probe A. Probe B is used for scattering and detection

excites the HE–plasmon mode, and the excitation efficiency depends on the direction of the incident light polarization.

In order to check whether the output light power on the aperture has been enhanced due to the edged structure, its spatial distribution on the probes with and without the edge were compared for d_f=100 nm. Note that the EH_{11} mode is not guided inside the probe with d_f=100 nm because it is smaller than its cutoff diameter (450 nm). A probe with a sharpened core with a 30-nm-thick gold coating was used as the probe B in order to increasing the scattering efficiency for sensitive detection [8]. Figure 4.8 shows the cross-sectional profiles of the measured distributions. Curves A and B are for edged probes, with the directions of the incident light polarization being orthogonal to each other. Curve C is for a probe without an edge. One can see that values of curve A is ten times larger than that of curve C. The full width at the half-maximum of curve A is 150 nm, which is comparable to that of the HE–plasmon mode (=120 nm) for d_f=100 nm estimated by the mode analysis. These results indicate the effective excitation of the HE–plasmon mode by the edge at the foot of the tapered core.

4.3 Multiple-Tapered Probes

Although the technique described in the previous section can be applied to probes with d_f>30 nm, it should be noted that a further increase in throughput requires some tailoring of the probe structure. This is because the guiding loss in the metallized tapered core is still high even if the HE–plasmon mode is excited. Since the guiding loss in the core is due to the loss in the metal cladding, the easiest way to decrease the loss is to shorten the length of the tapered core.

4.3.1 Double-Tapered Probe

In order to shorten the enormous absorption region, chemical etching is a powerful technique because it allows reliable design and fabrication of an optimum structure for the tapered core. First in this subsection we demonstrate the strong dependence of the throughput on the cone angle θ and hence on the resultant length l of a tapered core. For probes with different cone angles, throughput is evaluated for a wide range of aperture diameters (80 nm$\leq d_f \leq$900 nm). On the basis of these results, we optimize the shape of the probe, taking into account the experimental utilities [9]; for example, by increasing the length of the tapered core, we can avoid contact between the cladding and the bumpy surface structure of a sample.

We briefly describe the fabrication technique of the probe by a selective etching process, whose details have been described in Chap. 3. The cone angle θ can be controlled by the buffering conditions of the etching solution, which

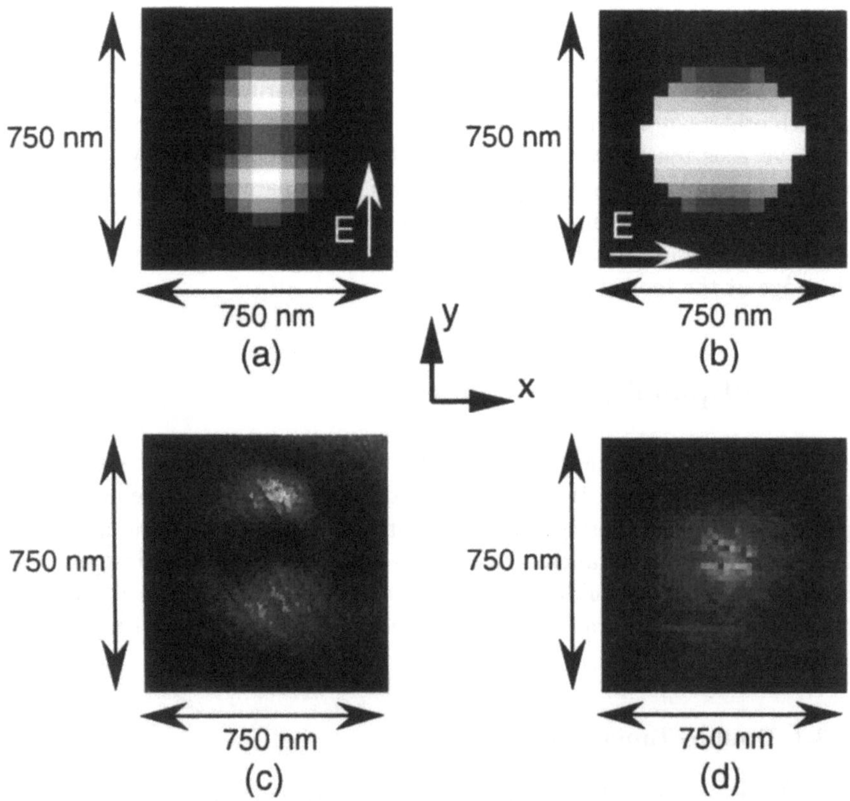

Fig. 4.7. Spatial distribution of the output light power on the top of an edged probe with d_f=500 nm. **a** Calculated result of the EH_{11} mode. **b** Calculated result of the HE–plasmon mode. The vector $\boldsymbol{E}$ in these figures represents the direction of polarization of the incident light. **c, d** Measured results. The direction of polarization of the incident light for these figures are orthogonal to each other

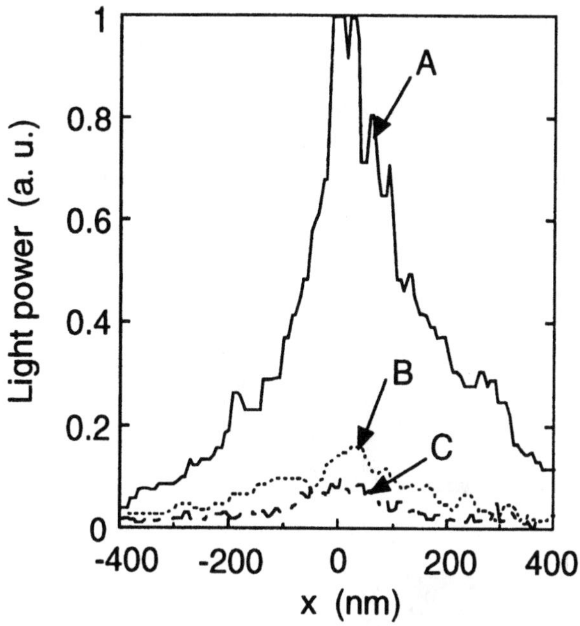

Fig. 4.8. Measured cross-sectional profiles of the output light power. Curves A and B are for the edged probe, where the directions of polarization of the incident light are orthogonal to each other. Curve C is for the probe without an edge

is adjusted by the volume ratio X of NH_4F, maintaining that of HF to H_2O at 1:1. Here, the composition of the solution is expressed as X:1:1. In order to evaluate the dependency of throughput on the cone angle, two types of probe are prepared using a one-step etching technique with a solution of X=10 (probe A) and X=2.7 (probe B). The cone angle θ and the length l of the tapered core of probe A are 20° and 6 μm, respectively, and those of probe B are 50° and 2.5 μm, respectively.

Using a sputter-coating method, the exterior surface of the etched core is coated with a gold film of 200 nm thickness, which is seven times the skin depth at 633 nm wavelength. For the fabrication of a small aperture, the selective resin coating (SRC) method is employed (see Chap. 3). Acrylic resin dissolved in an organic solution is coated as a guard layer over the sides of the metallized tapered core, leaving the top of the core free of the resin. Here, the surface tension of the resin is used. On removing the gold film from the top of the core with the commonly used KI–I_2 etching solution, a small protruding-type aperture (with the foot or aperture diameter d_f, c.f. Fig. 4.1) is fabricated. The aperture diameter d_f can be varied by controlling the etching time in the KI–I_2 etching solution and the coating condition, which can be adjusted by changing the cladding diameter and the viscosity of the resin solution.

To estimate the throughput, the light from a He–Ne laser (633 nm wavelength) of 130 μW power is coupled into the fiber probe with a coupling efficiency of 60%. The far-field light ejected from the aperture is collected with a 0.4 numerical aperture (NA) objective lens (see inset in Fig. 4.9). The output light power is measured with an optical power meter. The geometrical size of the aperture is estimated with a scanning electron microscope after the throughput measurement.

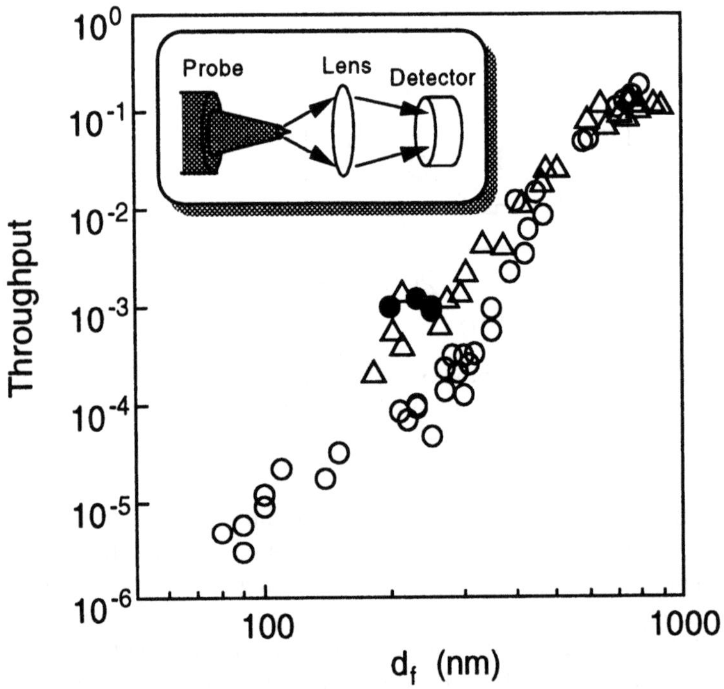

Fig. 4.9. Measured relations between the aperture diameter d_f and the throughput for probe A with a single tapered core with cone angle 20° (*open circles*), probe B with a single tapered core with cone angle 50° (*open triangles*), and a double-tapered probe (*closed circles*)

Figure 4.9 shows the measured value of the throughput as a function of the aperture diameter d_f for probes A (open circles) and B (open triangles) of different cone angles. When the diameter d_f is larger than the wavelength of He–Ne laser light in glass, λ_c (~400 nm), the throughput of probes A and B take almost the same values when the aperture diameters are the same. This result means that the efficiency of delivering light into the region $d_f > \lambda_c$ is not so strongly dependent on the cone angle θ and the length l of the tapered core, which is 5 μm long at most. On the other hand, in the region $d_f < \lambda_c$, the difference between the two types of probes is remarkable due to the influence of the metal coating. When d_f=200 nm, the throughput of probe A is lowered by ten times to that of probe B. Although the throughput is dependent on the spatial mode characteristics in the tapered core, the difference in guiding lengths between probes A and B can also contribute considerably to the throughput. The guiding lengths between core diameters of 400 nm (=λ_c) and 200 nm (=d_f) are 570 nm for probe A and 210 nm for probe B (see Fig. 4.10a and b). It is theoretically estimated that, in the case of the EH_{11} mode, the absorption of light by the metallic film coating drastically influences the region where the core diameter is smaller than λ_c. In such a strong loss region, a difference in the guiding length of as much as 360 nm results in a decline of the throughput by one order of magnitude. We can conclude that, in order to increase the throughput, it is reasonable to shorten the strong loss region in the tapered core by increasing the cone angle.

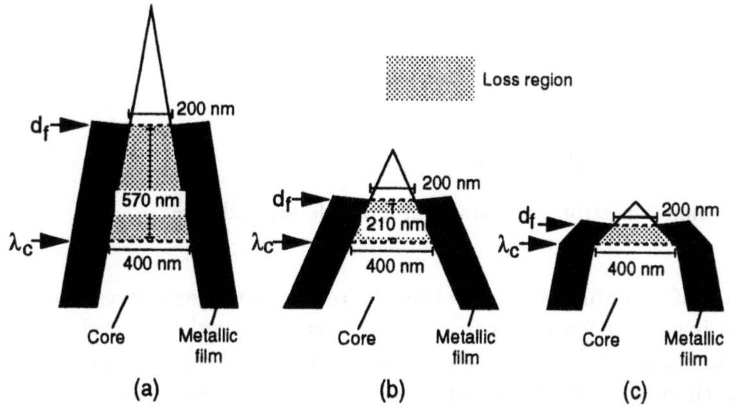

Fig. 4.10. Schematic diagrams of the cross-sectional profiles of probe A, with a single tapered core with cone angle of 20°, probe B, with a single tapered core with cone angle of 50°, and a double-tapered probe. Meshed area represents the loss region

However, a short tapered core with a large cone angle leads to contact between its cladding and the bumpy surface of the sample during the actual scanning operation. To avoid this inconvenience, it is necessary to lengthen the tapered core while maintaining a high throughput. For this purpose, we have successfully developed a two-step etching process for the fabrication of a double-tapered probe (Fig. 4.10c) with high reproducibility as detailed in Sect. 3.3.1. In the first step, using a solution with a composition of X=1.8, a short tapered core with a large cone angle of 150° is fabricated. Second, a long tapered region, which delivers light to the cutoff diameter, is obtained with a solution having X=10. As shown in scanning electron micrograph in Fig. 4.11, the resultant cone angle is about 90°. The fabrication technique for the aperture is the same as that described above. We again measured the throughput of some of the fabricated double-tapered probes with d_f=200 nm. These results are also plotted in Fig. 4.9 (closed circles). A throughput ten times higher than that of probe A (open circles) was achieved. This comparison implies that one of the most important factors in determining the throughput is the length of the loss region in the narrow tapered core.

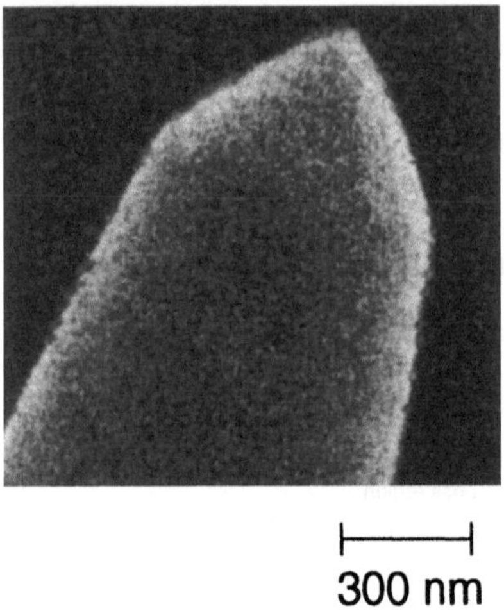

Fig. 4.11. Scanning electron micrograph of a double-tapered core

In the case of a double-tapered probe, the realization of apertures smaller than 100 nm is not yet feasible owing to the large cone angle required. However, owing to the limitations in measurement sensitivity, high throughput is more crucial than realizing a very small aperture for the detection of a weak signal in the advanced spectroscopy of semiconductors. The capabilities of double-tapered probes in such experiments are demonstrated in Chap. 9.

4.3.2 Triple-Tapered Probe

We have shown in the previous section that by using a double-tapered structure in the probe, the throughput can be increased more than ten times. However, a further problem with the double-tapered probe is the deterioration of resolution caused by the larger apex diameter of the double-tapered core owing to its large cone angle. In order to solve this problem, the triple-tapered probe shown in Fig. 4.12 has been proposed and fabricated. The diagram in the bottom right-hand corner of this figure shows the result of this fabrication, which is a probe with a very sharp third taper on top of the double-tapered core. Owing to this sharp taper, the generation and localization of a very high spatial Fourier frequency component of the optical near field is expected. Its fabrication consists of five steps. (Another fabrication technique for a triple-tapered probe is described in Sect. 3.5.1.)

1. Chemical etching to sharpen the core and reduce the cladding diameter.
2. Irradiating the FIB to form a very sharp tapered core on a flat floor, which is used as the third taper.
3. Chemical etching to form the first and second tapers at the foot of the third taper.
4. Coating with a gold film.
5. Removing the gold film from the top of the third taper in order to form an aperture.

The chemical etching for steps 1 and 3 are the established techniques described in Chap. 3. Step 5 was used to form a flat aperture in order to compare the throughput with that of the other apertured probes described in previous sections. However, it is also possible to maintain the top of the third taper after removing the gold film with a KI–I_2 etching solution in order to obtain a protruded triple-tapered probe, which can realize a very high resolution while maintaining a high throughput.

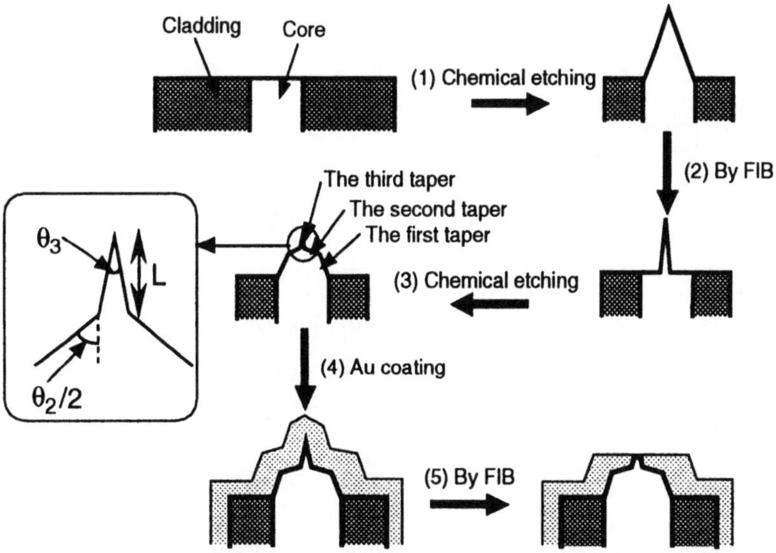

Fig. 4.12. Schematic diagram of the fabrication of a triple-tapered probe with an aperture at a flat top

Figure 4.13a and b show scanning electron micrographs of the results of step 2, which are obtained by using a Ga^+ ion beam of 30 nm and 1 μm diameter, respectively. A comparison between the two figures suggests that the 1-μm-diameter ion beam is as effective as the 30-nm ion beam in forming the third taper, and that an expensive FIB machine for tight focusing may not be necessary. Figure 4.13c and d show scanning electron micrographs of the triple-tapered core formed by step 3. A schematic illustration of a triple-

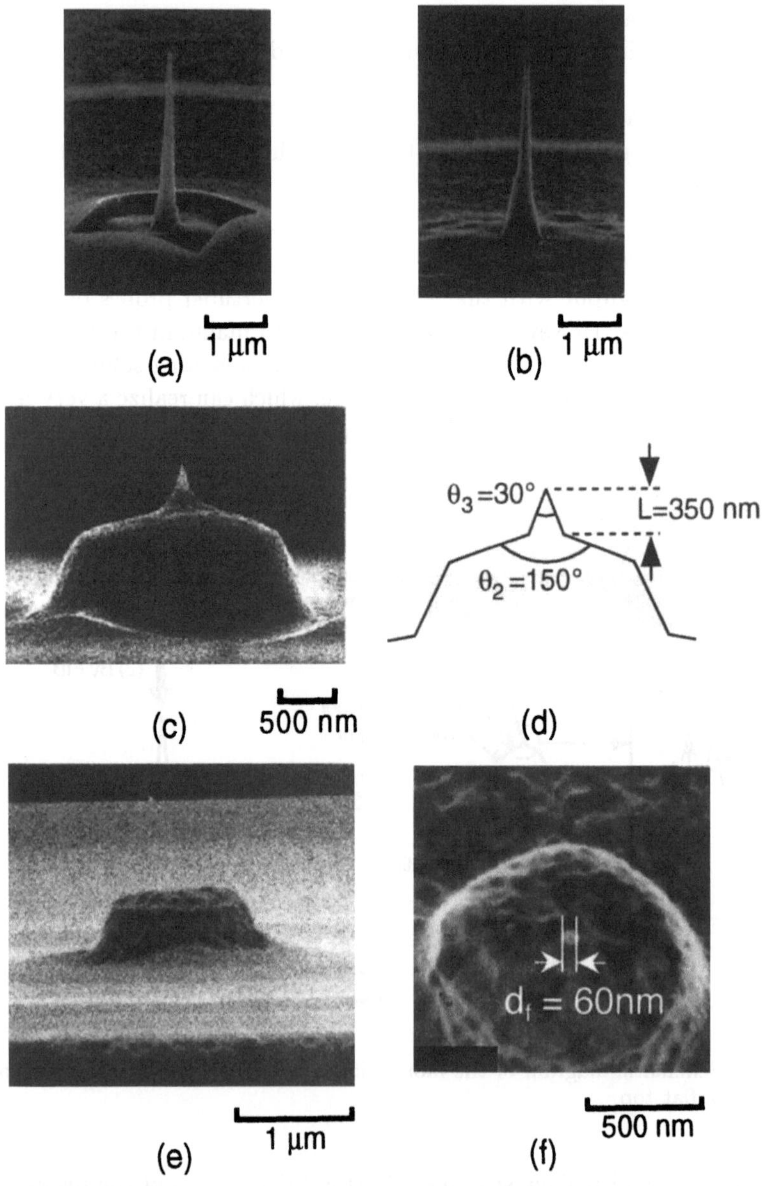

Fig. 4.13. Scanning electron micrographs of a triple-tapered probe. **a**, **b** Results of step 2 obtained by using a 30-nm- and 1-μm-diameter Ga^+ ion beam, respectively. **c**, **d** Profile of the triple-tapered core formed by step 3 and a magnified image of the third taper. **e**, **f** Side view and magnified top view of the probe after steps 4 and 5.

tapered core is given in the inset in Fig. 4.12, where θ_2 and θ_3 represent the cone angles of the second and third tapers, respectively. The length of the third taper is represented by L. Although Fig. 4.13c shows θ_2=150°, the value of θ_2 realized by this step can be controlled within the range 90–150° with an error of ±5°. Figure 4.13d is a magnified picture of the third taper, with θ_2=30° and L=350 nm and a fabrication error of 3° and 30 nm, respectively. It also shows the nanometric apex diameter of the third taper. Figure 4.13e and f show the side and magnified top views of the probe, respectively, as fabricated by steps 4 and 5. The value of d_f is 60 nm and the thickness of the gold film is 300 nm.

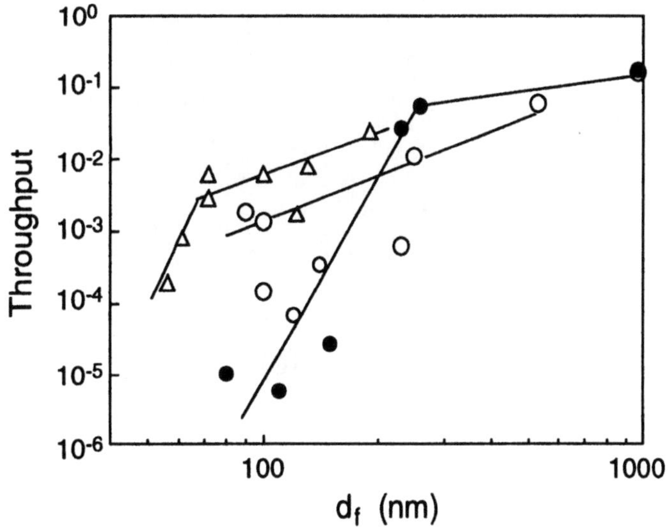

Fig. 4.14. Measured relations between the aperture diameter d_f and the throughput. *Closed circles*, *open circles*, and *open triangles* represent the values for a conventional probe, the edged probe in Sect. 4.2.2, and a triple-tapered probe, respectively

Figure 4.14 shows the measured results of the dependency of the throughput on d_f. The closed circles, open circles, and open triangles represent the values for the conventional probe, the edged probe in Sect. 4.2.2, and the triple-tapered probe, respectively. This figure confirms that the triple-tapered metallized fiber probe has a throughput 1000 times higher than that of a conventional probe for 60 nm$\leq d_f \leq$100 nm. In other words, a throughput as high as 1×10^{-4}–1×10^{-2} has been realized over this range of values for d_f.

The realization of such an extremely high throughput happens for several reasons.

1. The overall profile of the first and second tapers is similar to a convex lens, which allows the guided light to be focused from the fiber to the foot of the third taper.

2. The third taper is much shorter than the total length of a conventional fiber probe, which reduces the guiding loss.
3. Figure 4.14 also shows that the relation between d_f and the throughput of the triple-tapered probe is similar to that of the edged probe for 80 nm$\leq d_f \leq$300 nm. This similarity reveals the possible excitation of the HE–plasmon mode even in the triple-tapered probe. This is attributed to the edge formed between the second and third tapers due to the drastic change in the cone angle.

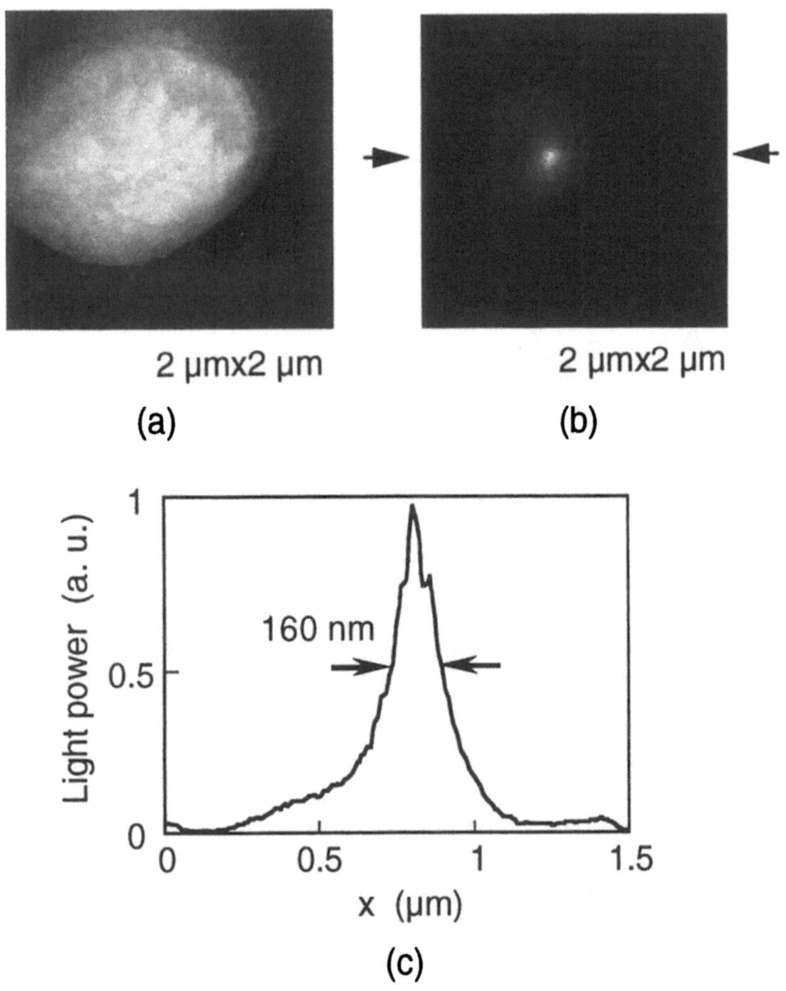

Fig. 4.15. Measured spatial distribution of the output light power on the planar aperture of a triple-tapered probe with d_f=60 nm. **a** Topographic image of the aperture surface measured simultaneously by a shear-force technique. **b** Two-dimensional profile of the distribution. **c** Cross-sectional profile of **b**

Figure 4.15a shows the measured spatial distribution of the output light power on the planar aperture (d_f=60 nm) at the top of a triple-tapered probe. The experimental set-up for this measurement was similar to that used in Fig. 4.6. Figure 4.15b shows a topographic image of the aperture surface measured simultaneously by the shear-force technique. By comparing these two figures, the generation of light at the center of the aperture is confirmed. Figure 4.15c shows the cross-sectional profile of the distribution derived from Fig. 4.15a. Its full width at half-maximum is 160 nm, which is the convolution between the aperture diameter (60 nm) and the size of probe B in Fig. 4.6. A remarkably sharp peak at the center of this curve is attributed to localization of the optical near field by the third taper on the top of the core.

References

1. U. Dürig, D. W. Pohl, F. Rohner, J. Appl. Phys. **59**, 3318 (1986)
2. L. Novotony, D. W. Pohl, B. Hecht, Opt. Lett. **20**, 970 (1995)
3. L. Novotony, C. Hafner, Phys. Rev. E **50**, 4094 (1994)
4. D. Marcuse, *Light Transmission Optics*, Chap. 10 (Van Nostrand Reinhold Company, New York, 1972)
5. T. Yatsui, M. Kourogi, M. Ohtsu, Appl. Phys. Lett. **71**, 1756 (1997)
6. T. Pangaribun, K. Yamada, S. Jiang, H. Ohsawa, M. Ohtsu, Jpn. J. Appl. Phys. **31**, L1302 (1992)
7. S. Mononobe, M. Naya, T. Saiki, M. Ohtsu, Appl. Optics. **36**, 1496 (1997)
8. Y. Inoue, S. Kawata, Opt. Lett. **19**, 159 (1994)
9. T. Saiki, S. Mononobe, M. Ohtsu, N. Saito, J. Kusano, Appl. Phys. Lett. **68**, 2612 (1996)

Chapter 5

Functional Probes

5.1 Introduction

This chapter describes functional probes which have a functional material on the tip of an optical fiber probe, especially those fabricated by a new fixation method proposed here. Figure 5.1 depicts a functional probe constructed from a metal-coated fiber probe. The size of the functional material is micro- or submicro-meter, which is equal to or less than an optical wavelength. However, it is worth noting that nanometric size is expected in the future with improved technology to be fully compatible with near-field optics.

Comparison between functional probes and conventional probes described in Chaps. 3 and 4 allows one to recognize the significance and applications of functional probes. Functional probes can be regarded as optical fiber sensors directed into miniaturization. Appropriate applications of functional probes are those which cannot be achieved by conventional probes, such as energy transfer, optical nonlinearity and chemical/biological sensing. Local investigation of conditions and environments as well as the optical field, including the optical near field, is possible.

Pioneering studies of functional probes have been carried out by Kopelman and co-workers [1–11]. By using a photoinitiated polymerization technique as the fixation method, submicrometer chemical sensors were demonstrated with metal-coated optical fiber probes produced by a heat-pulling pro-

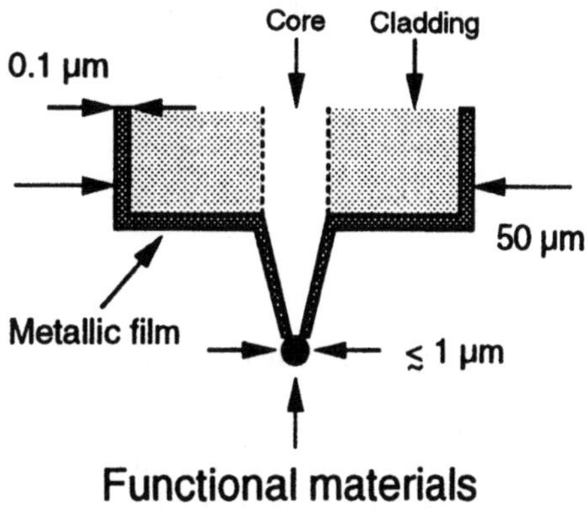

Fig. 5.1. Schematic diagram of a functional probe. A functional material is selectively attached to the tip of the probe

cess [1, 2]. In preliminary experiments using capillary-type probes, functional probes with a point light source and high throughput were reported [3–5].

Methods of fabricating the functional probes are considered in terms of fixation method, material selection, and practical use, which are closely correlated with each other. Since fixation methods are so undeveloped that they restrict material selection, practical use is the last consideration. In the next section, we describe a newly developed fixation method using a micropipette.

5.2 Methods of Fixation

The basic requirement for a functional probe is the selective fixation of a functional material on the probe tip, because selective fixation makes it simple to interpret the optical response of the probe. With a nonselective fixation method such as the dip-coating method, the whole surface of the probe is covered by the functional material. The resulting functional probe cannot indicate a local response because of diffusion effects in the functional materials, such as Brownian motion or carrier propagation.

A suitable method for selective fixation, proposed by Kopelman and coworkers, is photoinitiated polymerization due to UV light coupled at the other side of the probe. The size of a fixed polymer is determined by the aperture diameter of the metal-coated probe [1, 2]. Although the fixation method has excellent selectivity, it means that a limited number of materials can be fixed to photoinitiated polymers.

We propose a fixation method which uses a micropipette with a micromanipulator system in order selectively to fix the functional material onto the tip of the fiber probe. This newly developed fixation method enables one to attach various different kinds of material. Figure 5.2 shows three types of tip shape for the functional probes, which can all be constructed by the micropipette fixation method. In addition, this method can be applied to heat-pulled probes as well as the chemically etched probes.

Figure 5.3 shows a schematic diagram of the micromanipulator system, which consists of two electrically controlled micromanipulators and a conven-

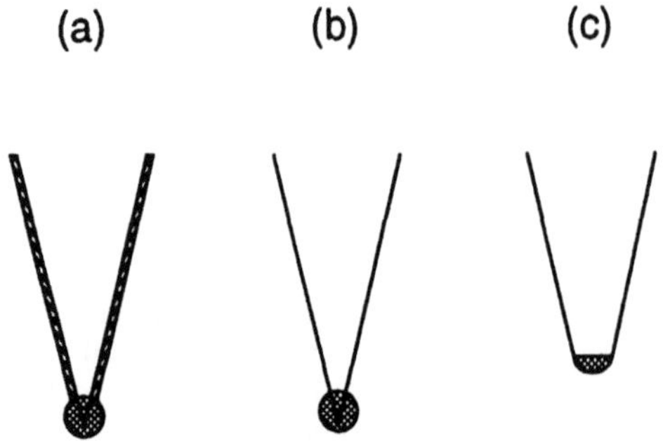

Fig. 5.2. Schematic diagram of tip shapes of functional probes constructed from **a** metal-coated fiber probe, **b** a bare fiber probe and **c** flattened top fiber probe

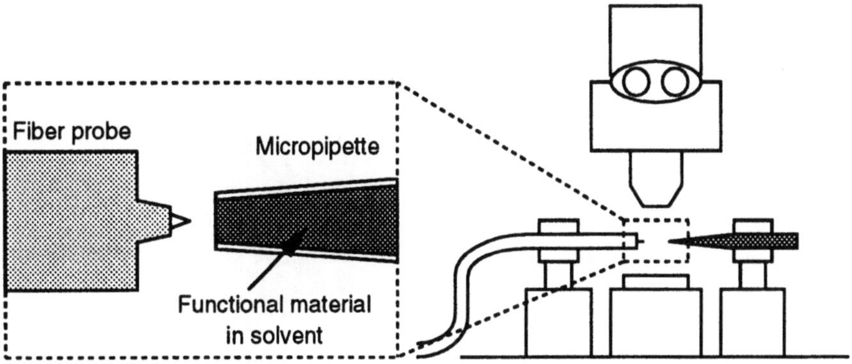

Fig. 5.3. Schematic diagram of the micromanipulator system used for the micropipette fixation method. The system consists of two micromanipulators and an optical microscope. Fixation is performed under observation through the optical microscope

tional optical microscope. A fiber probe is held on the holder of one micromanipulator while a micropipette is set on the holder of the other. Figure 5.4 illustrates a typical fixation procedure for the functional probes in Fig. 5.2a and b. The functional material adheres to the glass surface of the probe tip due to adsorption. Fixation sizes are controllable in the range of 1 to 3 μm. The upper limit of the fixation size can be expanded to more than 10 μm when heat-pulled probes are used. On the other hand, the lower limit of the fixation size can be decreased to less than 0.3 μm when the flattened top probe shown in Fig. 5.2c is used. The functional material is fixed by colliding the tip of the micropipette with the flattened top of the probe, as shown in Fig. 5.5a–c. Fixation size is limited not by the diffraction limit, but by the tip size of the micropipette. Figure 5.5d shows a fabricated functional probe with a fixation size less than the optical wavelength. The functional material

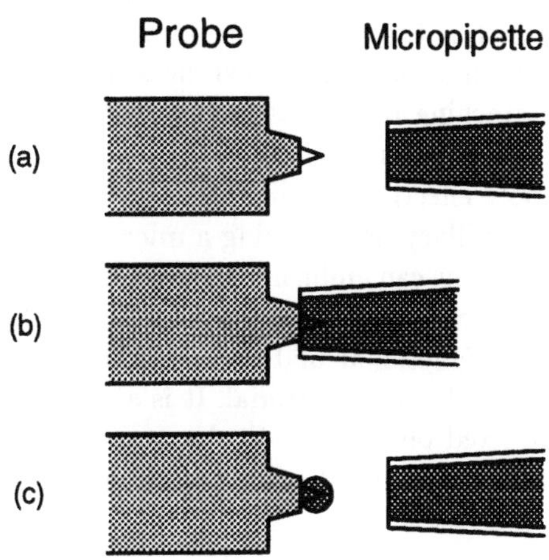

Fig. 5.4. Schematic illustration of the process for attaching a functional material to a probe tip by the micropipette fixation method. **a** The micropipette and the probe are aligned coaxially. **b** The micropipette is made to approach the probe until the probe tip is immersed in the functional material. **c** The probe with the functional material is removed from the micropipette

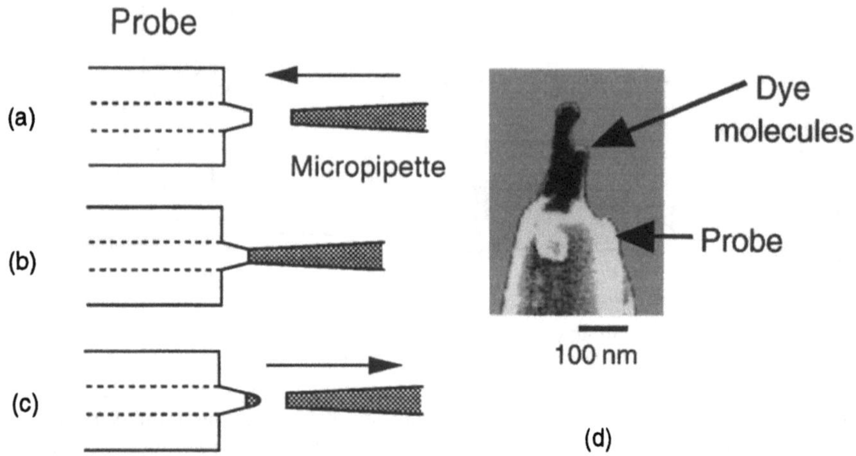

Fig. 5.5. Schematic illustration of the fixation process, and a SEM micrograph of a flattened-top probe. **a** The micropipette and the probe are aligned coaxially. **b** The micropipette is made to approach the probe until the tip of the micropipette collides with the flattened top of the probe. **c** The probe with the functional material is removed from the micropipette. **d** SEM micrograph of a functional probe tip with rhodamine 6G dye molecules of a total size less than the optical wavelength

is rhodamine 6G dye molecules, which are dissolved in an ethanol solvent, injected into a micropipette, and fixed by the micropipette fixation method.

5.3 Selecting a Functional Material

Since the fixation method selects the material to immobilize on the probe tip, we have to consider the requirements for the materials imposed by the micropipette fixation method. First, the materials must be liquid or dissolved in solvents in order to fill a micropipette. Second, the materials must be fixed on the probe surface by adsorption, which is the result of the intermolecular force between the glass surface and the matter. Polar substances are more appropriate than nonpolar ones for attaching to a glass surface with a polarity which is the same as that of the silanol group (Si–OH), because then the dipole-dipole interaction works more effectively. Third, the capillary effects (e.g., capillary condensation and capillary rise) of using a micropipettes should be taken into account, because they can influence chemical properties such as viscosity, concentration, and components of the material in the micropipette. Changing the viscosity of functional materials by the use of solvents can control the size and shape of the fixed material. It is also worth considering that fine particles can be fixed on a probe tip by means of an aggregation of colloidal particles dissolved in volatile solvents at the tip of the micropipette.

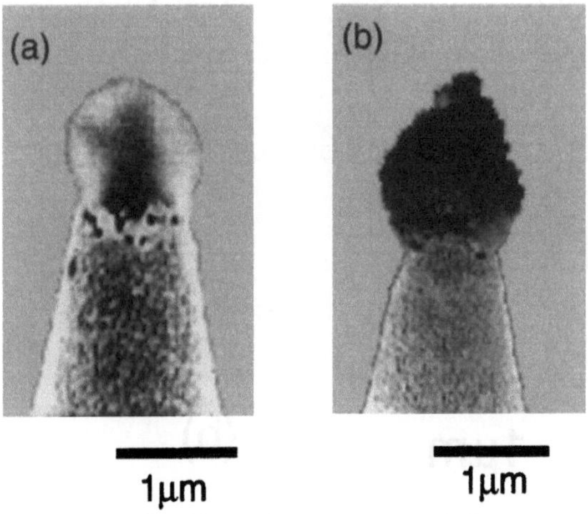

Fig. 5.6. SEM micrographs of functional probes. The functional materials are a PDA polymers and b GaAs fine particles

Figure 5.6a and b show examples of functional probes [12]. In Fig. 5.6a, the polymer is polydiacetylene (PDA) 3ECMU, an optical nonlinear material. Its solvent is 1,1,2,2-tetrachloroethane. In Fig. 5.6b, the fine particles are GaAs powder, which is prepared by crushing a GaAs wafer in a mortar for about an hour. GaAs powder in an ethanol solvent behaves like colloidal particles.

5.4 Probe Characteristics and Applications

This section describes the characteristics of functional probes and their possible applications, such as a point light source and a chemical sensor.

5.4.1 Dye-Fixed Probes

A dye-fixed probe is excellent for practical use in efficient fluorescence although it has the problem of photobleaching. It has promising application as light-emitting probe for use as a point light source. A more attractive application is a energy-transfer probe, which can achieve high spatial resolution using energy transfer from the donor dyes of a sample to the acceptor dyes of the probe. In this subsection, light-emitting probes are demonstrated with dye-fixed probes.

A dye-fixed probe of subwavelength size works as a point light source. With this probe, near-field optical microsopy measurements have been performed in collection mode [13]. A sample of a compact disc (CD) with a grating structure of 2 μm, a pit diameter of 0.3 μm and pit space of 0.1 μm is illuminated under total reflection by the beam of a 488 nm cw Ar^+ laser. The dye molecules on the probe are excited by an optical near field generated on the CD surface, and radiate fluorescence of around 600 nm wavelength.

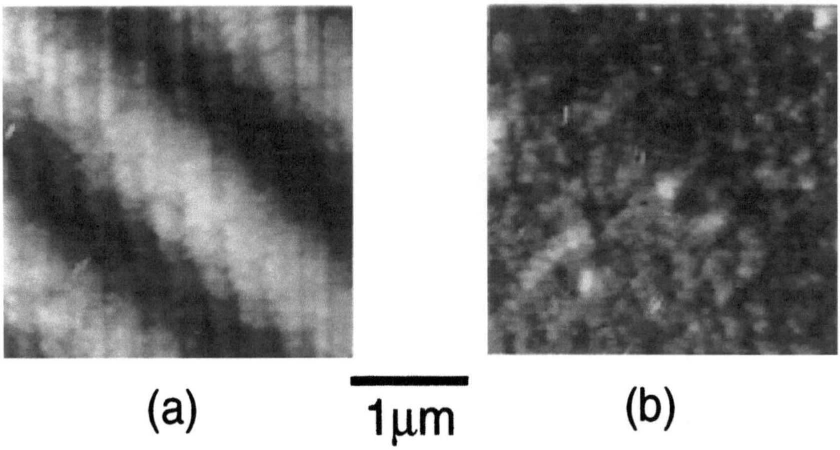

Fig. 5.7. NOM measurement of a CD surface with a light-emitting probe. a Shear-force image. b Fluorescence image

This propagates through the optical fiber of the probe, passes through a notch filter which rejects the 488-nm-wavelength light, and finally reaches a photomultiplier tube operated in photon-counting mode. The sample–probe separation is controlled by the shear force technique, and the fluorescent image is obtained by scanning the probe. Figure 5.7a and b show a shear-force topographic image of the CD surface and a fluorescence image of the CD surface, respectively, which are measured simultaneously. The spatial homogeneity of the fluorescent image in Fig. 5.7b reveals that the dye-fixed probe does not have any photobleacing problems within a measurement time of about 20 min. Compared with the shear-force topographic image, the fluorescent image has low contrast of the grating structures, but high contrast of the pit patterns. Low contrast at a large scale and high contrast at a small scale are due to the operation of the constant-distance mode by the shear-force feedback (see Sect. 6.2.2.1 for details).

5.4.2 Chemical Sensing Probes

Dyes such as phenolphthalein have been used as a fluorescent indicator of pH in analytical chemistry. This idea suggests that by using fixing materials like the fluorescent indicator, we can obtain chemical sensing probes which will enable us to measure the spatial distribution and temporal evolution of the ion concentration in a cell. In this subsection, we describe chemical sensing probes fixed with a plasticized poly(vinyl chloride) (PVC) membrane, which has been extensively used as a lipophilic membrane in ion-selective sensors operated in an optical mode [14–35] as well as in an electrical one. Much effort has been focused on the miniaturization of optical sensors using the plasticized PVC [9, 10].

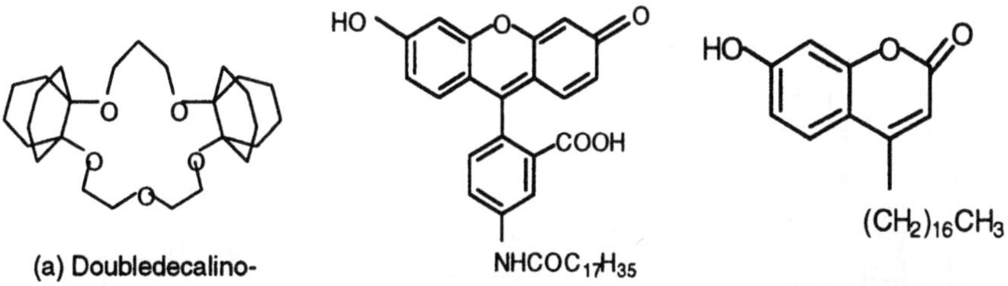

(a) Doubledecalino-16-crown-5 (DD16C5)

(b) 5-octadecanoyl-amino-fluorescein

(c) 4-heptadecyl-7-hydroxycoumarin

Fig. 5.8. Chemical structures of reagents. **a** DD16C5 is a Na^+-selective neutral ionophore. **b** 5-octadecanoyl-amino-fluorescein is an indicator dye for the measurement of fluorescence intensity while **c** 4-heptadecyl-7-hydroxycoumarin is a one for that of fluorescence spectral shift

An ion-selective membrane operated in fluorescent mode is prepared by incorporating ion-selective ionophores and indicator dyes into plasticized PVC. Here, the ionophore is a compound which selectively forms a complex with a certain ion. The sensing mechanism is based on an ion-exchange principle given by

$$S_O + AH_O + i^+ = SiA_O + H^+ \tag{5.1}$$

where S is an ion-selective neutral ionophore, AH is a lipophilic anionic dye of the indicator, i^+ is the ion to be sensed, SiA is an associated species, H^+ is a hydrogen ion and subscript "o" represents the organic phase. Ion concentration is measured by changes in the intensity or spectrum of fluorescence of the indicator dyes. The amount of the change is determined by a ratio of the deprotonation of anionic dyes, AH, in Eq. 5.1.

Figure 5.8 shows reagents of a neutral ionophore and indicator dyes in plasticized PVC membrane for two types of demonstration of chemical sensing probes. One is a measurement of fluorescence intensity using a lipophilic fluorescein of 5-octadecanoyl-amino-fluorescein, while the other is a measurement of fluorescence spectral shift using a lipophilic 7-hydroxycoumarin of 4-heptadecyl-7-hydroxycoumarin. Doubledecalino-16-crown-5 (DD16C5) [36] is a crown ether that forms an 1:1 complex with Na^+ by including it in the cavity of DD16C5. Adhesion of the membrane to the glass surface of a hydrophilic optical fiber is enhanced by replacing PVC with a copolymer of 90wt% vinyl chloride, 6wt% vinyl alcohol, and 4wt% vinyl acetate (OH–PVC). The plasticizer is bis(2-ethylhexy)sebacate (DOS). The plasticized PVC consists of 33wt% OH–PVC and 66wt% DOS. Fractional ratios of 5-octadecanoyl-amino-fluorescein and 4-heptadecyl-7-hydroxycoumarin are 3wt% and 0.6wt%, respectively, relative to the plasticized PVC. The neutral ionophore of DD16C5 is about 200 mol% relative to each of the indicator dyes. A solvent of tetrahydrofuran (THF) is used for mixing reagents.

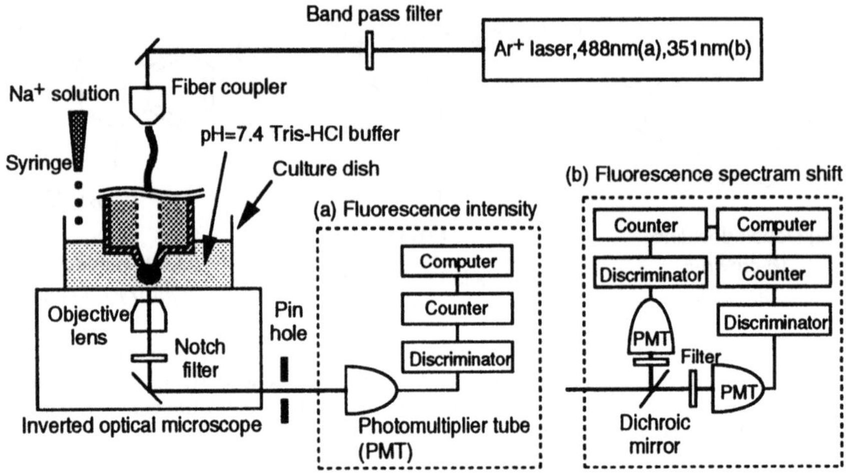

Fig. 5.9. Experimental set-ups for the measurement for a fluorescence intensity and b fluorescence spectral shift. The chemical sensing probe is placed in a pH=7.4 buffer solution and positioned over an objective lens of an inverted optical microscope. Fluorescence from the sensing membrane is collected and detected by the photon-counting method. Na^+ concentration is changed by adding droplets of Na^+ solution with a syringe into the culture dish

Figure 5.9 shows the experimental set-ups for fluorescence measurements. A single-mode cw Ar^+ laser light is coupled to an optical fiber at the other side of a chemical sensing probe and illuminates its sensing membrane. The excitation power is about 1 μW and the excitation wavelengths are 488 nm and 351 nm for lipophilic fluorescein and lipophilic 7-hydroxycoumarin, respectively. The chemical sensing probe is immersed in a solution of pH=7.4 Tris-HCl buffer in a culture dish on the table of the inverted optical microscope. The Na^+ concentration is changed from 0 to about 0.03 M by adding a NaCl drop at a time with a syringe into the culture dish. Fluorescence from dyes in the sensing membrane is collected with an N.A.=0.3 objective lens in the inverted optical microscope and detected by the photon-counting method after eliminating the laser light with a holographic notch filter. To measure fluorescence intensity using lipophilic fluorescein, a fluorescence of around 550 nm is directed to a photomultiplier tube (PMT), as shown in Fig. 5.9a. In contrast, to measure fluorescence spectral shift using lipophilic 7-hydroxycoumarin, a fluorescence of 380–450 nm is divided into two beams at 400 nm, near the isosbestic point of fluorescence spectra, with a dichroic mirror. The two divided fluorescences are separately detected by PMTs, as shown in Fig. 5.9b. It should be noted that a fiber with a Ge-doped core cannot be used for the probe because fluorescences of lipophilic 7-hydroxycoumarin and Ge-doped core overlap at around 400 nm when they are excited by a UV light of 351 nm. For 351 nm excitation, the probes fabricated from pure silica core fiber are used instead of the etched probes.

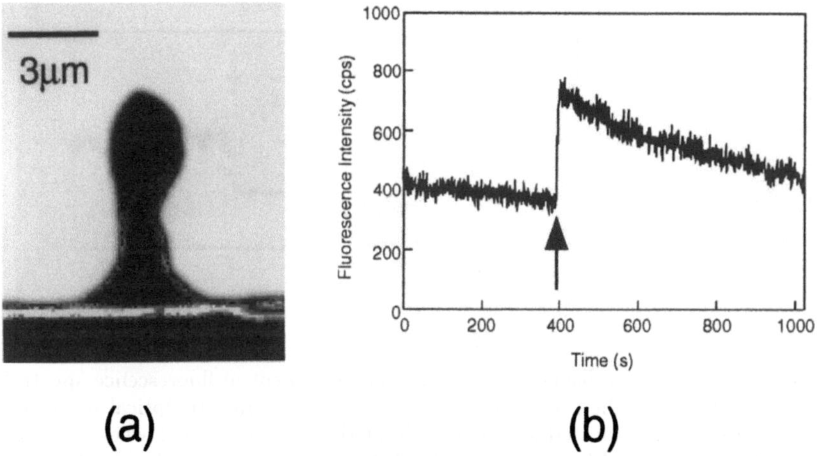

Fig. 5.10. A chemical sensing probe for the measurement of fluorescence intensity. **a** SEM micrograph of the probe tip. The fixed size is ~ 3 μm. **b** An optical response to a change of Na^+ concentration from 0 to ~ 0.03 M. The arrow represents the time at which a droplet of Na^+ solution is added to the buffer solution in the culture dish

Figure 5.10a and b show a SEM micrograph of a chemically etched probe with lipophilic fluorescein, and the measured temporal variation of the fluorescence intensity induced by a change in the Na^+ concentration, respectively. To acquire the signal representing the spectral wavelength shift, we used a heat-pulled probe with lipophilic 7-hydroxycoumarin, a SEM micrograph of which is shown in Fig. 5.11a. Figure 5.11b shows temporal variations of two fluorescence intensities. One is a longer wavelength than the isosbestic point of the fluorescence spectra, and the other is shorter. Figure 5.11c shows the temporal variation of the ratio between the two fluorescence intensities in Fig. 5.11b. Thus, this ratio in Fig. 5.11c is proportional to the magnitude of the spectral wavelength shift. An advantage of using fluorescence intensity ratio is that it rejects the contribution of photobleaching, which can be seen in Figs. 5.10b and 5.11b. The response time to a change in Na^+ is 10 s in both Figs. 5.10b and 5.11c.

By replacing the nontailed ionophore of DD16C5 with a tailed one of tetradecyl-doubletetramethyl-16-crown-5 (C14-DTM16C5) [36], the stability of the optical signals is improved [37]. This can be explained by the fact that the tail of the ionophore acts like an anchor that prevents the ionophore which includes Na^+ from leaching into the aqueous phase due to attachment to plasticized PVC. Details will be described elsewhere.

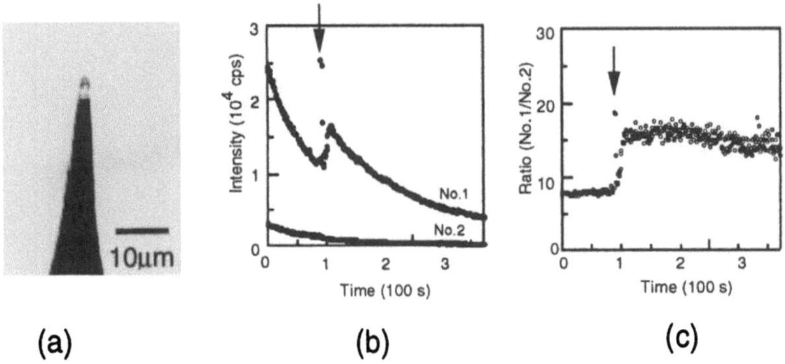

Fig. 5.11. A chemical sensing probe for the measurement of fluorescence spectral shift. **a** SEM image of the probe tip. The fixed size is ~ 5 μm. **b** Optical responses of two fluorescences separated approximately at the isosbestic point when the Na^+ concentration is changed from 0 to ~ 0.03 M. The fluorescence of the No. 1 profile has lower energy than that of the No. 2 profile. **c** A time-resolved profile of the fluorescence intensity ratio derived from profiles in **b**. Arrows in **b** and **c** represent the time at which a droplet of Na^+ solution was added. The peaks at the arrows in **b** and **c** are due to a flashlight

5.5 Future Directions

Although functional probes have such simple structures that materials can be selectively fixed to the probe tip, they have problems in practical use which are not yet solved. Developing a new fixation method is an essential task for the future so that various kinds of functional material with various sizes and shapes can be fixed. With an established fixation method, knowledge about materials and their related technology will play a major role in resolving many issues about the functionality of functional probes. This knowledge will cover near-field optics, optical fiber sensors, analytical chemistry, biological chemistry, colloidal chemistry, surface chemistry, polymer science, solid state physics, optical microscopy, fluorescence spectroscopy, laser spectroscopy and so forth. In particular, knowledge about spectroscopy for measuring weak optical signals is vital.

Biological applications of functional probes are especially interesting. An in vivo investigation of cells and neurons could be carried out using chemical sensing probes by measuring the spatial distribution and time evolution of biological materials. Chemical structure and arrangement could also be determined using energy-transfer probes by simultaneously measuring topographical and energy-transfer images of biological substances whose specific groups are labeled with dye molecules.

References

1. W. Tan, Z.-Y. Shi, S. Smith, D. Birnbaum, R. Kopelman, Science **258**, 778 (1992)
2. W. Tan, Z.-Y. Shi, R. Kopelman, Anal. Chem. **64**, 2985 (1992)
3. R. Kopelman, K. Lieberman, K. Liberman, Mol. Cryst. Liq. Cryst. **183**, 333 (1990)
4. R. Kopelman, A. Lewis, K. Lieberman, J. Lumin. **45**, 298 (1990)
5. K. Lieberman, S. Harush, A. Lewis, R. Kopelman, Science **247**, 59 (1990)
6. Z. Rosenzweig, R. Kopelman, Anal. Chem. **67**, 2650 (1995)
7. Z. Rosenzweig, R. Kopelman, Anal. Chem. **68**, 1408 (1996)
8. M. Shortreed, R. Kopelman, M. Kuhn, B. Hoyland, Anal. Chem. **68**, 1414 (1996)
9. M. Shortreed, E. Bakker, K. Kopelman, Anal. Chem. **68**, 2656 (1996)
10. M. Shortreed, E. Monson, R. Kopelman, Anal. Chem. **68**, 4015 (1996)
11. A. Song, S. Parus, R. Kopelman, Anal. Chem. **69**, 863 (1997)
12. K. Kurihara, M. Ohtsu, Technical Digest, The Pacific Rim Conference on Lasers and Electro-Optics, ThK3, July 1997, Makuhari, Japan, pp. 148–149
13. K. Kurihara, K. Watanabe, M. Ohtsu, Technical Digest, The Eleventh International Conference on Optical Fiber Sensors, Fr2-3, May 1996, Sapporo, Japan, pp. 694–697
14. K. Suzuki, K. Tohda, Y. Tanda, H. Ohzora, S. Nishihama, H. Inoue, T. Shirai, Anal. Chem. **61**, 382 (1989)
15. S. S. S. Tan, P. C. Hauser, N. A. Chaniotakis, G. Suter, W. Simon, Chimia **43**, 257 (1989)
16. W. E. Morf, K. Seiler, B. Lehmann, C. Behringer, K. Hartman, W. Simon, Pure & Appl. Chem. **61**, 1613 (1989)
17. K. Seiler, W. E. Morf, B. Rusterholz, W. Simon, Anal. Sci. **5**, 557 (1989)
18. K. Suzuki, H. Ohzora, K. Tohda, K. Miyazaki, K. Watanabe, H. Inoue, T. Shirai, Anal. Chem. Acta **237**, 155 (1990)
19. Y. Kawabata, R. Tahara, T. Kamichika, T. Imasaka, N. Ishibashi, Anal. Chem. **62**, 1528 (1990)
20. Peter Holý, W. E. Morf, K. Seiler, W. Simon, J.-P. Vigneron, Helv. Chim. Acta **73**, 1171 (1990)
21. J. N. Roe, F. C. Szoka, A. S. Verman, Analyst **115**, 353 (1990)
22. W. E. Morf, K. Seiler, B. Rusterholz, W. Simon, Anal. Chem. **62**, 738 (1990)
23. S. Ozawa, P. C. Hauser, K. Seiler, S. S. S. Tan, W. E. Morf, W. Simon, Anal. Chem. **63**, 640 (1991)
24. S. S. S. Tan, P. C. Hauser, K. Wang, K. Fluri, K. Seiler, B. Rusterholz, G. Suter, M. Krüttli, U. E. Spichiger, W. Simon, Anal. Chim. Acta **255**, 35 (1991)
25. K. Seiler, K. Wang, M. Kuratli, W. Simon, Anal. Chim. Acta **244**, 151 (1991)

26. S. J. West, S. Ozawa, K. Seiler, S. S. S. Tan, W. Simon, Anal. Chem. **64**, 533 (1992)
27. M. Lerchi, E. Bakker, B. Rusterholz, W. Simon, Anal. Chem. **64**, 1534 (1992)
28. E. Bakker, W. Simon, Anal. Chem. **64**, 1805 (1992)
29. K. Seiler, W. Simon, Anal. Chim. Acta **266**, 73 (1992)
30. H. He, H. Li, G. Mohr, B. Kovács, T. Werner, O. S. Wolfbeis, Anal. Chem. **65**, 123 (1993)
31. E. Bakker, M. Lerchi, T. Rosatzin, B. Rusterholz, W. Simon, Anal. Chim. Acta **278**, 211 (1993)
32. G. J. Mohr, O. S. Wolfbeis, Anal. Chim. Acta **316**, 239 (1995)
33. I. Murkovic, A. Lobnik, G. J. Mohr, O. S. Wolfbeis, Anal. Chim. Acta **334**, 125 (1996)
34. M. R. Shortreed, S. L. R. Bakker, R. Kopelman, Sensors and Actuators B **35–36**, 217 (1996)
35. E. Wang, L. Ma, L. Zhu, C. M. Stivanello, Anal. Lett. **30**, 33 (1997)
36. K. Suzuki, K. Sato, H. Hisamoto, D. Siswanta, K. Hayashi, N. Kasahara, K. Watanabe, N. Yamamoto, H. Sasakura, Anal. Chem. **68**, 208 (1996)
37. K. Kurihara, M. Ohtsu, T. Yoshida, T. Abe, H. Hisamoto, K. Suzuki, Europt(r)ode IV, pI12, March 1998, Münster, Germany, pp. 49–50

Chapter 6

Instrumentation of Near-Field Optical Microscopy

6.1 Operation Modes of NOM

With the probes whose fabrication was described in Chap. 3 based on the criteria concepts described in Chap. 2, the near-field optical microscope (NOM) can generally be operated under two different modes, called the collection-mode (c-mode) NOM and the illumination-mode (i-mode) NOM. Figure 6.1 gives a schematic illustration of the two different modes of operation. In c-mode NOM (Fig. 6.1a), the light is incident on the sample at total internal reflection. The three-dimensional optical near field generated and localized on the sample surface is scattered by a probe, and part of the scattered field is collected through the same probe. There exist optical near fields from the planar background and from the sample. However, no optical near field from the background will be picked by the probe when probe parameters such as apex diameter and foot diameter are subwavelength and comparable to the size of the sample feature. The principle of operation of i-mode NOM (Fig. 6.1b) is similar, except that the probe acts as a generator of the optical near field which illuminates the sample surface. The field scattered by

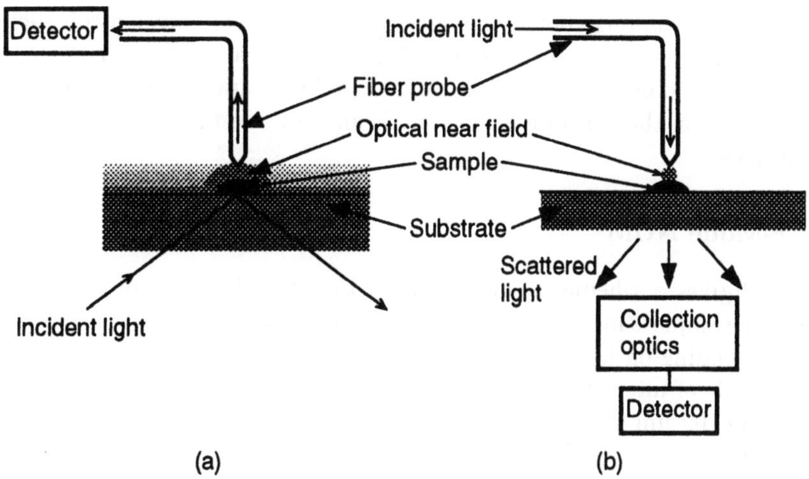

Fig. 6.1. Schematic illustration of **a** the c-mode and **b** the i-mode NOM

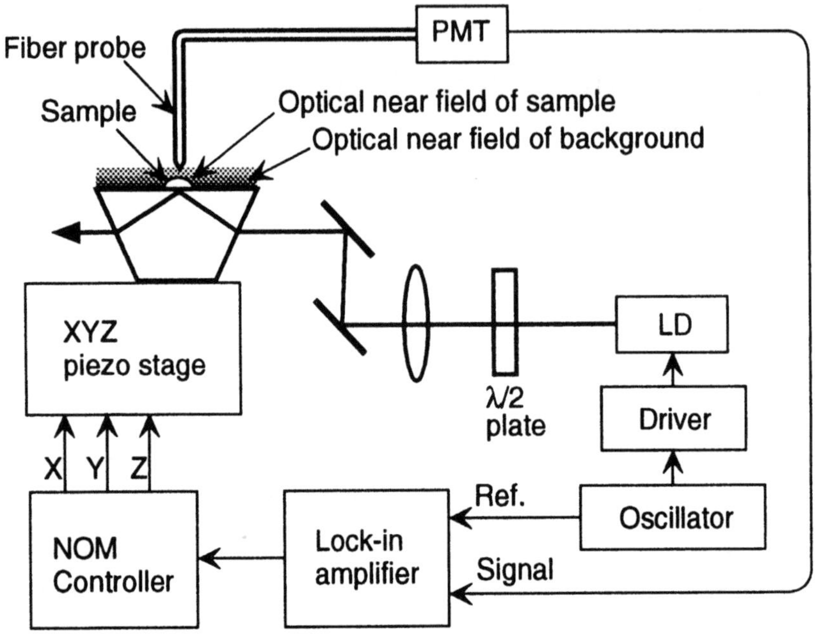

Fig. 6.2. Schematic illustration of c-mode NOM

the sample is collected by conventional optical elements. In either c-mode or i-mode, probe parameters such as apex diameter, foot diameter (for these definitions, see Chap. 2), and sample–probe separation determine the detection efficiency, the obtainable resolvable smallest feature of the sample, and the image contrast.

In this chapter we describe the features of these modes of operation. Further, this chapter gives a detailed discussion on the second essential element of NOM, i.e., the scanning control. Different methods of scanning control, such as the shear-force feedback technique and the optical near-field intensity feedback technique to obtain a constant sample–probe separation, are discussed. Some reviews on the constant-height mode and the constant-distance mode are also given.

6.1.1 c-Mode NOM

Figure 6.2 gives a schematic illustration of the optical system of a c-mode NOM. The sample is mounted on an *xyz* piezo stage. The piezo parameters are chosen depending on features such as the lateral and longitudinal variations of the sample of interest. Here, the piezo stage has a total lateral scan span of 100 μm$\times$100 μm so that it is suitable for scanning biological specimens, for example. The z stage has a span of 5 μm. A right–angle or dove prism is used for generating the optical near field on the sample surface. The sample is mounted on the prism with a sandwich of index-matching oil. Here, care

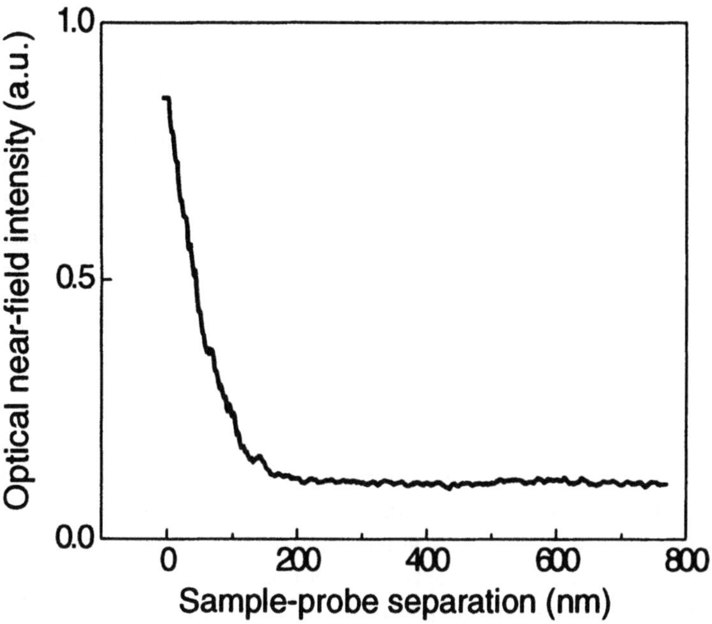

Fig. 6.3. Variation of the optical near-field intensity as a function of the sample–probe separation

should be taken to avoid unnecessary scattering from the prism surface, such as from dust, or scratches, as strong scattered fields will lead to a large background signal. The probe is also mounted on an *xyz* stage so that it can be moved freely over the different areas of the sample. It is also necessary that both the prism–sample and the fiber probe are mounted rigidly so that no external vibration affects the system.

The light from a laser diode (LD) of wavelength 680 nm was used as the source. The light is incident on the prism in such a way that it is totally internally reflected and the optical near field is generated on the sample surface. This optical near field is localized in three-dimensional space and it follows the spatial variations of the sample, as shown by the shaded region in Fig. 6.2. The probe starts to brighten when it approaches closer to the sample. This corresponds to the fact that the optical near field is scattered by the probe. The scattered field contains information about the sample features, and part of the scattered field gets in through the same probe by propagating through the fiber. By having a highly sensitive detector such as a photomultiplier tube at the other end of the probe, information about the sample features can be extracted. In this system, the sample substrate is moved to get a two-dimensional (2D) sample feature distribution. Typical values for the detected power of the scattered light are of the order of a few nanowatts to a few hundred picowatts. Further, by modulating the light from the LD and detecting the signal at the modulated frequency with a lock-in amplifier, it will be possible to perform a highly phase sensitive detection of the signal over the background.

Figure 6.3 shows the variation of the optical near-field intensity as a function of the sample–probe separation. As the probe approaches the sample surface, the optical near-field intensity increases sharply. This monotonous variation of the optical near-field intensity can be used as a feedback control signal to maintain a constant sample–probe separation while performing the 2D raster scanning of the sample. The 2D raster scanning and the feedback operation can be performed with a commercially available controller. The modes of scanning and scanning control will be discussed in Sect. 6.2.

6.1.2 i-Mode NOM

Figure 6.4 shows a schematic illustration of the experimental system of an i-mode NOM. In this system, light from a laser (here, an Ar ion laser is used as the light source) is coupled into the fiber probe to illuminate the sample. The sample is mounted on an *xyz* piezo stage to make the raster scan possible. The light coming out of the probe contains both the radiating far fields and the optical near fields. Their ratio depends strongly on probe characteristics such as apex diameter and foot diameter: the smaller these parameters, the weaker will be the radiating far fields. Hence, on approaching the sample surface very close to the substrate, i.e., within a few nanometers, the optical near field localized around the probe gets scattered by the fine features of the sample. With an objective lens with a high numerical aperture, most of the scattered field is collected to be detected by a photomultiplier tube. Here, by using techniques such as labeling with an absorbing or fluorescent dye, it is possible selectively to investigate or identify the sample features.

In both c-mode and i-mode, the sample is scanned by the probe at a sample–probe separation of around 10 nm. There are basically three ways of maintaining this order of separation during the raster scanning. One is a free running operation without regulating the sample–probe separation. This is called the constant-height mode. In other words, the probe is kept at a constant vertical position and the sample is moved only in the x and y directions to obtain an image. Although this kind of operation could result in the probe bumping into the protruding features of the sample, hence leading to possible damage to the sample and the probe, it can be implemented successfully if some a priori information is available about the sample. We will demonstrate the results in Sect. 6.2.1. Another way to maintain the sample–probe separation is called the constant-distance mode. This can be acheived in two ways: One is to use the monotonous variation of the optical near-field intensity as a feedback signal [1, 2]. The other is through the use of an auxiliary signal which simultaneously maps the topographical variations of the sample, such as the shear force [3, 4]. The third possibility is atomic force which is sometimes used for feedback control [5]. We will describe the

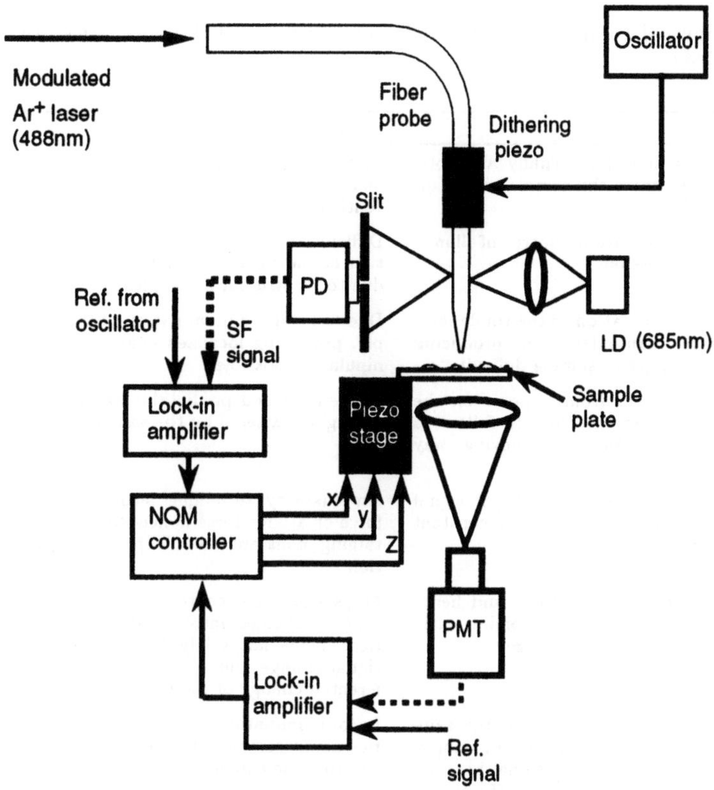

Fig. 6.4. A schematic illustration of i-mode NOM

control techniques using the optical near-field intensity and the shear force in detail in Sect. 6.2.

6.1.3 Comparative Features of Modes of NOM

In this section, a comparative evaluation of the features of c-mode and i-mode NOMs has been obtained empirically based on the authors' experience through performing experiments. Table 6.1 gives the results of the empirical comparison of the different features and characteristics of the modes.

Based on the features of NOMs in Table 6.1, choice of the operation mode of NOM depends on the sample, the type of image, such as the near-field optical image distribution and the near-field spectral characteristics. In the following Chaps. 7–11, both modes have been used extensively based on the sample features and the type of information required.

Table 6.1. Empirical comparison of the different features and characteristics of c- and i-mode NOMs

No.	c-mode	i-mode
1	Sample illuminated externally and optical near field from the sample picked up by the probe	Sample illuminated locally by the probe and the scattered light collected by conventional optics
2	Due to 1, polarization control of illumination light possible	Difficult to predict the state of polarization of the light coming from the probe due to its structural complexity
3	Due to the illumination of the whole sample surface, manipulation or producing localized changes in sample difficult	Due to localized illumination of the sample, producing localized changes or manipulation possible
4	Optical near field from a point on the surface of the sample mostly follows a monotonous variation when moving away from the surface	No definite and predictable variation of the light scattered by the sample exists
5	Due to 4, an inherent feedback signal is available for maintaining a constant sample–probe separation	An auxiliary technique such as shear-force or atomic force is needed for maintaining constant sample–probe separation
6	Maps equi-intensity contour, and hence the contrast of the image is coming solely from the optical signal change	Maps equi-force contour. This leads to contrast changes introduced into the optical image not solely from the optical signal change, but coming from the differential change of the probe motion
7	Due to whole illumination of the sample, for a large scattering type of sample, the background scattered signal becomes large, reducing the strength of the optical near-field signal	Local illumination of sample avoids worries about sample scattering characteristics to some extent
8	Fluorescent labeling possible. However, due to the illumination of the whole sample area, possibility of bleaching of the dye occurs	Fluorescent labeling of the sample makes it possible to do discriminatory inspection and spectroscopic studies
9	Labeling of the sample with absorbing dye may introduce large roughness variations, leading to large scattering	Labeling of the sample by absorbing dye enhances the contrast of the image
10	Due to the availability of inherent optical near-field intensity feedback mechanism, no lateral dithering of the probe is required	In the auxiliary mechanism of shear-force feedback, the probe is dithered laterally, leading to damage of the sample
11	Because of optical near-field intensity feedback, operation can easily be extended to liquid	Shear-force employed for feedback control makes it difficult to extend the operation in liquid. Capillary forces and large Q of resonance lead to poor sensitivity of shear-force
12	Image interpretation is relatively easy as the sample is scanned by keeping the optical near-field intensity constant	Image interpretation not straightforward as the image contrast is not solely from an optical signal change, but rather from the shear-force induced contrast changes in the image
13	As an operational precaution, it is necessary to avoid largely scattering samples. Moreover, it is difficult to image opaque and semi-transparent samples	No restriction on the sample. Both opaque and transparent sample possible

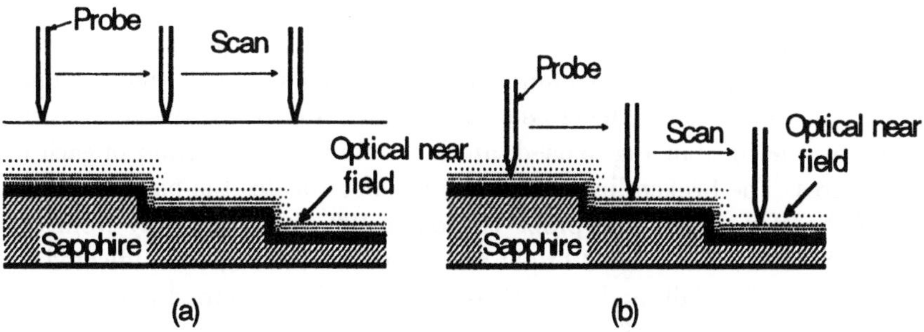

Fig. 6.5. A schematic illustration of (a) constant-height mode and (b) constant-distance mode

6.2 Scanning Control Modes

It is necessary to have the probe in the near-field region while making the 2D raster scan of the sample with the probe in both c- and i-mode NOMs. In other words, the sample–probe separation should be of the order of a few to a few tens of nanometers in order for it to be possible to image the finer features of the nanometric sample. This is because the separation has to be as close as possible to the height variation of the smallest feature in the sample. By imaging an ultrasmooth sapphire surface (cf. Chap. 7), the constant-height and constant-distance modes of imaging are demonstrated in the following sections.

6.2.1 Constant-height Mode

Figure 6.5a is a schematic illustration of this mode of scanning. Here, the sample under observation is an ultrasmooth sapphire surface. The sample is obtained by annealing a mirror-polished sapphire surface at 1000–1400°C. During the process, misorientation of the atomic planes leads to atomic steps with a step height in the range 2–4 nm and a step width of 1.3 μm [6]. The sample is moved in the x and y directions in a raster fashion while positioning the probe at some vertical position corresponding to a point along the distance–intensity curve (see Fig. 6.3). The resulting image is a two-dimensional map of intensities as a function of the lateral position in the sample. This kind of scanning is fast, as no feedback control is implemented. Further, it also excludes the possibility of introducing any artificial contrast in the image. To perform this kind of scanning, however, some kind of a priori information about the height-feature variation of the sample is required in order not to bump into the sample, and hence not to damage both the sample and the probe.

Figure 6.6a gives a shear-force topographic image of the ultrasmooth sapphire surface. Figure 6.6b–d show the results of constant-height mode images obtained by scanning the same region of the surface at sample–probe separations of around 100 nm, 50 nm, and 5 nm, respectively. In order not to bump into different steps, a scanning area is selected so that it contains only one step on the surface. The cross-sectional variation across the step, as indicated by the lines in Fig. 6.6a–d, are illustrated at the bottom of each of these figures. The broken line in the illustration indicates the position of the edge of the step. When the probe is around 100 or 50 nm away from the surface, there exists a blurred bright region around the edge, as indicated by the arrows in Fig. 6.6b and c which is represented by a broad curve in the illustrated cross section. In contrast, on going close to the step height (i.e., around 5 nm), the bright region becomes very sharp, and is followed by a dark region. The dark region corresponds to the falling edge of the step. The bright narrow region corresponds to the region before the edge of the step. In each of figures b–d, the arrow at the top indicates the position of the step as estimated from the shear-force topographic image. This also agrees with the expected result [7–11]. It is expected that at the edge, due to surface discontinuity, there should exist a highly localized three-dimensional optical near field, with the field extent determined by the step height. As the step height is less than 5 nm, the optical near field is also localized within the volume determined by this height. Therefore, on bringing the probe as close as one step height or closer, the region near the edge of the step appears sharp and bright. These results show that it is possible to obtain high-contrast images under constant-height mode provided that the scanning height is very close to the height variation of the sample under observation.

6.2.2 Constant-Distance Mode

Figure 6.5b shows a schematic illustration of the constant-distance mode. The probe is scanned so as to follow the topography of the sample while performing raster scanning of that sample.

6.2.2.1 Shear-force Feed Back

One of the techniques generally used in following the topographical variations of the sample is the shear-force technique. Here, the differential force or frictional force is detected by dithering the probe at its resonance frequency at some amplitude, and the amplitude of dithering is monitored optically with an LD and a photodetector, as shown in Fig. 6.4 [3, 4]. Due to the presence of van der Waals force in the proximity of the surface, the resonance frequency is changed or shifted [12]. Therefore, the dithering amplitude of the resonance frequency decreases when the probe approaches closer to the sample surface. Figure 6.7 shows the variation of the dithering amplitude as a function of the sample–probe separation. The inset shows the resonance frequency spectrum of the probe. Raster scanning is done by dithering the probe at a constant amplitude and hence constant sample–probe separation. In spectroscopic applications, the use of an additional laser beam in detecting the vibrations of

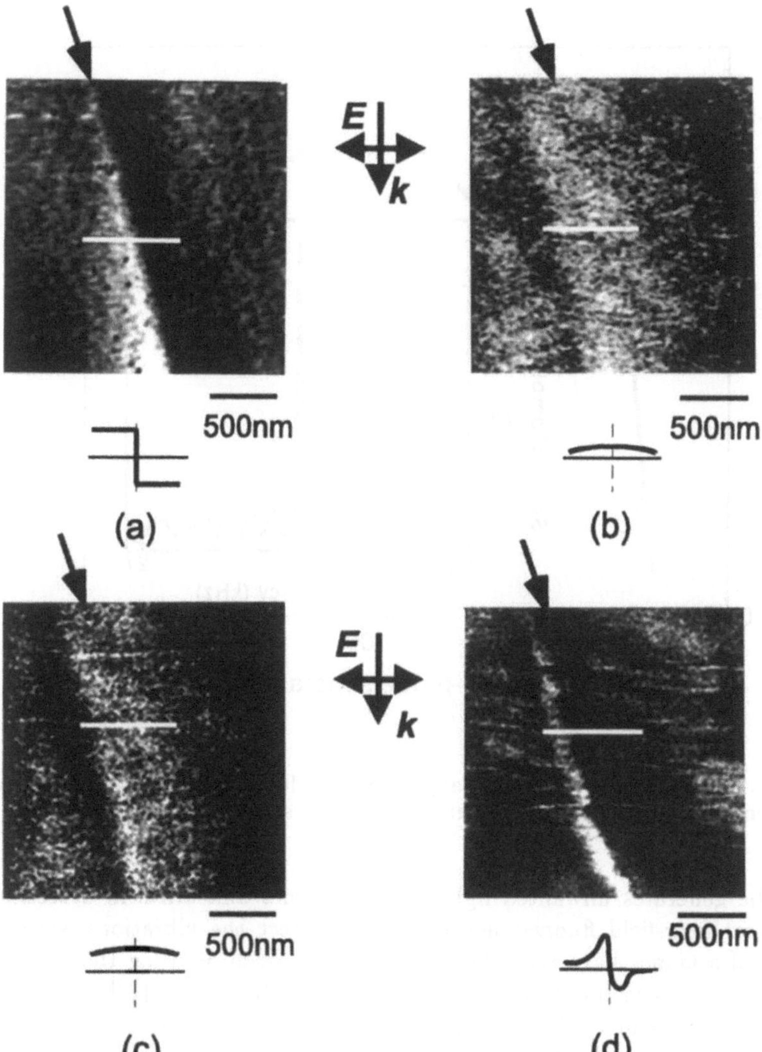

Fig. 6.6. **a** Shear-force topographic image of an ultrasmooth sapphire surface. The pixel size is 10 nm×10 nm. **b–d** c-mode NOM images of the same region obtained under constant-height mode at sample–probe separations of around 100 nm, 50 nm, and 5 nm, respectively. The arrow at the top of each image indicates the position of the step edge as estimated from the shear-force topographic image. Corresponding schematic sketches at the bottom of each image represent the variation across the line shown in each image. In these sketches, the *broken line* represents the edge of the step. The bright, sharp region in *d* corresponds to the region just before the edge of the step, and it appears as a blurred bright region in **b** and **c**. The falling edge of the step appears dark in **d**. The arrows in **b–d** indicating the bright region, which approximately corresponds to the edge of the step. The arrows drawn between **a** and **b**, and between **c** and **d** represent the directions of the wave vector $\boldsymbol{k}$ and the electric field vector $\boldsymbol{E}$ of the incident light

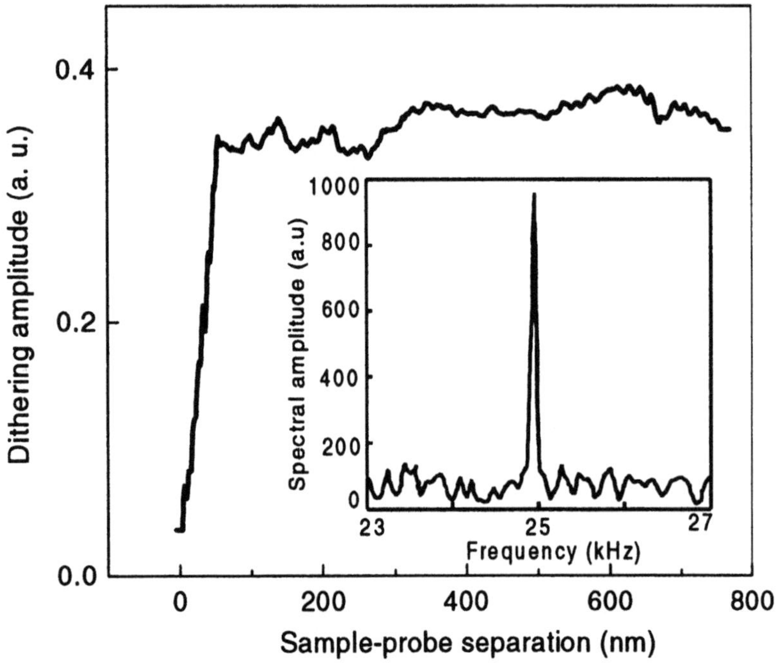

Fig. 6.7. Variation of the dithering amplitude as a function of the sample–probe separation. The inset shows the resonance frequency spectrum

the probe generates unwanted light and introduces difficulties in detecting the optical near-field fluorescence signal. To detect the vibrations, several nonoptical schemes have recently been proposed, such as using the capacitive change between the probe and a nearby capacitance lead [13], using the impedance change in a balanced bridge with an oscillating tip on one arm [14], or using a quartz crystal tuning fork with the fiber probe attached to one of the prongs and detecting the voltage between the electrodes attached to the prongs [15–18].

In general, the image obtained is a map of constant force, and it can be considered equivalent to the topography of the sample. It has generally been used as a feedback control signal mechanism in some of the experiments in the following chapters owing to its versatility. In using shear force for sample–probe separation feedback control, the lateral dithering of the probe has got nearly two degenerate degrees of freedom, and this will sometimes lead to the presence of two closely spaced resonance peaks instead of a single peak in the resonance spectrum of the fiber probe. Such a spectral characteristic will introduce instability in the feedback control loop. A simple way to solve such

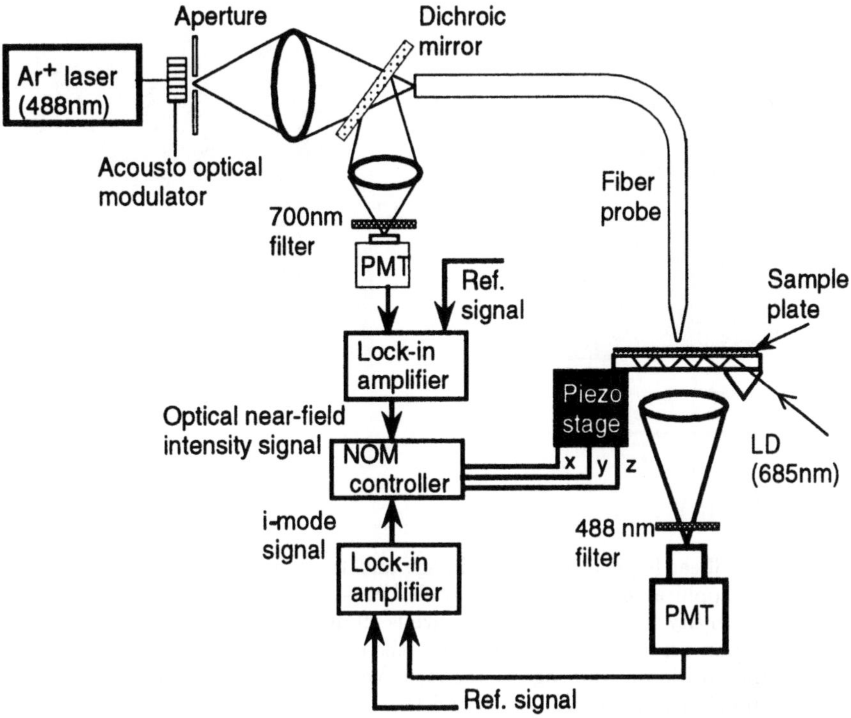

Fig. 6.8. Schematic illustration of an i-mode NOM under optical near-field intensity feedback

a problem is by using a four-sectioned piezo and redistributing the voltage between the terminals in order to match the applied voltage to one of the degenerate modes [19].

However, recently it has been found that with shear force as the auxiliary feedback for separation control, a cross-talk problem arises, which is also referred to as an "artifact problem" (see Sect. 1.4) [20]. In other words, the sample–probe separation control process through the use of the shear-force technique can give rise to undesirable features in the optical image that are highly correlated with the shear-force topographical features.

6.2.2.2 Optical Near-Field Intensity Feedback

The other technique for monitoring the equivalent topography of the sample is through the use of optical near-field intensity as a feedback signal. Figure 6.8 shows a simple experimental modification for generating the optical near field in an i-mode NOM. The easiest way to generate the optical near-field signal is by using a prism to make a light of a different wavelength to incident at an angle greater than the critical angle of total internal reflection. Here, the sample plate is replaced by a parallel plate onto which a small right-angled prism is optically bonded. The sample plate and the parallel plate are sandwiched with a layer of index-matching oil. A three-dimensionally

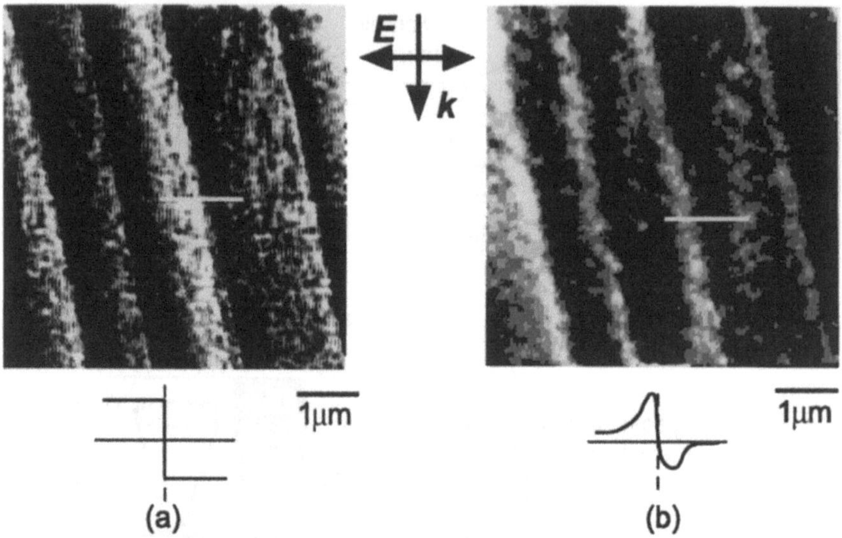

Fig. 6.9. **a** Shear-force topographic image and **b** c-mode NOM image of the ultrasmooth sapphire surface obtained under constant-distance mode with optical near-field intensity feedback. The pixel size is 39 nm×39 nm. A schematic diagram of the variation across the step is shown at the bottom of each image. *Broken lines* indicate the position of the edge of the step. Vectors drawn between **a** and **b** indicate the directions of propagation $\boldsymbol{k}$ and the electric field $\boldsymbol{E}$ of the incident light. Based on a qualitative comparison of **a** and **b**, the region just before the edge of the step in **a** appears as a bright strip in **b** followed by a dark region corresponding to the falling edge of the step

localized optical near field is generated on the surface of the sample. The localization of this optical near field is dependent on the features of the sample and not on the wavelength of the light [21]. As can be seen from Fig. 6.3, as the sample–probe separation gets shorter, the probe starts to scatter the localized optical near field, and hence to be able to pick up some of this scattered field.

The 2D raster scanning is done by the proper choice of a particular intensity value, and hence the sample–probe separation. In this case, the resulting image is a map of the constant intensity or the optical near-field topography. Figure 6.9a shows the shear-force topographic image of the sapphire surface, and b shows the corresponding image obtained by keeping the optical near-field intensity constant. The schematic sketches of the variation across a single step region are shown at the bottom of each image. In the sketches, the broken line indicates the position of the edge of the step. In b, the region just before the edge of the step appears as a sharp, bright strip, and the region corresponding to the falling edge of the step appears dark in the constant-intensity image. This agrees with the expected results based on theory [7–11]. Around the edges of the ultrasmooth sapphire surface, a

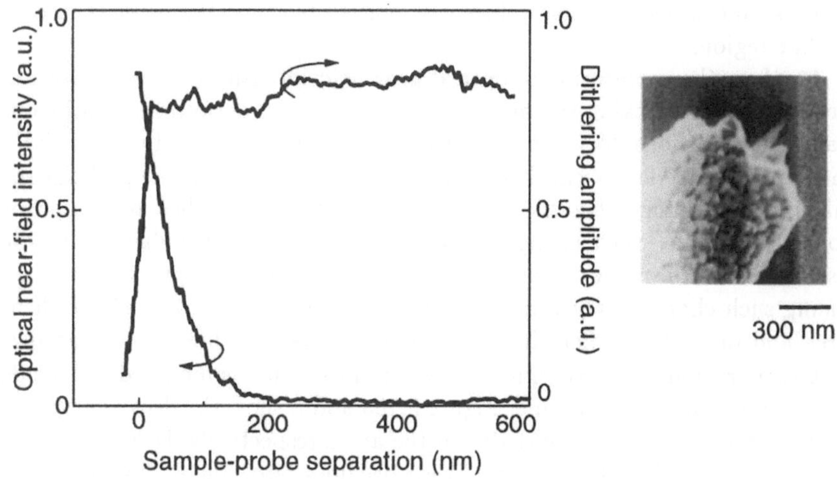

(a)

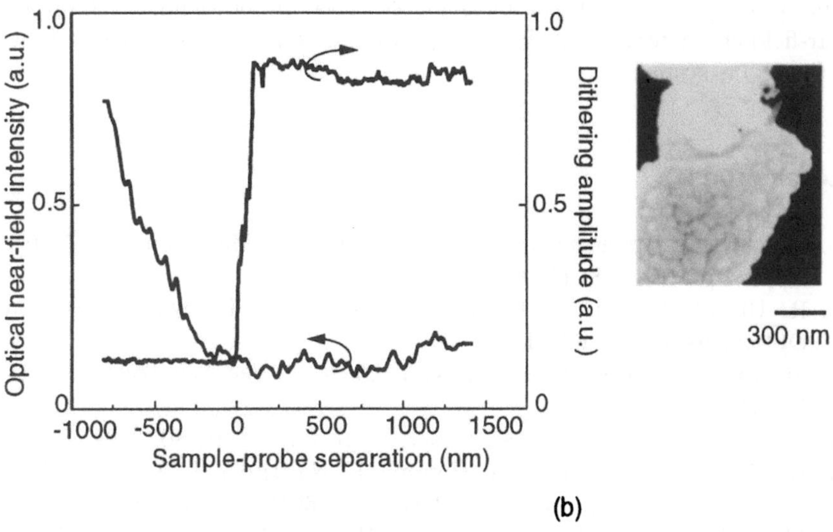

(b)

Fig. 6.10. Variation of the dithering amplitude and the optical near-field intensity as a function of the sample–probe separation for **a** a clean probe and **b** a probe with dirt at the apex region strong, highly localized optical near field is present compared with the flat region on either side of the edge. This field gets scattered by the probe, and part of it is detected as a finite intensity value. As the probe is moved up and down in the vertical direction to maintain this intensity constant, it is obvious that larger intensity values will be detected near the edges than over the flat regions.

Further, the monitoring of optical near-field intensity can be used as an error signal for checking the quality of the probe. On performing continuous scanning operations, for example, with the generally employed i-mode NOM under shear-force feedback, there is a strong probability that the probe shape will undergo changes because of dirt often becomes attached to the probe from the sample. Such changes will introduce an overall reduction in the resolution of the image. A simple technique [22] has been proposed for deducing such changes, which involves the simultaneous recording of the shear force and optical near-field intensity decay curves as a function of sample–probe separation. The experimental system in Fig. 6.8 is modified to include a shear-force detection scheme. Figure 6.10a and b show the decay curves for a clean probe and a probe with dirt at the apex, respectively. Defining zero as the point where the dithering amplitude becomes zero, Fig. 6.10b shows that an increase in the optical near-field intensity occurs a long way after the zero point for the probe with dirt compared with the increase for a clean probe. This dramatic change in the morphology of the probe changes the optical near-field characteristics of the probe, leading to a poor resolution capability of the NOM system.

References

1. M. Naya, R. Micheletto, S. Mononbe, R. Uma Maheswari, M. Ohtsu, Appl. Opt. **36**, 1681 (1997)
2. R. Uma Maheswari, S. Mononobe, H. Tatsumi, Y. Katayama, M. Ohtsu, Optical Review **3**, 463 (1996)
3. M. Vaez-Iravani, R. Toledo-Crow, Y. Chen , J. Vac. Sci. Technol. A **11**, 742 (1993)
4. E. Betzig, P. L. Finn, J. S. Weiner, Appl. Phys. Lett. **60**, 2484 (1995)
5. H. Muramatsu, N. Chiba, K. Homma, K. Nakajima, T. Ataka, S. Ohta, A. Kusumi, M. Fujihira, Appl. Phys. Lett. **66(24)**, 3245 (1995)
6. M. Yoshimoto, T. Maeda, T. Ohnishi, H. Koinuma, O. Ishiyama, M. Shinohara, M. Kubo, R. Miyura, A. Miyamoto, Appl. Phys. Lett. **67**, 2615 (1995)
7. A. Zvyagin, M. Ohtsu, Opt. Commun. **133**, 339 (1997)
8. A. Zvyagin, J. D. White, M. Ohtsu, Opt. Lett. **22**, 955 (1997)
9. R. Carminatti, A. Madrazo, M. Neito-Vesperinas, J.-J. Greffet, J. Appl. Phys. **82(2)**, 501 (1997)

10. K. Jang, W. Jhe, Opt. Lett. **21**, 236 (1996)
11. K. Kobayashi, O. Watanuki, J. Vac. Sci. Technol. B **14(2)**, 804 (1996)
12. C. J. Chen,*Introduction to Scanning Tunneling Microscopy*, (Oxford University Press, New York, 1993)
13. J. K. Leong, C. C. Williams, Appl. Phys. Lett. **66**, 1432 (1995)
14. J. W. P. Hsu, M. Lee, B. S. Deaver, Rev. Sci. Instrum. **66**, 3177 (1995)
15. K. Karrai, R. D. Grober, Appl. Phys. Lett. **66**, 1842 (1995)
16. Y. H. Chuang, C. J. Wang, J. Y. Huang, C. L. Pan, Appl. Phys. Lett. **69**, 3312 (1996)
17. W. A. Atia, C. C. Davis, Appl. Phys. Lett. **70**, 405 (1997)
18. A. G. T. Ruiter, J. A. Veerman, K. O. van der Werf, N. F. van Hulst, Appl. Phys. Lett. **71**, 28 (1997)
19. A. V. Zvyagin, J. D. White, M. Kourogi, M. Kozuma, M. Ohtsu, Appl. Phys. Lett. **71**, 2541 (1997)
20. B. Hecht, H. Bielefeldt, Y. Inoue, D. W. Pohl, L. Novotony, J. Appl. Phys. **81**, 2492 (1997)
21. T. Saiki, M. Ohtsu, K. Jang, W. Jhe, Opt. Lett. **21**, 674 (1996)
22. R. Uma Maheswari, S. Mononobe, M. Ohtsu, Appl. Opt. **31**, 6740 (1996)

Chapter 7

Basic Features of Optical Near-Field and Imaging

7.1 Resolution Characteristics

We characterize the basic features of optical near field and imaging using the system developed in Chap. 6. As an essential measure to represent the quality of the imaging, this section discusses the resolution. Although the sample is always three-dimensional, we evaluate the longitudinal and lateral resolutions separately for practical and experimental convenience. They are both influenced by various factors such as the probe parameters, sample–probe separation, and the polarization of the illuminating light. We discuss about the ways of evaluating these characterizing resolutions and their dependence on various factors.

7.1.1 Longitudinal Resolution

The longitudinal resolution of a near-field optical microscope (NOM) is characterized with a c-mode NOM. In order to do this characterization, the sample must be chosen carefully so that its surface variations are small enough to reject the scattered light. To satisfy this the criterion, the sample used is an ultrasmooth sapphire surface. The sample is fabricated by subjecting a mirror-polished sapphire surface to annealing treatments at temperatures between 1000 and 1400°C for several hours in air. During this process, the top-most surface takes the form of a staircase-like structure due to the misorientation of the atomic planes. A schematic model of this misorientation process, based on molecular dynamics calculations [1, 2] is shown in Fig. 7.1a. Figure 7.1b shows the shear-force topographic image of such a staircase-like structure. The step height is 4 nm and the step width is 1.3 μm, but both of these depend on the annealing conditions and the surface conditions, such as the crystallographic face of the sapphire [3]. The experimental system is the same as that shown in Fig. 6.2.

The sample is illuminated under total internal reflection, and the localized optical near field generated on the sample surface is picked up by a probe. In this experiment, the probe used is fabricated by the selective resin coating method described in Sect. 3.4.1, and has a protruding tip with an apex diameter d of < 10 nm and a foot diameter d_f of 30 nm. The thickness of the

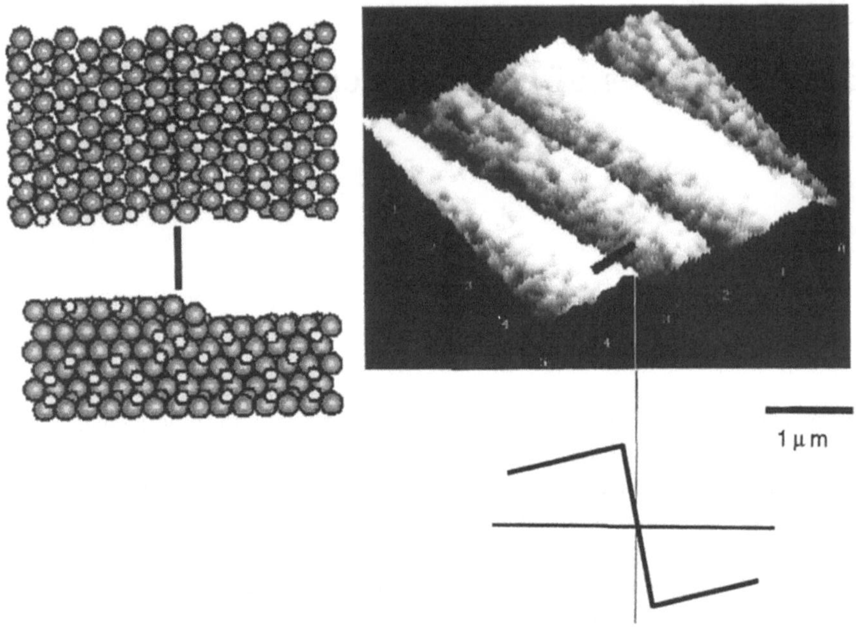

(a) (b)

Fig. 7.1. **a** Top (*left*) and side (*right*) views of the atom arrangements for the (0001) plane of sapphire simulated by molecular dynamics. The small and large balls indicate aluminum atoms and oxygen atoms, respectively. The line indicates the center of the step edge with flat terraces on either side. **b** Perspective view of the shear-force topographic image of the ultrasmooth sapphire surface. The step height is 4 nm and step width is 1.3 μm. A sketch of the variation across the line drawn is shown below the image

gold coating is around 150–200 nm. The observation was done under both constant-height and constant-distance modes, as described in Sects. 6.2.1 and 6.2.2, respectively. Figure 7.2 shows an image of the ultrasmooth sapphire surface obtained under constant-distance mode at a sample–probe separation of less than 10 nm. The sketch below the image represents the intensity variation across the line drawn in the image. The line corresponds to the position of the edge of the step. Hence, in the region around the edge of the step the image appears brighter, and immediately behind the step the image appears darker. When the sample–probe separation is increased to around 80 nm, only a flat image with no intensity variation appears. This shows the high discrimination sensitivity of the probe to optical near field from the ultrasmooth sapphire surface.

Further, based on NOM images obtained under constant-height mode shown in Fig. 6.6b–d, the optimum sample–probe separation has been found to be about the same as the step height. Figure 7.3a and b show a schematic

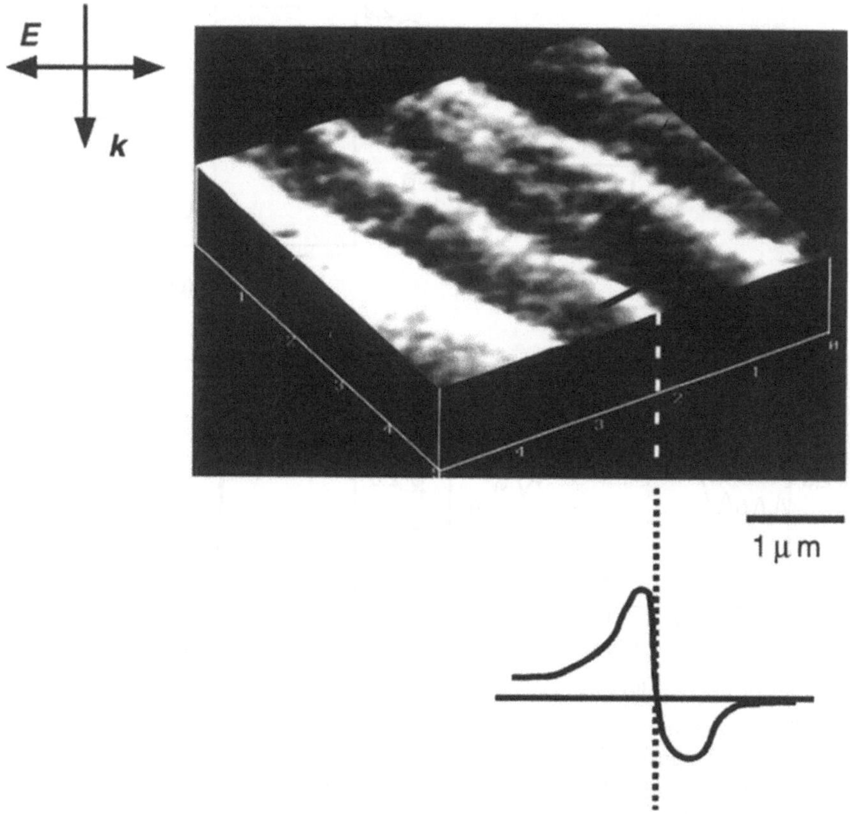

Fig. 7.2. Perspective view of the ultrasmooth sapphire surface at sample–probe separations of less than 10 nm obtained under constant-distance mode using the optical near-field intensity as the feedback signal. The sketch beneath the image represents the intensity variation across the line drawn in the image. The *broken line* corresponds to the position of the step. The peak corresponds to the region just before the edge of the step, and the valley corresponds to the falling edge of the step. The vectors in the left-hand corner indicate the directions of the propagation $\boldsymbol{k}$ and the electric field vector $\boldsymbol{E}$ of the incident light

cross-sectional view containing the step and shear-force topographic variation, respectively. Figure 7.3c and d show the corresponding one-dimensional intensity profiles of the constant-height mode images across a single step area at probe heights around 100 nm and 5 nm, respectively. The broken lines drawn across each part correspond to the position of the step. As seen in c, at around 100 nm, there appears a broad bright region around the step region. At around 5 nm, as seen from d, the bright region just before the step region gets sharper and brighter and the falling edge of the step region appears dark resulting in a valley. This is due to the fact that a strong optical near field is localized around the edge. This gets scattered, and is detected on bringing the probe as close as possible to the step height.

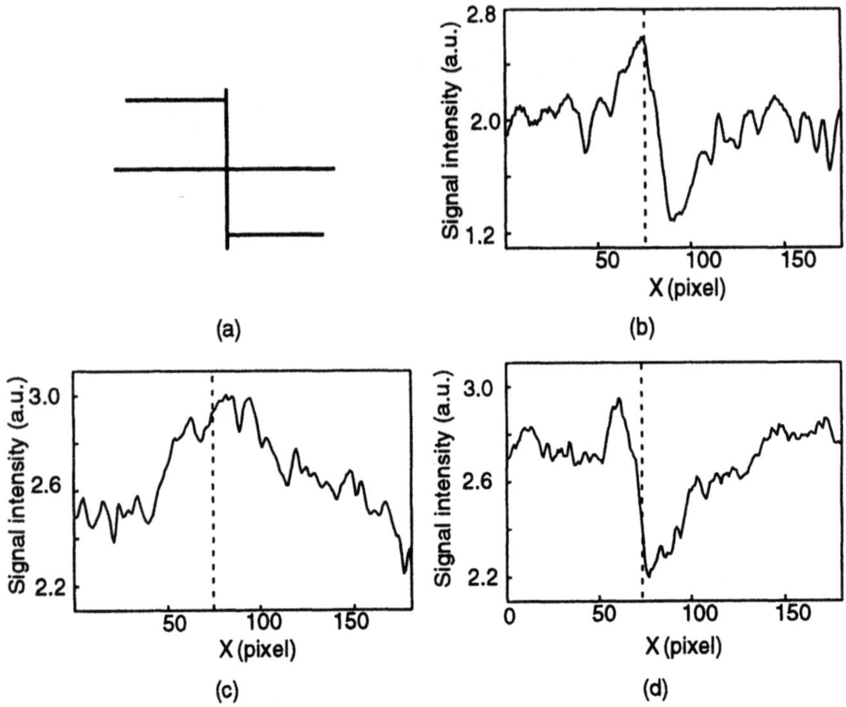

Fig. 7.3. **a** A one dimensional schematic cross-sectional view containing a single step. **b** One dimensional shear-force topographic profile. **c,d** Corresponding one-dimensional intensity variations across the step region of the images obtained under constant-height mode at heights of 100 nm and 5 nm, respectively. The *broken lines* drawn across **b–d** indicate the position of the step

Based on the above results, the longitudinal resolution of NOM could be estimated to be of the order of a few nanometers. Hence, the optical near field would be localized within a few nanometers for a step of height of a few nanometers. This can also be inferred from the fact that the optical near field, which localizes around the sample feature, is characterized by the extension of the sample feature as discussed in Chaps. 1 and 2.

7.1.2 Lateral Resolution

This section evaluates the lateral resolution using a system function of the NOM. Since the optical near field mediates the short-range electromagnetic interaction between the sample and the probe (see Sect. 1.3), multiple scattering efficiency depends on various factors such as sample–probe separation.[1] In

[1] "Multiple scattering" means that nanometric conformations of the interacting sample and probe are involved in the scattering process. Therefore, it is always a linear optical process. In other words, it means that the proportionality between the incident and scattered light powers is not violated by the nonlinear

other words, the operation condition of NOM, and consequently the image characteristics, depend on these factors. However, microscopic imaging always assumes a sufficiently weak interaction, i.e., no reaction from the probe to the sample. It also assumes a linear measurement without modifying the sample conformation and structure, even though the optical near field is destroyed by the probe.[2] This sufficiently weak interaction has been assumed not only for NOM, but also for other optical or even nonoptical measurement instruments.

It should also be noted that causality is always assured for the multiple scattering process [4] because the relation between the electric fields of the incident and scattered lights is a linear function of the relative positions of the sample and the probe. This means that a linear scattering matrix can be defined, including the information about the conformations and structures of the sample and the probe. Therefore, by assuming a sufficiently weak interaction, a system function can be defined and used for NOM which has been employed for other linear and causal systems. In other words, a system function can be defined as long as NOM is used as a measurement instrument. This sufficiently weak interaction can be assumed for practical NOM systems by limiting their use to a relevant spatial Fourier frequency range. This section evaluates an experimental system function by assuming a sufficiently weak interaction in order to derive design criteria for NOM.[3]

One method for evaluating a system function and the lateral resolution is through the use of power spectral analysis [5]. Using the power spectral density (PSD), it is possible to evaluate the system function. From this system function, the highest −3 dB cutoff spatial Fourier frequency could be defined as the lateral resolution of the system, where the amplitude of the system function is attenuated to half its maximum. An alternative way of defining the resolution could be the frequency at which the signal-to-noise ratio of the PSD is zero dB. In contrast, one more essential measure of imaging to represent the clarity of the sample image, could be defined as the ratio of the

---

components of the induced polarizations of the sample and the probe. Thus, it is different from a nonlinear optical process.

[2] This microscope should be a processing instrument if the sample conformation or structure is modified. Although it is important to use NOM as a processing instrument by increasing the optical power, such a strong interaction regime is not allowed for in microscopic imaging.

[3] Use of a system function is indispensable for electronic system engineers to design novel electronic amplifiers and related devices. A similar assumption has been made here. The nonlinear response characteristics of the amplifier vary its gain if the input electronic signal level is increased. This increase can modulate even the bias voltage, which leads to a change in the operation conditions. However, since causality holds between the input and output electronic signals, a system function can be defined and used by linearizing the response character and fixing the operation conditions under a small signal approximation. (This approximation corresponds to assuming sufficiently weak interaction in NOM, as described above.)

PSD at the frequency corresponding to the size of the smallest feature in the sample to the PSD at the frequency corresponding to the surface variation of the substrate.

In order to evaluate lateral resolution, i-mode NOM has been used. The experimental system is the same as that described in Sect. 6.1.2. As a sample, gold particles of diameter 20±0.5 nm in the colloidal form have been used. Colloidal gold particles are commercially available, and their size is well characterized with a transmission electron microscope (TEM). A drop of gold solution is dispersed on a glass substrate and allowed to dry at room temperature. The sample is then mounted on the piezo stage, which is equipped with in-built sensors to compensate for the effects of hysteresis during raster scanning. The probe used was fabricated by the selective chemical etching method to have a flattened apex, as shown in Fig. 3.7 [6], followed by coating with chromium and gold to a thickness of 10 nm and 200 nm, respectively to form a nanometric aperture. For a view of a scanning electron micrograph of the aperture, see Fig. 3.23.

Figure 7.4a shows the NOM image of clustered gold particles distributed randomly on a glass slide substrate. When obtaining this image, the sample–probe separation was maintained at less than 5 nm. It should be noted that the slope corresponding to the tail of the shear-force curve is small in this position (as shown in Fig. 6.7). In other words, the shear-force gain was kept sufficiently low, and the actual scanning was done under an almost free-running condition in order to avoid artifact effects from the shear-force feedback [7]. The pixel size is 7 nm×7 nm. In this image, the region marked I corresponds to gold particles forming large clusters consisting of overlapping particles. Figure 7.4b shows a magnified view of region II, which is in the clustered region I. Dark regions (indicated by the arrows) correspond to small clusters of gold particles.

Figure 7.5a shows the two-dimensional (2D) PSD of the image shown in Fig. 7.4a. The 2D PSD is radially symmetric; the spectral variation is averaged along the angular direction (θ) of the polar coordinate system. The variation of the radially averaged power spectral density (APSD) is shown in Fig. 7.5b. The APSD clearly shows a minimum at the spatial frequency (f_0) 5.6×10^7 m^{-1}. The theoretical APSD is calculated [8] based on a theoretical modeling equivalent to the random distribution of particles. Under large fluctuations of the spatial distribution of the particles, the APSD corresponds to the PSD of a single particle, which turns out to be the well-known Bessel function of first order. The effect of the random distribution appears to be the monotonous linear variation of the APSD in the low Fourier frequency range below the frequency f_1. Here f_1 is the cross-over frequency between the Bessel function for the single particle effect and the monotonously varying component of the random distribution effect. A characteristic of the Bessel function is that it has a first zero at a frequency α_0 which is proportional to the inverse of the size ρ of the particle given by $\alpha_0 = 1.22/\rho$. Using this

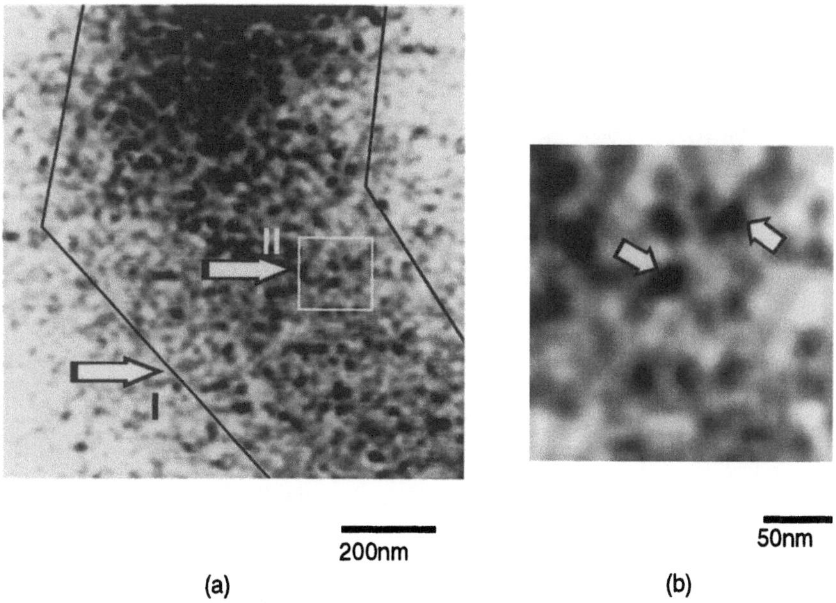

Fig. 7.4. a NOM image of gold particles. Region I is the region of clustered gold particles. b Magnified view of region II in a. The dark regions indicated by *arrows* correspond to either single particles or small clusters of particles

relation and f_0, the diameter of a single gold particle could be estimated to be 22 nm, which agrees with the nominal value evaluated by TEM.

Further, by defining the system function as the ratio of the experimental power spectrum to the theoretical power spectrum, an estimate of −3 dB could be obtained. The theoretical power spectrum is given by the Bessel function, with ρ being 20 nm, and the experimental one is given in Fig. 7.5b. The estimated system function takes the form $H(f/f_0) = 0.81(f/f_0)^{-0.16}$, being normalized to unity at $f = f_1$. From this equation, the −3 dB cutoff frequency could be determined as 1.2×10^9 m^{-1} which corresponds to a size of 0.8 nm. This gives an estimate for the lateral resolution capability of the NOM system. Such a high resolution could be attributed to the presence of the high-frequency optical near fields confined around the edges of the dielectric–metal boundary at the apex of the probe [9, 10].

7.2 Factors Influencing Resolution

Although resolution of the images obtained using the optical near field is highly influenced by many of the probe parameters, the present discussion is limited to the influence of the most essential parameters, i.e., the probe parameters and the sample–probe separation.

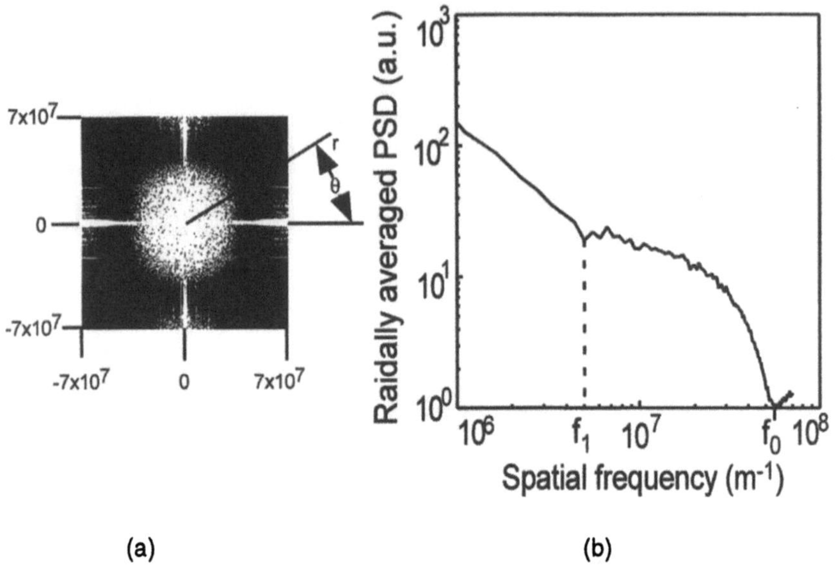

Fig. 7.5. a Two-dimensional power spectral densities (PSD) of the NOM image shown in Fig. 7.4a. b Radially averaged power spectral density (APSD) given as a function of spatial frequency f on a log-log scale. Here f_1 is the frequency below which APSD is dominated by the effects due to the random arrangement and clustering of the gold particles. For $f > f_1$, the ASPD resembles that due to a single particle. f_0 is the frequency at which the APSD takes a minimum value

7.2.1 Influence of Probe Parameters

Resolution is determined by the localization volume of the optical near field [9, 10]. In turn, this optical near field is determined by probe parameters such as the apex and the foot diameter of the probe [11]. Figure 7.6a and b show the c-mode NOM images of the flagellar filaments of *Salmonella* obtained under constant-height mode with probes having foot diameters d_f of 30 nm and 100 nm, respectively. The apex diameter d is less than 10 nm and the cone angle is 20° in each case. During the observations, the sample–probe separation was maintained at 15 nm. As can be seen, with a larger foot diameter, the filament indicated by a long arrow in Fig. 7.6a appears broader in Fig. 7.6b. The broadening is due to the probe with a larger foot diameter effectively picking up the field of low spatial frequency components.

7.2.2 Dependence on Sample–Probe Separation

Sample–probe separation plays a crucial role in the resolution of NOM imaging. As demonstrated in Sect. 7.1.1, for imaging an ultrasmooth sapphire surface with a height of 2–4 nm, the value of the sample–probe separation has to be made as close as possible to the step height. The lateral resolution is also strongly influenced by the relative values of the sample–probe

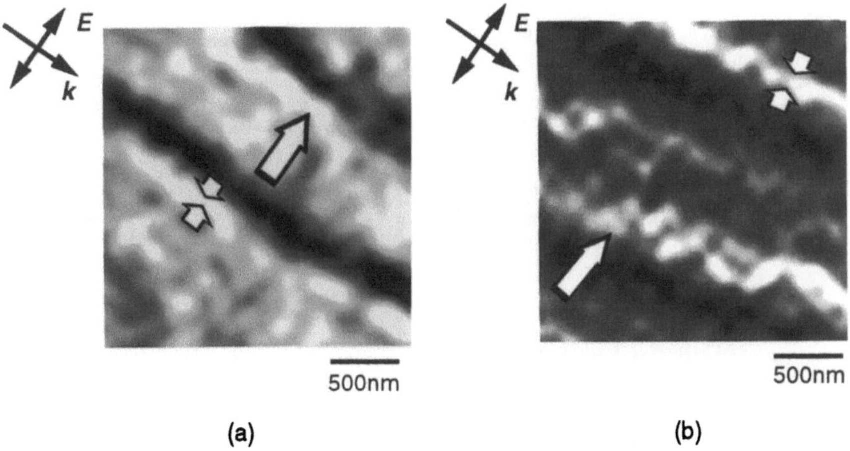

Fig. 7.6. c-mode NOM images of *Salmonella* flagellar filaments with probes having foot diameters d_f of **a** 30 nm and **b** 100 nm obtained under constant-height mode. Sample–probe separation was 15 nm. *Long arrows* in the images indicate a filament. Note the increase in the width of the filament, as indicated by the *short arrows*, with increased d_f. The full width at half-maximum of the filament, as indicated by *short arrows* in **a** is around 50 nm. The arrows at each top left-hand corner represent the directions of the wave vector $\boldsymbol{k}$ and the electric field vector $\boldsymbol{E}$ of the incident light

separation and the height of the sample structure [12]. Figures 7.6a and 7.7a show c-mode NOM images of flagellar filaments of *Salmonella* obtained under constant-height mode at the sample–probe separations of 15 nm and 65 nm, respectively. The probe had an apex diameter d of less than 10 nm and a foot diameter d_f of 30 nm. Owing to the increase in separation, the full width at half-maximum of the filament, as indicated by the short arrows in Figs. 7.6a and 7.7a gets larger from 50 nm to 150 nm. In order to study the dependence of the images on the sample–probe separation more quantitatively, the spatial Fourier spectra of the images in Figs. 7.6a and 7.7a were calculated. For an exact estimation, all the data containing a large number of flagellar filaments were used, and the results are shown in Fig. 7.7b. Comparing the two curves in the figure, it can be seen that the value of the high-frequency component decreases with increasing sample–probe separation. This is attributed to the fact that the size of the localization of the optical near field is of the order of that of the sample [9, 10].

7.3 Polarization Dependence

The c-mode has been used to study the dependence of NOM imaging on polarization because the polarization of the incident light can easily be controlled in the case of the c-mode. The polarization dependence has been investigated

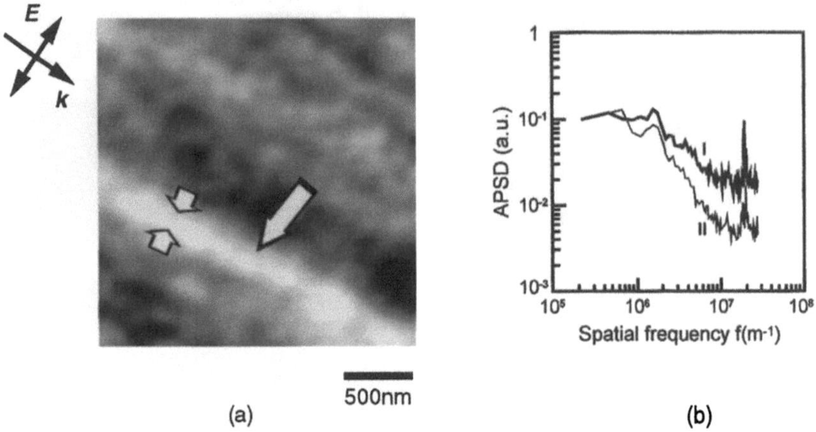

Fig. 7.7. **a** c-mode NOM image of flagellar filaments under constant-height mode for a sample–probe separation of 65 nm. The *long arrow* indicates a flagellar filament. The full width at half-maximum of the filament, as indicated by the *short arrows* increased from 50 to 150 nm. The probe had a foot diameter d_f of 30 nm. The arrows at the top left-hand corner represent the directions of the wave vector $\boldsymbol{k}$ and the electric field vector $\boldsymbol{E}$ of the incident light. **b** The spatial Fourier spectrum of NOM images for sample–probe separations of 15 nm (I), corresponding to Fig. 7.6a, and 65 nm (II), corresponding to Fig. 7.7a

with two different types of sample: one is an ultrasmooth sapphire surface, and the other is $LiNbO_3$ nanocrystals formed on this surface. $LiNbO_3$ is well known for its high optical nonlinearity in the bulk [13].

7.3.1 Influence of Polarization on the Images of an Ultrasmooth Sapphire Surface

The ultrasmooth sapphire surface used for characterizing the longitudinal resolution has also been used for investigating the influence of polarization. Figure 7.8a shows a shear-force topographic image. Figure 7.8b and c show the corresponding images obtained under constant-distance mode of the same region of the sample for s-and p-polarizations, respectively. It should be mentioned here that the detected light power in the case of p-polarization is at least ten times over than that for s-polarization. The position of the step edge is indicated by an arrow at the top of each part. Optical near-field intensity feedback was employed for the constant-distance mode.

Figure 7.9a–c show the cross-sectional profiles of the shear-force topographic and optical near-field intensity variations under s- and p-polarizations, respectively along the lines indicated in Fig. 7.8a–c. The broken line corresponds to the position of the step edge. The schematic sketches shown in the inset of Fig. 7.9b and c correspond to the expected cross-sectional intensity profiles across a single step region. The expected profiles were obtained

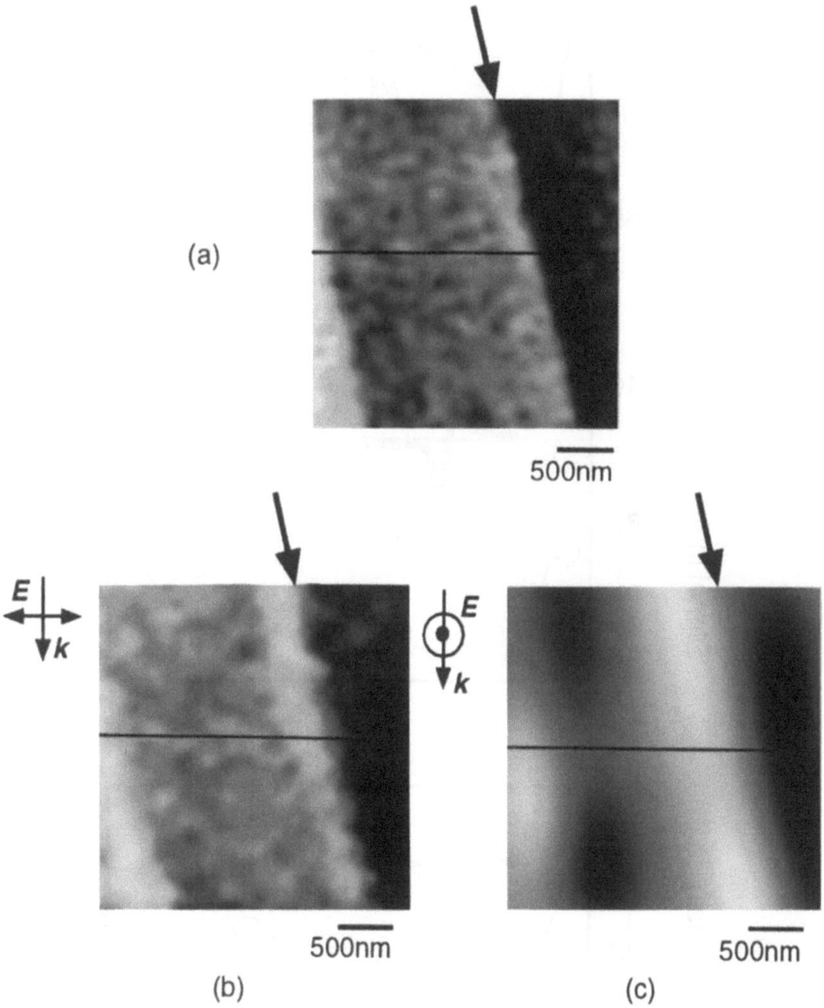

Fig. 7.8. **a** Shear-force topographic image of the sapphire substrate. **b**, **c** c-mode images obtained under constant-distance mode by optical near-field intensity feedback for s- and p-polarizations, respectively of the incident electric vector over the same region as **a**. **c** shows the image after filtering out the interference fringes of the incident light, which is observed because of the low detected light power. The arrow at the top of each figure indicates the step edge. The arrows at the left-hand corners represent the directions of the wave vector $\boldsymbol{k}$ and the electric field vector $\boldsymbol{E}$ of the incident light

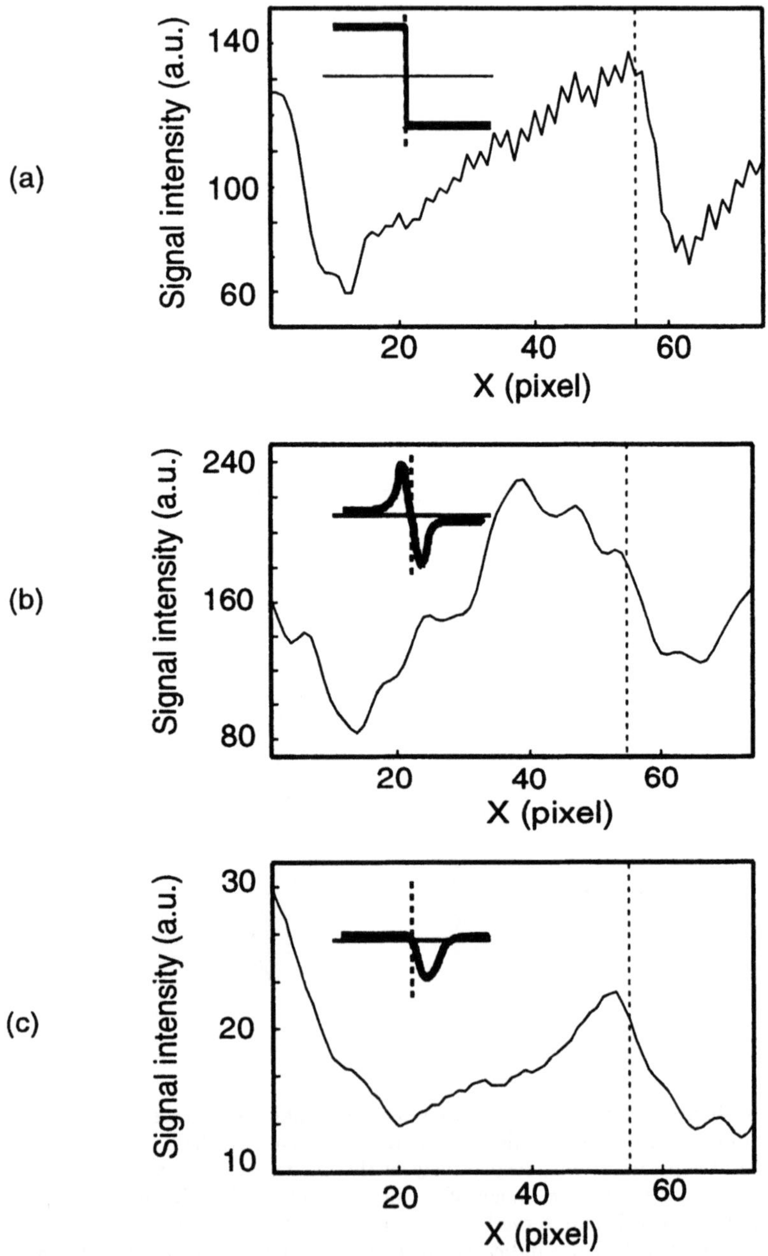

Fig. 7.9. a–c cross-sectional profiles of the shear-force topographic and optical near-field intensity variations under s- and p- polarizations, along the lines indicated in Fig. 7.8a–c, respectively. The inset schematic sketches are the theoretically expected cross-sectional intensity profiles across a single step region. The *broken line* corresponds to the position of the step edge

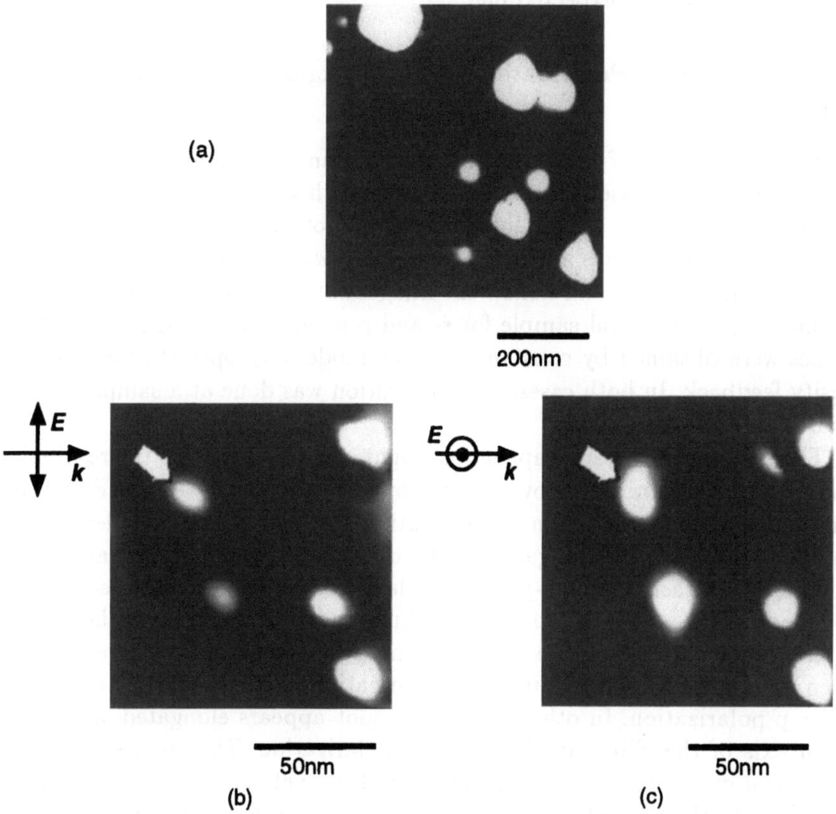

Fig. 7.10. **a** Atomic force micrograph of $LiNbO_3$ nanocrystals. **b,c** c-mode NOM images of nanometric $LiNbO_3$ nanocrystals obtained under constant-height mode by optical near-field intensity for s-polarized and p-polarized incident lights, respectively. The arrows near the top left-hand corner represent the directions of the wave vector $\boldsymbol{k}$ and the electric field vector $\boldsymbol{E}$ of the incident light. Note that the spots indicated by the *arrows* have elongated in different directions depending on the incident polarization

based on the coupled dipole mode presented in Sect. 2.1.2 and the theoretical models presented in Ref. [14, 15]. Under s-polarization, as seen from the profile obtained in Fig. 7.9b, the region nearest the edge appears bright, with the falling edge of the step appearing dark. It can be seen that the profile obtained almost agrees with that expected result. Under p-polarization, the region close to the falling edge of the step appears dark. The profile obtained is almost same as the expected one.

7.3.2 Influence of Polarization on the Images of $LiNbO_3$ Nanocrystals

Nanometric crystals of $LiNbO_3$ were grown on an ultrasmooth sapphire surface by the pulsed laser ablation technique with a KrF excimer laser [16]. Figure 7.10a shows an atomic force micrograph of the $LiNbO_3$ nanocrystals. A nanometric crystal has a length of 100 nm, a width of 80 nm, and a height of 10 nm. Figure 7.10b and c show the images obtained from the same area of the $LiNbO_3$ nanocrystal sample for s- and p-polarization, respectively. The images were obtained by constant-distance mode with optical near-field intensity feedback. In both cases, the observation was done at a sample–probe separation of around 10 nm.

The nanometric crystals appear as bright spots. Comparing the results in b and c, one spot (indicated by arrows) appears to be elongated in a direction perpendicular to the direction of propagation of the s-polarized incident light, while it appears to be elongated in the direction of propagation, and with much less contrast, under p-polarized illumination. In order to discuss the difference, we consider the spot as an ellipse and take the ratio of the major to the minor axes. It can be seen that the averaged ratio is always smaller when the spot is observed under s-polarization than when it is observed under p-polarization. In other words, the spot appears elongated along the minor axis of the ellipse in the case of s-polarization. This agrees with the theoretical calculations based on the optical extinction theorem of Zvyagin et al. [15]. In their results, in the case of s-polarization, a dielectric spherical sample appears to be elongated along the minor axis, and in the case of p-polarization, it appears to be elongated along the major axis.

References

1. K. Kawamura, *Introduction to Molecular Simulations*, Chaps. 7 and 8 (Kaibun-do, Tokyo, 1989)
2. A. Miyamoto, K. Takeichi, T. Hattori, M. Kubo, T. Inui, Jpn. J. Appl. Phys. **31**, 4463 (1992)
3. M. Yoshimoto, T. Maeda, T. Ohnishi, H. Koinuma, O. Ishiyama, M. Shinohara, M. Kubo, R. Miyura, A. Miyamoto, Appl. Phys. Lett. **67**, 2615 (1995)

4. M. Born, E. Wolf, *Principles of Optics* (Pregamon Press, Oxford, 1980)
5. J. W. Goodman, *Fourier Optics* (McGraw Hill, NewYork 1968)
6. R. Uma Maheswari, S. Mononobe, M. Ohtsu, IEEE J. Lightwave Technol. **13**, 2308 (1995)
7. B. Hecht, H. Bielefeldt, Y. Inoue, D.W. Pohl, L. Novotony, J. Appl. Phys. **81**, 2492 (1997)
8. R. Uma Maheswari, H. Kadono, M. Ohtsu, Opt. Commun. **131**, 133 (1996)
9. K. Jang, W. Jhe, Opt. Lett. **21**, 236 (1996)
10. T. Saiki, M. Ohtsu, K. Jhang, W. Jhe, Opt. Lett. **21**, 674 (1996)
11. M. Naya, S. Mononobe, R. Uma Maheswari, T. Saiki, M. Ohtsu, Opt. Commun. **124**, 9 (1996)
12. R. Uma Maheswari, S. Mononobe, M. Ohtsu, Appl. Opt. **31**, 6740 (1996)
13. P. N. Butcher, D. Cotter, *The elements of nonlinear optics*, (Cambridge University press, Cambridge, 1991)
14. K. Kobayashi, O. Watanuki, J. Vac. Sci. Technol. B **14(2)**, 804 (1996)
15. A. Zvyagin, M. Ohtsu, Opt. Commun. **133**, 339 (1997)
16. G. H. Lee, M. Yoshimoto, H. Koinuma, Applied Surface Science (1998)

Chapter 8
Imaging Biological Specimens

8.1 Introduction

Biological specimens have been investigated with optical microscopes such as the differential interference contrast microscope and the confocal microscope. Although they provide a way to monitor the biological process in real time [1], their resolution is limited by the diffraction of the light. They have also been studied by electron microscopes, which offer high resolution. However, electron microscopes have several disadvantages such as:

1. loss of surface information due to the use of special sample preparation techniques such as ultra thin slicing employed in transmission electron microscopy and surface coating used in the case of scanning electron microscopy;
2. destructive observation;
3. the requirement of a high vacuum.

One possible method avoiding diffraction is through the use of scanning probe microscopes [2–8], including the near-field optical microscope (NOM). With NOM, it is also possible to attach specific biomolecules with fluorescing dye labels in order to investigate the spatial distribution and functional properties of the molecules with a resolution well beyond the diffraction limit. In this chapter, we discuss the imaging of some biological specimens, namely, *Salmonella*, neurons, microtubules, fibroblasts, and DNA, in order to demonstrate the high-resolution capability of our NOM using fiber probes fabricated by the techniques described in Chap. 3.

8.2 Observation of Flagellar Filaments by c-Mode NOM

The *Salmonella* bacterium is a primitive single cell organism with a basal body, from which five or six flagellar filaments extend. The bacterium can translate by rotating these filaments at very high speed through a motor-like structure on its body. This structure is also known as a molecular motor, and its size is of nanometer order. Although the reason for such movements still remains a mystery [9], it is becoming important in the field of biology and micro-machine fabrication. A single filament has a typical length of around 2 μm and a diameter of 25 nm. As the diameter of the filament is well characterized, it can be employed as a reference sample for NOM imaging. Figure 8.1 shows a transmission electron micrograph of the filaments. For observation under c-mode NOM [10], the filaments have been separated from the body and fixed on a hydrophilized microscopic cover-glass substrate.

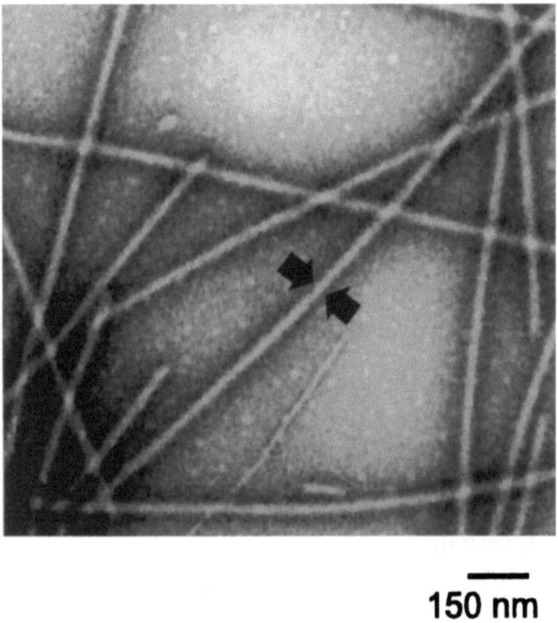

Fig. 8.1. A transmission electron micrograph of the straight-type flagellar filaments of *Salmonella*. The diameter of a single filament, as indicated by the *arrows*, is 25 nm

8.2.1 Imaging in Air

The sample substrate was mounted on the prism surface of the system for c-mode NOM, as shown in Fig. 6.2, under s-polarized incident light with the electric field parallel to the sample surface. The probe used was fabricated by selective chemical etching and selective resin coating methods in order to form a protrusion extending from the metal-coated probe (see Sect. 3.4.1). The apex and foot diameters (d and d_f, respectively) of the protrusion were 10 nm and 30 nm, respectively.

Figure 8.2 shows a perspective view of a segment of a flagellar filament obtained by c-mode NOM under the s-polarized incident light, i.e., its electric vector is parallel to the sample surface. The scanning was done under constant-height mode. The montonously increasing optical near-field intensity (see Fig. 6.3) was used to set the initial height. In order to compensate

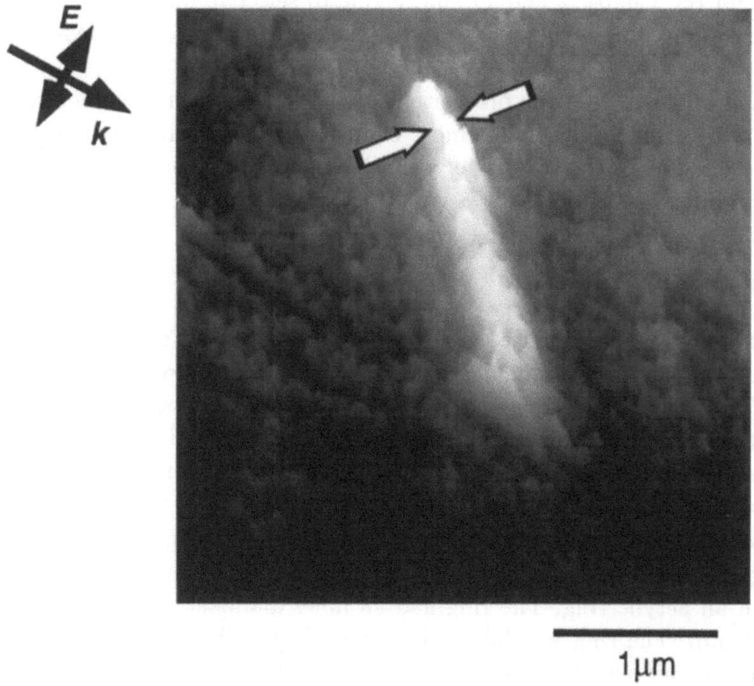

Fig. 8.2. NOM image of a flagellar filament with s-polarized incident light. The sample–probe separation was about 15 nm. The arrows at the top left–hand corner represent the direction of the wave vector $\boldsymbol{k}$ and the electric field $\boldsymbol{E}$ of the incident light. The full width at half-maximum of the bright region corresponding to the filament, as indicated by the *arrows*, is 50 nm

for the effects of tilt in the substrate, the time constant of the servo loop has been carefully adjusted to follow only slow topographical variations. Since the response time of the controller was less than 0.1 s and the scanning frequency was 2 Hz, the feedback control does not respond to sample features smaller than 1 μm. This means that the NOM system is sufficiently sensitive to pick up high spatial frequency components of the optical near field in order to image the fine profile of the sample. The pixel size is 19 nm$\times$19 nm. The sample–probe separation was less than 15 nm, and was estimated in the following way. After finishing the scan, the probe was brought towards the sample by applying a voltage corresponding to a distance of 15 nm. At this position, when the scanning was repeated, we observed streaks in the image due to the probe scratching the sample surface.

The bright region in this figure corresponds to the filament, and the full width at half-maximum of this region, as indicated by the arrows, is 50 nm [10]. This value is comparable to that obtained using a transmission electron microscope, and indicates the high-resolution capability of the NOM system. This is essentially because of the size-dependent feature of the optical near field (cf. Chaps. 2 and 7), and is due to selective detection of high spatial frequency components of the optical near field by the protruded-type probe.

It should be noted that the pixel size of 19 nm in this experiment is larger than the size of the probe and almost comparable to the diameter of the filament. By reducing the pixel size, a much smaller value for the width, or a higher resolution, is expected.

8.2.2 Imaging in Water

In the case of biospecimens, it becomes necessary to conduct observations in a liquid environment, i.e., in the natural environment of the biospecimen. The c-mode NOM makes it easier to extend observation in a liquid environment owing to the inherent possibility of using the optical near-field intensity feedback. The optical near-field intensity for constant-distance mode operation means that the probe is free from the strong viscous drag of the liquid, and hence is suitable for imaging the surface of a soft sample [11].

A flagellar filament sample fixed on a hydrophilized glass substrate was fitted with an acrylic ring. The ring had an inner diameter of 10 nm and a height of 2 nm. The ring was filled with water during the observation. The ring makes the surface of the water flat and its surface tension uniform over the whole surface. The probe was inserted into the water at a contact angle of 90°, allowing the probe to be immersed without bending.

Figure 8.3 shows the image of the filaments obtained in water. The pixel size is 10 nm×10 nm. During the scan, the sample–probe separation was estimated to be less than 30 nm. The bright segments represent fragments of the filament. The full width at half-maximum of the fragment, as indicated by the arrows, is 50 nm. On comparing this value with the nominal value estimated from TEM observations in vacuum, the difference is only around 25 nm, which indicates the potential of NOM in conducting successful high-resolution observations in liquid. The reason for the difference is the difference in the surrounding conditions around the filaments.

8.3 Observation of Subcellular Structures of Neurons by i-Mode NOM

Surface information about neural interconnections such as synaptic contacts is important to our understanding of the mechanism of formation of neural interconnections. Figure 8.4 shows a schematic representation of a neuron, which consists of a cell body and neural processes. Inside a branching neural process, there are many microtubules made up of a protein called tubulin [14]. They are responsible for the transport of proteins and intercellular vesicles (packets of protein substances). The end of the neural process is called a

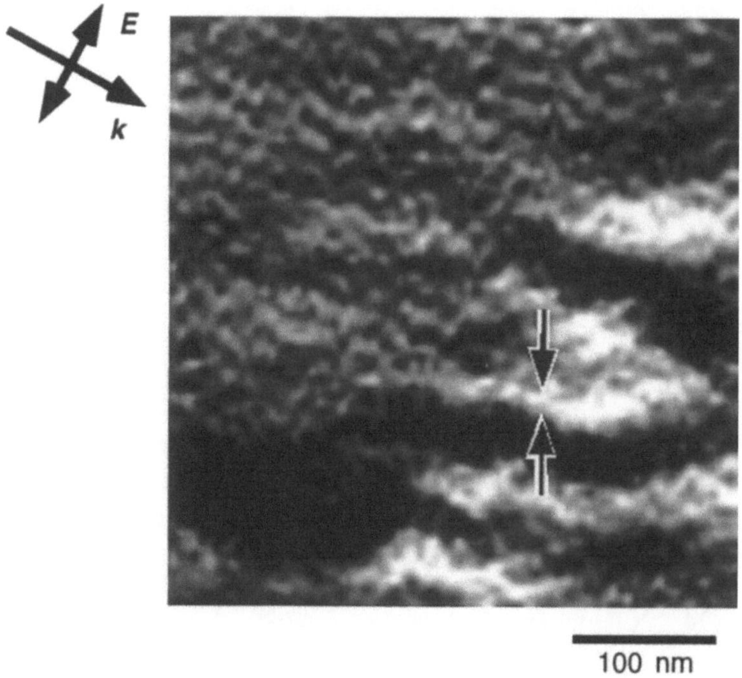

Fig. 8.3. Perspective view of the image of a filament in water obtained by c-mode NOM. The arrows top left represent the directions of wave-vector $\boldsymbol{k}$ and electric field vector $\boldsymbol{E}$ of the incident light. The full width at half-maximum of the bright region corresponding to the filament, as indicated by the arrows, is 50 nm

growth cone, and it has a palm-like lamellipodium. Inside the growth cone, there are many tubulin and actin filaments forming a network-like structure [15]. This growth cone is thought to be stretching itself to make connections with other cells. Although an optical microscope makes it possible to perform studies on the growth cone [12], the resolution is limited to the order of wavelength. On the other hand, it is difficult to investigate a soft membrane with a contact-mode atomic force microscope owing to the presence of strong forces.

As the neuron samples are relatively thick, it is difficult to image them by c-mode NOM. Also, when the specimens are stained with absorption dyes, large absorption will lead to very weak signals with c-mode NOM. In order to overcome these difficulties, the observation of neurons has been done with i-mode NOM [13].

8.3.1 Imaging in Air Under Shear-Force Feedback

The neurons used here were taken from the tissue of the hippocampal region of the brain of a Wistar rat. The functions of this region are related to learning

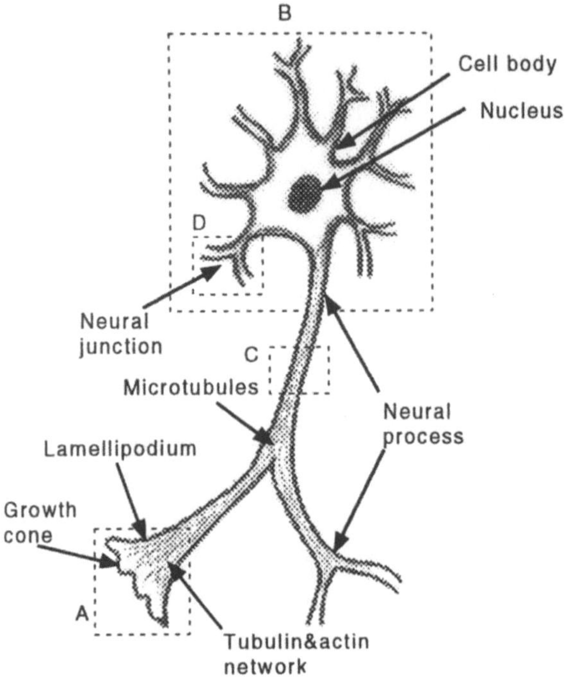

Fig. 8.4. A schematic representation of a neuron with its cell body and neural processes. Areas A, B, C, and D were imaged, and are shown in Figs. 8.5, 8.6, 8.7, and 8.9, respectively

and memory. After the neuron tissues had been cultured for 30 days, they were fixed with para-pharmaldehyde (4%) and glutal aldehyde phosphate (2%) buffered saline solution and rinsed with distilled water. Finally, they were dried on a microscopic cover-glass substrate. We prepared two kinds of samples, one with a dye label (toludine blue) with an absorption band around 600 nm wavelength, and one with dye.

The probe used in the experiments is the same as the one shown in Fig. 3.23 with no protrusion and a foot diameter of less than 30 nm.

8.3.1.1 Imaging of Neurons Without Dye Labeling

Figure 8.5a shows the NOM image of a neural process obtained under constant-distance mode by shear-force feedback with a sample–probe separation of around 40 nm. The filled arrows indicate the neural processes and the open arrow indicates the growth cone. The edges of the neural processes are seen with good contrast.

Figure 8.5b shows a magnified view of the region of the flat lamellipodium as indicated by a square in Fig. 8.5a. During the scan, the sample–probe separation was less than 20 nm. The honeycomb-like structure seen in this

(a)

(b)

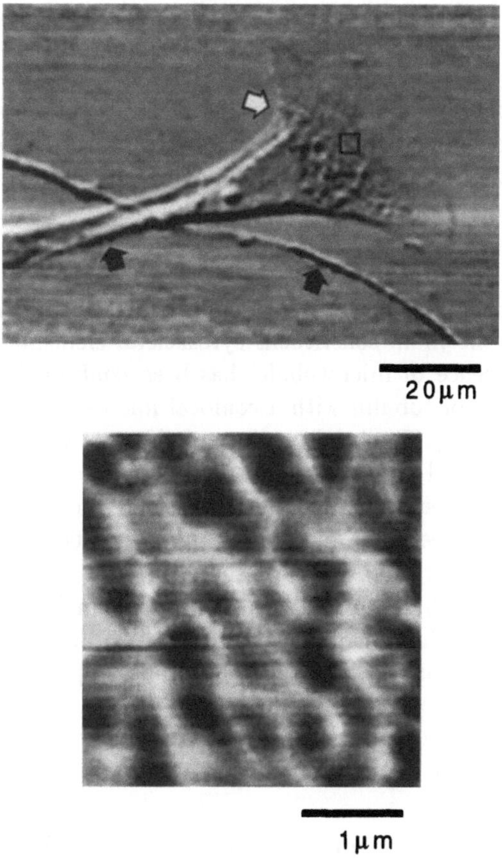

Fig. 8.5. a NOM image of an unlabelled neural process with extending growth cone. *Filled arrows* indicate the neural processes and the *open arrow* indicates the growth cone. b The magnified scan over the lamellipodium of the growth cone indicated by a square in a. The network like structure consisting of tubulin and actin filaments is formed during the drying of the sample

figure is the network of cytoskeleton consisting of actin and tubulin filaments. The network-like structure was formed during the drying of the sample.

8.3.1.2 Imaging of Neurons Labeled with Toluidine Blue

Figure 8.6a shows the NOM image obtained from a neuron sample labeled with toluidine blue under a sample–probe separation of less than 50 nm. The filled arrows indicate neural processes and the open arrow indicates the cell body. The image shows enhanced contrast due to the presence of strong absorption variations. Figures 8.6b and c show the simultaneously measured NOM image and shear-force topographic image, respectively, of the magnified view of the region indicated by the long arrow in the neural process in Fig. 8.6a. A sample–probe separation of less than 10 nm was maintained by shear-force feedback. In the NOM image of Fig. 8.6b, the fringe-like structures is because the structures lie just underneath the cell membrane on the surface. Such an observation is possible owing to the accessibility of the optical near field.

However, no clear structure could be seen in the shear-force topographic image in Fig. 8.6c because the shear force is sensed only by variations on the surface. Hence, NOM excludes the necessity of removing the cell membrane with detergent which is done to make observations with electron microscopy and atomic force microscopy. In other words, light could resolve structures which lie below the cell membrane.

The fringe-like structures seen in Fig. 8.6b could be identified as microtubules because they are the main constituent cytoskeletal elements of the neural processes. The presence of microtubules has been confirmed by immunohistochemical staining of tubulin with a confocal microscope (H. Tatsumi, personal communication). The bright regions of different widths correspond to microtubules observed either in the form of bundles or as a single tube. The dark regions with a still smaller width, seen on either side of the bright regions, are voids between the microtubules. Figure 8.6d shows the cross-sectional optical near-field intensity variation across the bright region indicated by the line in Fig. 8.6b. Its full width at half-maximum is estimated to be 26 nm, which agrees with the nominal diameter of 25 nm of a single microtubule measured by a scanning electron microscope [16]. This agreement demonstrates the high-resolution capability of the present NOM, which could be due to the field enhancement effects at the boundary between the dielectric and the metal coating of the probe [17].

8.3.2 Imaging in Water Under Optical Near-Field Intensity Feedback

Neurons isolated from the brain cortex of a Wistar rat were fixed on a microscopic cover glass and cultured for 14 days. They were then chemically fixed with 4% paraphamaldehyde for 30 min, followed by washing three times in phosphate buffered solution (PBS). Prepared samples were stored in PBS at 4°C to prevent the deterioration of the cells. The PBS was used during the actual observations in an aqueous environment.

8.3.2.1 Imaging in Air

An experimental set-up of i-mode NOM with optical near-field intensity feedback was used (as shown in Fig. 6.8). A shear-force feedback scheme as illustrated in Fig. 6.4 was also included for qualitative comparison. Figure 8.7 shows the NOM image of a section of the neural process corresponding to region C in Fig. 8.4 obtained under shear-force feedback. The long arrow indicates bundles of microtubules (neurofilaments) in the neural process.

Figures 8.8a and b show images obtained under shear-force and optical near-field intensity feedback, respectively, of the region indicated by a

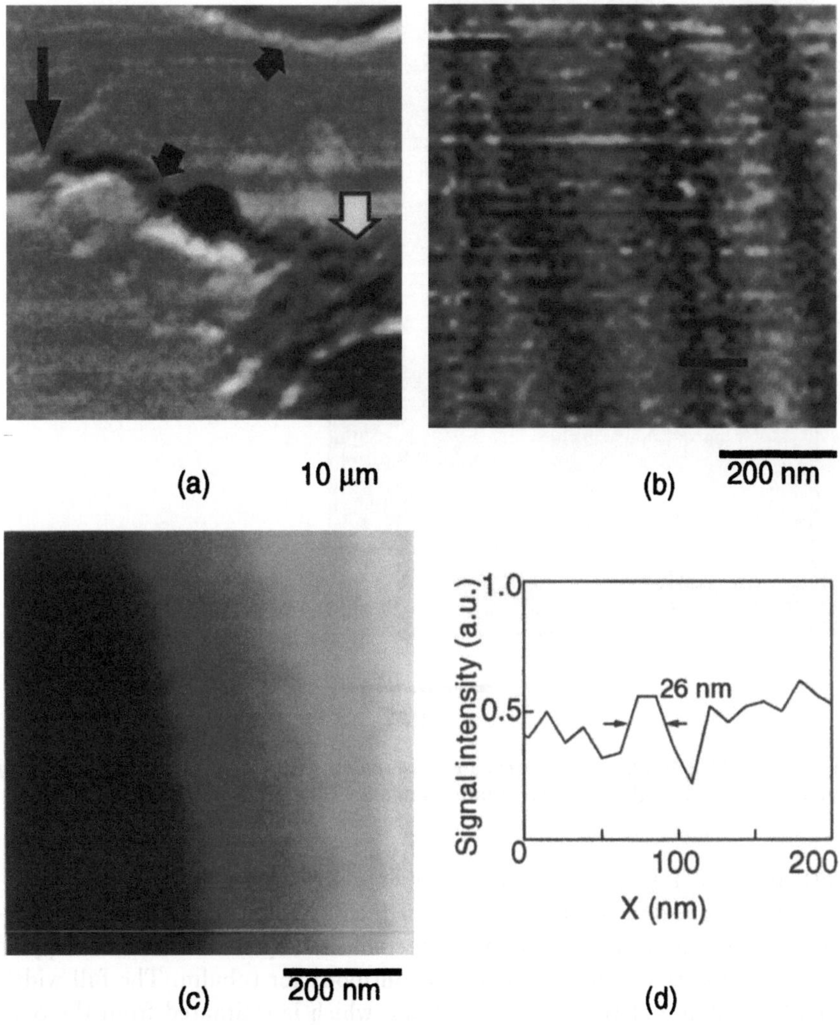

Fig. 8.6. **a** NOM image of a neuron labeled with toluidine blue. The wide *filled arrows* indicate the neural process and the *open arrow* indicates the cell body. **b**, **c** NOM and shear-force topographic images respectively, of the region indicated by a *long arrow* in **a**. The fringe-like structures in **b** are microtubules. In the shear-force topographic image, the bright regions correspond to valleys and the dark regions correspond to hills. **d** The cross-sectional optical near-field intensity variation across the tube as indicated by the *line* in **b**

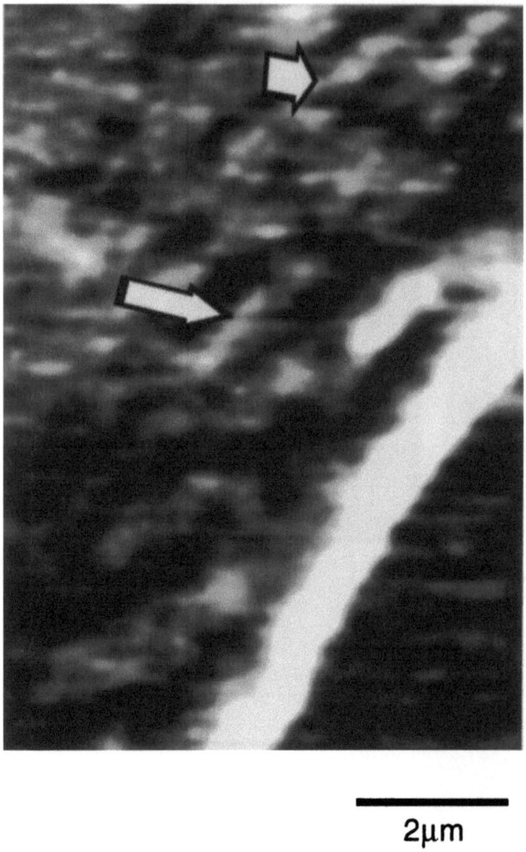

Fig. 8.7. NOM image of a neural process under shear-force feedback. The *long arrow* indicates bundles of microtubule filaments. The pixel size is 78 nm×78 nm

short arrow in Fig. 8.7. The arrows in Figs. 8.8a and b indicate bright regions which are due to large bundles of microtubules. The presence of microtubules in neural processes has been confirmed by confocal microscopy of these neurons stained with monoclonal antibody for tubulin. The full width at half-maximum of the bundle is 70 nm, which is estimated from the one-dimensional intensity variation across the line in Fig. 8.8b. The results show that optical near-field intensity feedback could very well be used in sample–probe separation control of i-mode NOM to image the neuron sample. By this technique, the artifact problem arising from using shear-force feedback could also be avoided [20].

8.3.2.2 Imaging in PBS

The sample plate was attached with a plastic ring so as to hold the liquid. The ring also makes the surface tension uniform all over the whole liquid surface. After the ring attached to the plate is placed over the parallel plate with a

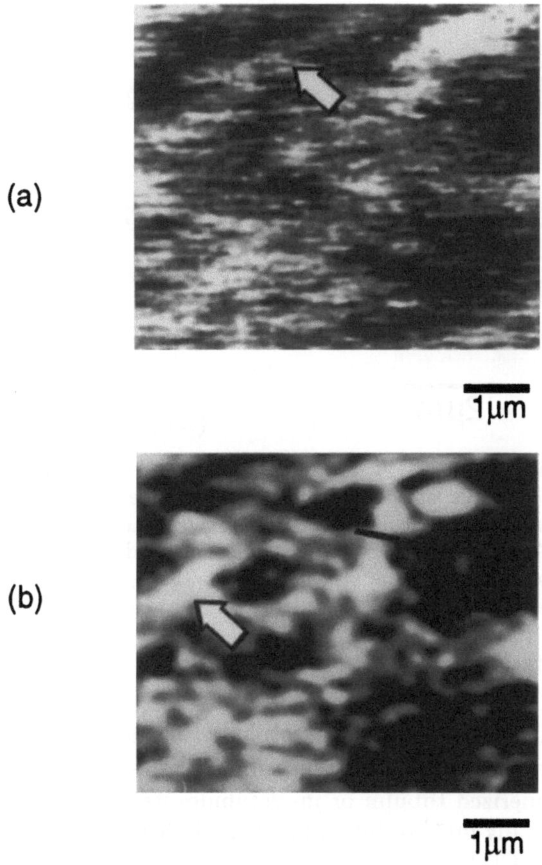

Fig. 8.8. NOM image of the region indicated by a short arrow in Fig. 8.7 under **a** shear-force feedback and **b** optical near-field intensity feedback. The pixel size is 39 nm×39 nm. *Arrows* indicate the bundles of microtubulin present within the neural process. The full width at half-maximum of the bundle is 70 nm, which is estimated from the one-dimensional intensity variation across the line shown in **b**

sandwich of index matching oil, scanning is performed in a constant-distance mode with optical near-field intensity feedback.

Figure 8.9a shows the i-mode NOM image obtained by scanning region D in Fig. 8.4. A schematic trace of the image in a indicating the parts is shown in Fig. 8.9b. This region corresponds to the junction of the neural processes extending from the cell. The open arrows in these figures indicate two neural processes. These figures also show the junction which is indicated by a filled arrow. These results indicate that it is possible to operate the i-mode NOM under optical near-field intensity feedback for operation in liquid [21].

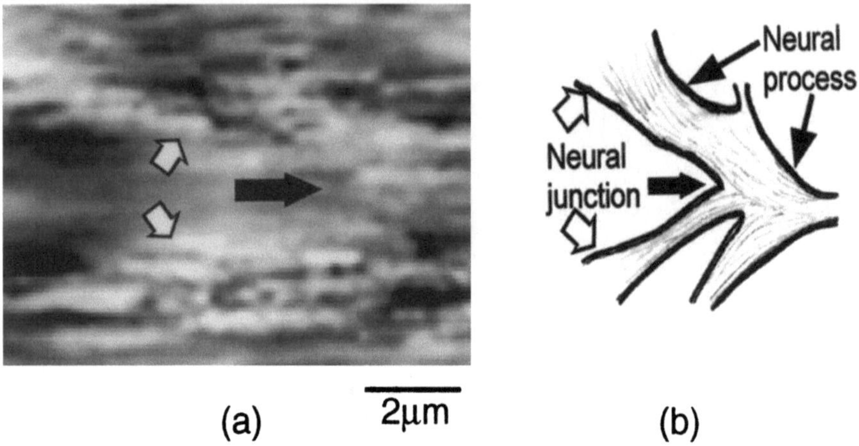

Fig. 8.9. a NOM image showing the junction of two neural processes extending from the cell which was obtained under optical near-field intensity feedback. The pixel size is 78 nm×78 nm. *Open arrows* indicate the neural processes, and the *filled arrow* indicates a junction. b A schematic trace of the image in a indicating the junction and the neural processes

8.4 Imaging of Microtubules by c-Mode NOM

The preparation of tubulin and the growth of microtubules was done by following the technique of Valters et. al. [22]. The tubulin molecules were isolated from a fresh pig brain. Polymerized tubulin or microtubules were obtained from an assembly solution of 1 mg ml^{-1} of tubulin in 20 mM (N-morpholino) ethanosulfonic acid, 80 mM NaCl, 1 mM ethylenebis (oxyethylenenitrilo) terraacetate, 0.5 mM $MgCl_2$, 0.36 mM GTP (guanosine 5'-triphosphate) of pH 6.4 after incubation at 37°C for 30 min. The stabilized structure was diluted to control the density of microtubules. The stabilized microtubules were placed on a poly-lysine coated cover-glass substrate and wicked off after a few minutes. Figure 8.10 shows a transmission electron micrograph of the microtubule sample. The dark region in the middle of a strand corresponds to the microtubule and the bright regions on either side correspond to the metal coating done for observation. The width of the dark region, as indicated by the arrows, is 25 nm.

The experimental set-up of c-mode NOM, as shown in Fig. 6.2, was used [23], where a shear-force feedback system was added to get a topographic image for comparison with the NOM image. The sample was mounted on the prism. The incident light was controlled by a half-wave plate to be s-polarized, i.e., its electric field vector is parallel to the sample surface. The probe used was fabricated by the selective chemical etching and selective resin coating methods, which is the same as that shown in Fig. 3.22. The protruded probe has an apex diameter d and a foot diameter d_f of 10 nm and 30 nm, respectively.

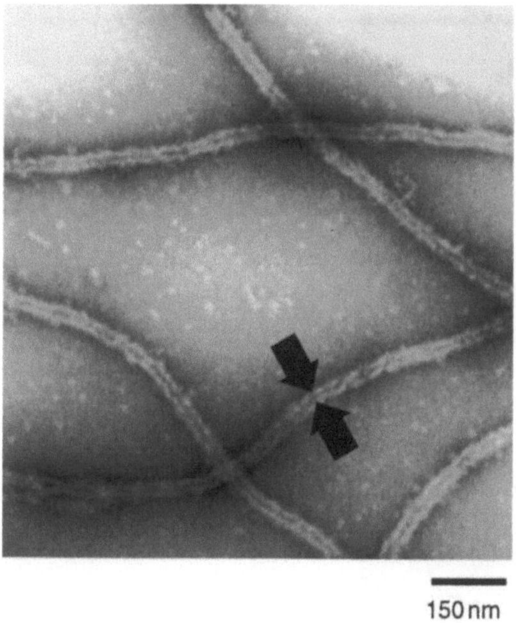

Fig. 8.10. A transmission electron micrograph of the microtubule sample

Figure 8.11a shows a shear-force topographic image of a microtubule obtained by maintaining the sample–probe separation at approximately 5 nm. The long narrow bright structure corresponds to a microtubule, and the bright round structures correspond to protein aggregates. The height and the full width at half-maximum of the microtubule are 16 nm and 60 nm, respectively. Figure 8.11b shows the magnified NOM image of the rectangular region in Fig. 8.11a obtained under constant-height mode at a sample–probe separation of 30 nm without using shear-force feedback. The arrows indicate the directions of the propagation vector $\boldsymbol{k}$ and the electric field vector $\boldsymbol{E}$. Both the microtubule and the protein aggregate appear darker with bright edges in Fig. 8.11b. In other words, the NOM image appears inverted and exhibits edge enhancement. This agrees with the results of the theory based on the optical extinction theorem [24]. The diameter of the microtubule was 40 nm, which was as estimated from the full width at half-maximum at the position indicated by arrows in Fig. 8.11b.

8.5 Imaging of Fluorescent-Labeled Biospecimens

One of the major advantages of NOM over its contemporary scanning probe microscopes is the possibility of labeling the biological samples with some specific fluorescent dye and then performing fluorescence imaging. This sec-

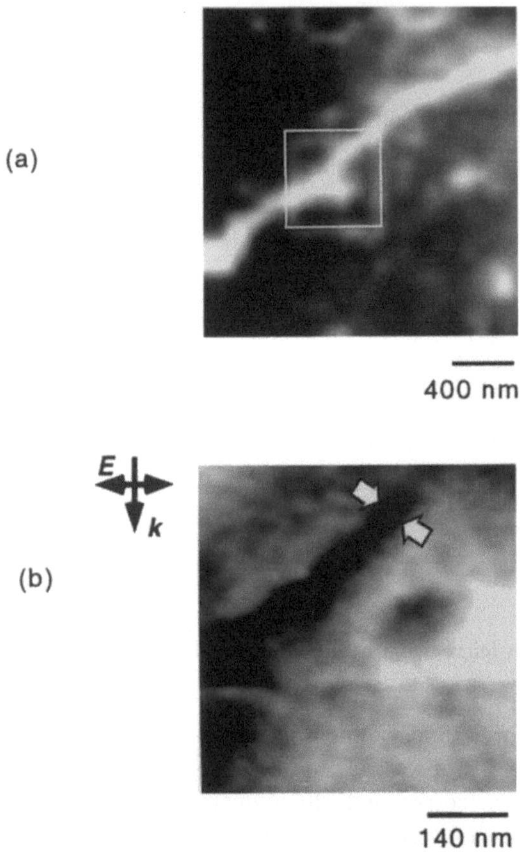

Fig. 8.11. a Shear-force topographic image of the microtubule (long and narrow) and protein aggregates (round), which appear white. b Magnified NOM image of the rectangular region in a taken under constant-height mode. The arrows top left indicate the directions of the propagation vector $\boldsymbol{k}$ and the electric field vector $\boldsymbol{E}$. The full width at half-maximum of the dark line structure is approximately 40 nm at the position indicated by the *arrows*

tion discusses the imaging of fluorescence from fibroblasts of cells taken from a normal rat kidney by using an i-mode NOM.

The fibroblasts were obtained from the subcloned cell line of a normal rat kidney. The cell line was grown in Eagle's minimum essential medium (MEM), washed with concentrated H_2SO_4, warmed, dried, and then transferred to a cover-glass plate. After 24 h, the cell line was fixed with a 3% parapharmaldehyde and PBS solution at 25°C for 30 min. After washing with 0.1% triton x-100 detergent, a hole was made in the surface lipid layer. The sample was placed in 0.2 M glysin for around 15 min to quench the fluorescence from the paraform. Next, the cell line was made to react with fluorescein (a dye emitting fluorescence at around a wavelength (λ) of 500 nm) attached to an

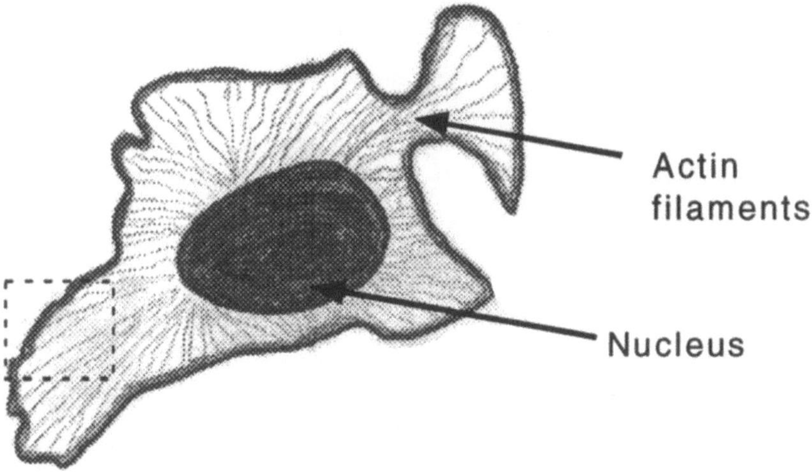

Fig. 8.12. A schematic representation of a fibroblast cell with a nucleus at the center and actin filaments extending radially. The *square* drawn in broken lines indicates the region in which the probe of the NOM was scanned

anti-rabbit antibody which was taken from goat. During this step, blocking is done using a 10% goat serum/PBS in order to avoid nonspecific attachment of the antibodies. Finally, the cell line with specific labeling done for actin filaments is put in distilled water and rinsed for short periods in alcohol and acetic acid consecutively.

Figure 8.12 shows a schematic representation of a fibroblast cell. There is a nucleus at the center with actin filaments radially extending over the cell. The square drawn in broken lines indicates the region in which the probe of the NOM was scanned. Fluorescence was detected using the photon counting method. The fluorescent dye-labeled fibroblast sample was mounted on the piezo stage of the i-mode NOM system with shear-force feedback for constant-distance mode. Light from an Ar^+ laser through the probe was used for excitation. The fluorescent light emitted from the sample was collected by a microscopic objective lens and detected by an avalanche photodiode. A bandpass filter with a center at $\lambda = 550$ nm and a notch filter, rejecting $\lambda = 488$ nm (rejection factor $= 1 \times 10^{-6}$), were used for the efficient detection of fluorescence.

Figure 8.13a and b show the simultaneously obtained shear-force topographic image and the fluorescence intensity distribution, respectively. The bright streak indicated by an open arrow in Fig. 8.13b corresponds to the actin bundles. The filled arrows in the two figures show the places where clusters of dye are attached to the actin bundles. Information about the cellular dynamics are expected in the near future through improvements in

(a)

(b)

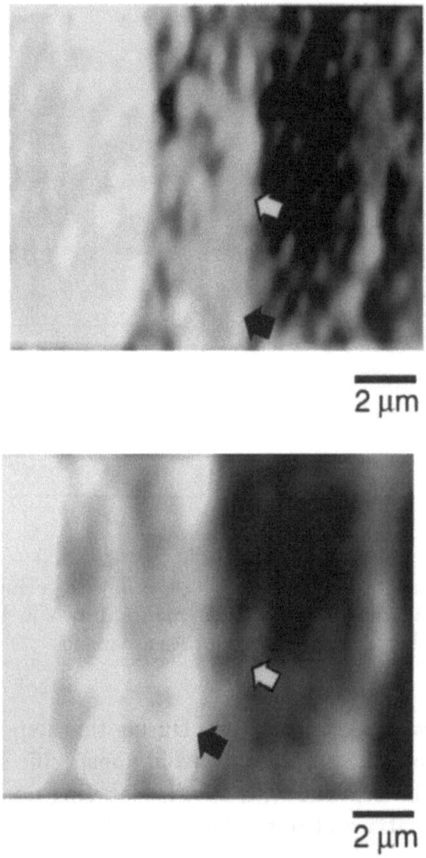

Fig. 8.13. a Shear-force topographic image and b the fluorescence intensity image obtained simultaneously across the region indicated by a square in Fig. 8.12. The *open arrow* indicates the actin bundles and the *filled arrow* indicates the clusters of dye attached to the actin bundle

sample preparation and the system for conducting in vivo observations of the cell.

8.6 Imaging DNA Molecules by Optical Near-Field Intensity Feedback

DNA (deoxyribonucleic acid) molecules are the main constituents of the cells of a living organism for they contain all the information coded in the formation of that organism [15]. The sample consists of double-stranded plasmid DNA with a ring structure (pUC18, 2868 base pairs) diluted in distilled water and fixed on an ultrasmooth sapphire surface. The ultrasmooth sapphire surface was used as the substrate because of its flatness and stability in air and

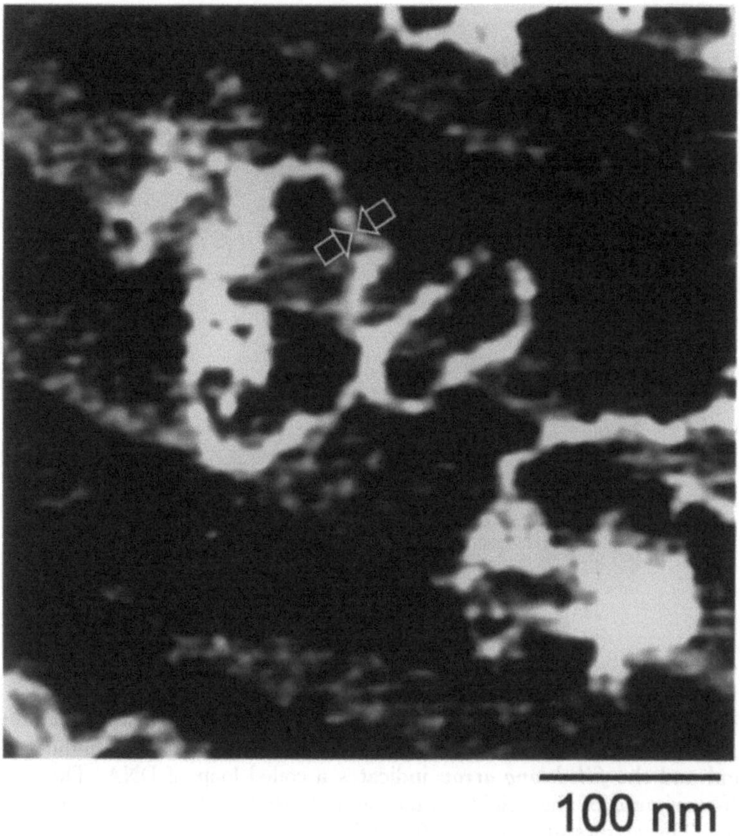

Fig. 8.14. An atomic force image of DNA on a sapphire substrate. The width of a single strand, as indicated by the *arrows*, is around 10 nm

water [25]. Further surface height variations, as discussed in Ref. [26], can be controlled to within less than 1 nm, and the lateral widths of the terraces of the steps could also be adjusted. As the ultrasmooth sapphire surface itself is hydrophobic, it is necessary to make it hydrophilic for a stronger adsorption of DNA to the surface. This was done by treating the surface with sodium diphosphate for few minutes and then washing in distilled water. The concentration of the DNA solution was 5 ng μl^{-1}, and a quantity of 2 μl was dispersed on the surface and blown dry with compressed air.

The DNA filaments on the surface are in the form of either a single strand or a coiled loop of length around 400 nm. The width is around 2–4 nm and the height is around 4 nm. Figure 8.14 shows the image of DNA on the ultrasmooth sapphire surface obtained using a noncontact-mode atomic force microscope. The observed width of a single strand, as indicated by the arrows, is estimated to be around 10 nm.

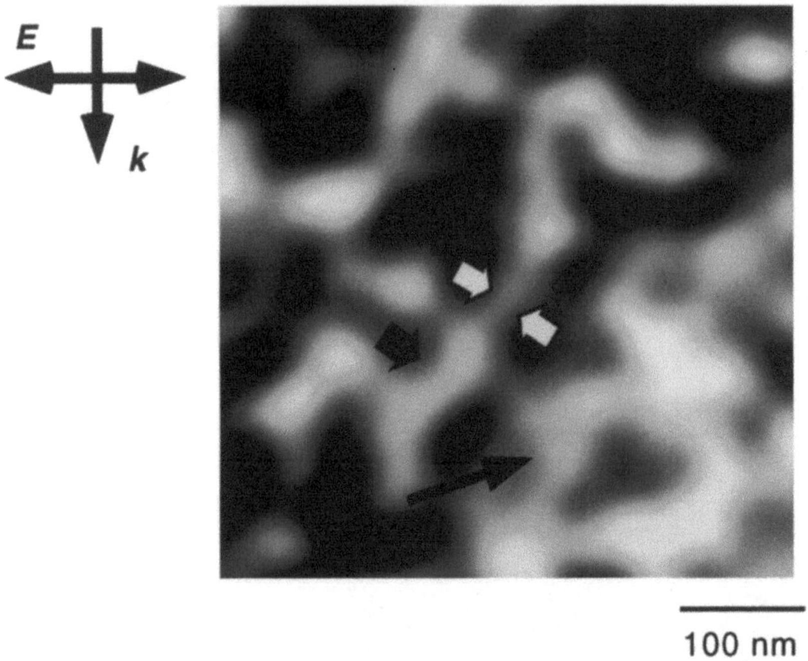

Fig. 8.15. c-mode NOM image of DNA obtained under constant-distance mode with optical near-field intensity as the control signal. The *filled short arrow* indicates a single strand and the *filled long arrow* indicates a coiled loop of DNA. The full width at half-maximum of the bright strand, as indicated by the *open arrows*, is around 20 nm. The arrows at the top left-hand corner of the figure indicate the propagation $\boldsymbol{k}$ and electric field $\boldsymbol{E}$ vectors of the incident light

The experimental system of c-mode NOM shown in Fig. 6.2 was used to image the DNA sample. The DNA sample is mounted on the prism and the highly localized optical near field is picked up by the probe. The probe used in this experiment is optimized to have both high sensitivity and high resolution. The probe is pencil-shaped, which is produced by etching (see Sect. 3.4.1) and made to have a very small nanometric protrusion with a thin layer of silver coating at its apex to enhance the sensitivity. Figure 8.15 shows the NOM image under constant-distance mode obtained by optical near-field intensity feedback. The pixel size is 5 nm×5 nm. Arrows at the top left-hand corner of the figure indicate the propagation $\boldsymbol{k}$ and electric field $\boldsymbol{E}$ vectors of the incident light. Both single strands (indicated by a filled short arrow) and coiled loops (indicated by a filled long arrow) of DNA could be observed in this figure. As seen, the individual strands are clearly resolved to have a full width at half-maximum of around 20 nm, as indicated by the open arrows. This high-resolution imaging of a DNA strand is attributed to the efficient pick-up of the high-frequency components of the optical near field by the probe. Recently, by improving the performance of our NOM system, the observed width of a single strand has been reduced to 4 nm when the pixel size was 2 nm × 2 nm. Details of this result will be published elsewhere.

References

1. R. J. Cherry, *New techniques of optical microscopy and microspectroscopy, Topics in Molecular and structural biology* Vol. 15 (Macmillan Press Ltd., London, 1995)
2. H. G. Hansma, J. Vesenka, C. Siegerist, G. Kelderman, H. Morrett, R. L. Sinsheimer, V. Elings, C. Bustamante, P. K. Hansma, Science **256**, 1180 (1992)
3. R. Weisendanger, H.-J. Guntherodt (eds.), *Scanning tunneling microscopy II* (Springer-Verlag, Berlin 1992)
4. E. Betzig, J. K. Tautman, Science **257**, 189 (1992)
5. W. P. Ambrose, P. M. Goodwin, J. C. Martin, R. A. Keller, Science **265**, 364 (1994)
6. R. Dunn, G. R. Holtom, L. Mets, X. S. Xie, J. Phys. Chem. **98**, 3094 (1994)
7. R. J. Pylkki, P. J. Moyer, P. E. West, Jpn. J. Appl. Phys. **33**, 3785 (1994)
8. E. Betzig, R. J. Chichester, F. Lann, D. L. Taylor, Bioimaging **1**, 129 (1993)
9. S. Kato, H. Okino, S. -I. Aizawa, S. Yamaguchi, J. Mol. Biol. **219**, 471 (1991)
10. M. Naya, S. Mononobe, R. Uma Maheswari, T. Saiki, M. Ohtsu, Opt. Commun. **124**, 9 (1996)
11. M. Naya, R. Micheletto, S. Mononobe, R. Uma Maheswari, M. Ohtsu, Appl. Opt. **36**, 1681 (1997)
12. H. Tatsumi, H. Sasaki, Y. Katayama, Jpn. J. Physiology **43**, S221 (1993)
13. R. Uma Maheswari, H. Tatsumi, Y. Katayama, M. Ohtsu, Opt. Commun. **120**, 325 (1995).
14. J. G. Nicholls, A. R. Martin, B. G. Wallace, *From neuron to brain* (Sinauer Assoc., Massachusettes, 1992)
15. B. Alberts, D. Bray, J. Lewis, M. Raff, J. D. Watson, *Molecular biology of the cell* (Garland publishing, New York, 1994)
16. H. Hartwig, J. Cell Biology **118**, 1421 (1992)
17. M. Specht, J.D. Pedaring, W. M. Heckl, T. W. Hänsch, Phys. Rev. Lett. **68**, 476 (1992)
18. K. Jang, W. Jhe, Opt. Lett. **21**, 236 (1996)
19. T. Saiki, M. Ohtsu, K. Jhang, W. Jhe, Opt. Lett. **21**, 674 (1996)
20. B. Hecht, H. Bielefeldt, Y. Inoue, D. W. Pohl, L. Novotony, J. Appl. Phys. **81**, 2492 (1997)
21. R. Uma Maheswari, S. Mononobe, H. Tatsumi, Y. Katayama, M. Ohtsu, Optical Review **3**, 463 (1996)

22. W. Valter, W. Fritasche, A. Schaper, K. J. Bohm, E. Unger, T. N. Jovin, J. Cell Science **108**, 1063 (1995)
23. A. V. Zvyagin, M. Ohtsu, Opt. Commun. **133**, 328 (1997)
24. A. V. Zvyagin, J. D. White, M. Ohtsu, Opt. Lett. **22**, 955 (1997)
25. K. Yoshida, M. Yoshimoto, K. Sasaki, T. Ohnisni, T. Ushiki, J. Hitomi, M. Sigeni, Biophysical Journal **74**, 1654 (1998)
26. M. Yoshimoto, T. Maeda, T. Ohnishi, H. Koinuma, O. Ishiyama, A. Shinohara, M. Kubo, R. Miyura, A. Miyamoto, Appl. Phys. Lett. **67**, 2615 (1995)

Chapter 9

Diagnosing Semiconductor Nano-Materials and Devices

9.1 Fundamental Aspects of Near-Field Study of Semiconductors

9.1.1 Near-Field Spectroscopy of Semiconductors

In this chapter, we describe the near-field photoluminescence (PL) spectroscopy of semiconductor micro/nano structures fabricated on opaque substrates. In such experiments, several specific techniques are necessary compared with the standard measurements shown in other chapters, from the viewpoints of probe structures, optical configurations, signal detections, and so on. In particular, the protruded probe (see Sect. 4.1 and Fig. 4.1) with a small foot diameter is not suitable for such measurements from the aspects of sensitivity and contrast. To date, even without high resolution, the experimental results obtained with an apertured probe (see Fig. 4.1) have been much more informative than data obtained using a protruded probe. Almost all the experiments in this chapter were performed using an apertured probe with foot diameters, d_f, from 100 nm to 1 μm.

Near-field optical techniques have made a remarkable contribution to imaging and spectroscopy in the diagnostics of semiconductor devices, including laser diodes, photodetectors, and light emitting diodes [1–7]. In the study of semiconductors, the near-field optical microscope (NOM) has been employed in two different ways. First, spatially resolved spectroscopy has been successfully implemented to determine the optical structures and responses of photonic devices and materials, tracing their surface topography [8–18]. Second, NOM can be used as a probe for local excitation and collection to achieve single particle spectroscopy, i.e., the observation of individual PL spectra of inhomogeneously broadened systems such as the islands of quantum wells [19], quantum wires [20–22] and quantum dots [23–25].

Recent rapid progress with pulsed laser sources and photodetectors [26] enables us to combine the near-field method with conventional time-resolved spectroscopy or nonlinear spectroscopy [27–31]. These techniques provide physical insights into the ultrafast dynamic processes of excited carriers and elemental excitations. Furthermore, in the last few years, NOM operation at low temperature has been an indispensable technique for the fundamental investigation and understanding of the intrinsic optical properties of semiconductors [32–35].

In these advanced experiments, we encounter serious problems which are particular to the optical investigation of semiconductors. One of the problems is concerned with the conversion of electromagnetic modes generated by a small aperture. Since semiconductors have large refractive indices, the evanescent modes are easily coupled into the propagating modes due to the interaction with semiconductor surfaces. This conversion results in reduced NOM resolving power. In order to retain the evanescent modes in the optically dense semiconductors, d_f should be much smaller than the wavelength of the excitation light, which causes low excitation efficiency. The diffusion of photoexcited carriers brings about another problem. In the detection of PL images and spectra, an illumination–collection hybrid mode is an essential technique in order to prevent the deterioration of the resolution, which is determined by the carrier diffusion length.

9.1.2 Optical Near Field Generated by a Small Aperture and Its Interaction with Semiconductors

The modal analysis based on Fourier optics reveals that the optical field through a small aperture is composed of propagating modes and evanescent modes [36]. Such a mode distribution in tangential wavevector $k_{||}$ (a component projected onto the flat aperture plane) space is illustrated in Fig. 9.1. The degree of occupation in the evanescent modes, which is characterized by the cutoff wavevector (k_c) of the distribution, is determined by d_f and the refractive index n of the circumstance. When the probe is placed in the free space, the modes with $k_{||}$ smaller than k_0 ($=2\pi/\lambda$) can propagate in the direction of ϕ, where ϕ is determined by $k_0 \sin\phi = k_{||}$. On the other hand, when $k_{||}$ is larger than k_0, such modes with pure imaginary ϕ cannot propagate into the free space and then behave as evanescent modes, which localize near the probe.

Since typical semiconductors, such as Si and GaAs, have large refractive indices of around n=3.5–4, the near-field profile from the probe is largely modified in the proximity of the sample. When the probe is positioned close to the semiconductor surface, some of the evanescent modes present in the region $k_{||}<nk_0$ are coupled into propagating modes in the material, and only some of the modes in $k_{||}>nk_0$ remains evanescent. In the spectroscopy of semiconductors, d_f should be carefully chosen to obtain high resolution beyond the diffraction limit [37, 38].

To characterize the optical field through the probe, a near-field photocurrent (PC) technique has frequently been employed [37–39]. For a photodetector of near-field light, we use a lateral p–n junction with a design (inset in Fig. 9.2) which is described in detail in Sect. 9.2.1. Figure 9.2 shows the near-field PC line scans in the vicinity of the p–n junction. The dependence of the resolution on d_f can clearly be seen. When d_f is equal to 200 nm, the

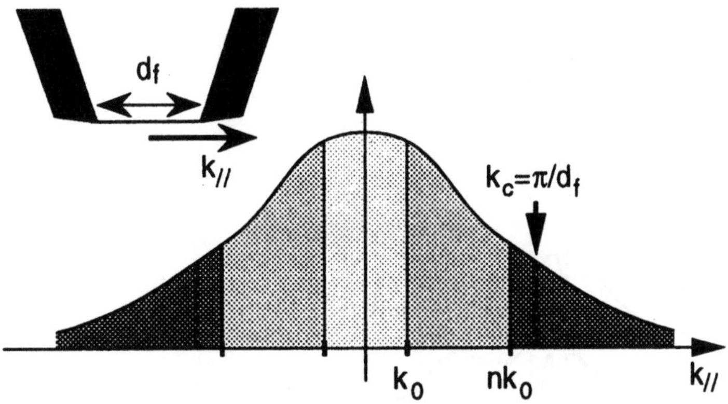

Fig. 9.1. Density of optical modes versus tangential wavevector in the plane of the aperture

signal width of 1.6 μm is much larger than d_f. Since the penetration depth of the excitation light (λ=830 nm) is about 1 μm, such a poor resolution originates from the propagation of light into the crystal and the resultant absorption at the p–n interface. On the other hand, in the case of d_f=100 nm, much narrower profiles, with a width of 0.6 μm, are obtained compared with those at d_f=200 nm. This comparison demonstrates that a large part of the excitation light through the probe with d_f=100 nm remains evanescent in the GaAs. Even in this case, however, the p–n junction is not imaged with the resolution determined by d_f. The diffusion of photoexcited carriers reduces the resolving power of NOM.

The bottom line scan in Fig. 9.2 shows the case of d_f=200 nm with λ=488 nm excitation. Almost the same resolution as that with d_f=100 nm and 830 nm excitation is achieved even by using such a large d_f. In this case, although all the light through the probe couples into the propagating modes, the shallow penetration depth of 80 nm due to the strong absorption contributes to the local excitation being equal to the evanescent excitation.

The probe-size-dependent feature of evanescent modes can be demonstrated from another aspect: determination of the decay length of the optical near field in the vertical direction [40, 41]. By measuring the PC intensity as a function of sample–probe separation z, we obtain the separate contributions of the evanescent modes and propagating modes to the near-field profile transmitted through the probe. Measurements with d_f=100, 150, and 200 nm are shown in Fig. 9.3a. Compared with d_f=150 and 200 nm, the signal with d_f=100 nm rises abruptly in the proximity of the sample surface. These results reflect the difference in evanescent mode occupancies for each probe; when d_f is small, the modes with larger $k_{||}$ are occupied and the resultant decay length of the signal becomes shorter. To estimate the contribution ratio of the evanescent modes to the propagating modes for d_f=100 nm, the position dependence of the decay length of the PC signal is measured, as

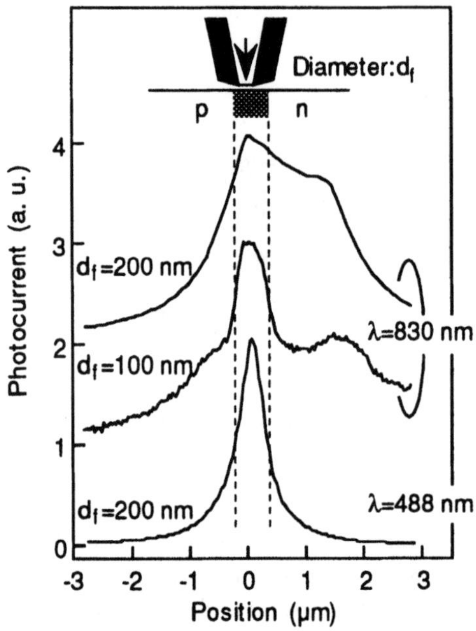

Fig. 9.2. Near-field photocurrent line scans using probes with d_f=100 and 200 nm at an excitation wavelength of λ=830 nm. The bottom line scan is the result of the measurement with d_f=200 nm and λ=488 nm excitation

shown in Fig. 9.3b. When the probe is positioned beside the active region ($x=\pm0.5$ μm), only the propagating modes generate the signal. When it is put just above the p–n junction (x=0 μm), on the other hand, not only the propagating modes but also the evanescent modes give rise to the PC. The significant difference in the signal behavior around $z\sim0$ nm reveals the contribution of the evanescent modes at x=0 μm. Such a remarkable difference is not observed at the same measurement for d_f=150 and 200 nm.

Based on these experimental data, it is concluded that in order to generate evanescent modes in GaAs, d_f should be smaller than 100 nm. This result is interpreted as indicating that the critical d_f to achieve evanescent excitation for semiconductors is less than half of the wavelength in the material: $d_f<\lambda/2n$. Since this criterion is transferred to $\pi/d_f<nk_0$, it can be said that the cutoff wavevector of the evanescent distribution lies at π/d_f (a similar conclusion appears in Ref. [37]). The dependence on the spread angle of the beam for propagating modes into GaAs on the excitation wavelength also supports this consideration, which is discussed in Sect. 9.2.5.

9.1.3 Operation in Illumination–Collection Hybrid Mode

In common with PL measurements in semiconductors, one very serious problem is the deterioration of spatial resolution due to the diffusion of pho-

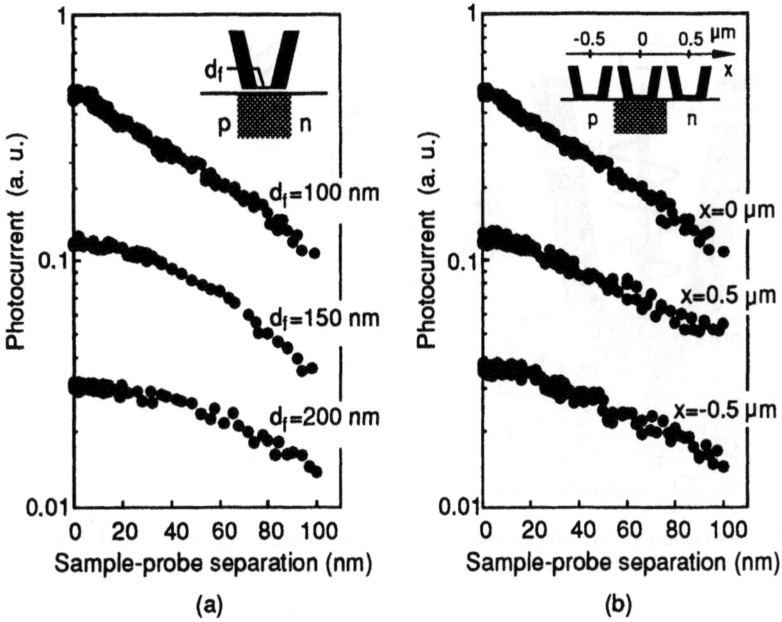

Fig. 9.3. a Near-field photocurrent intensity as a function of sample–probe separation for various d_f. b Similar plots for d_f=100 nm at the positions indicated in the inset

toexcited carriers prior to recombination. As well as the i-mode operation of NOM, there are several methods for local excitation of carriers in the region <100 nm: local electron injection methods, such as a high-energy electron beam from a scanning electron microscope (cathodoluminescence), and low-energy electron injection through a scanning tunneling microscope tip (tunneling luminescence). In the local excitation methods, however, we frequently encounter the situation where the photoexcited or injected carriers can migrate into the extended region (typically in the order of 1 μm) before recombination, depending on their diffusion coefficients and lifetime. In such a case, the spatial resolution is limited by the diffusion area as long as the PL signal is collected by a lens in the far-field configuration (Fig. 9.4). The only way to overcome this difficulty is the near-field collection of the PL signal through the probe, as illustrated in Fig. 9.4, where a resolution as small as d_f can be expected. (High resolution much smaller than d_f has been obtained in some cases as shown in Sect. 9.3.) Although the low collection efficiency of the probe has frequently been pointed out, recent progress in the design and fabrication of probes [42–46], as described in Chapt. 4, can provide us with very advanced measurements in illumination–collection hybrid mode. A theoretical understanding of the mechanism of near-field collection is also important to improve the resolution and detection efficiency.

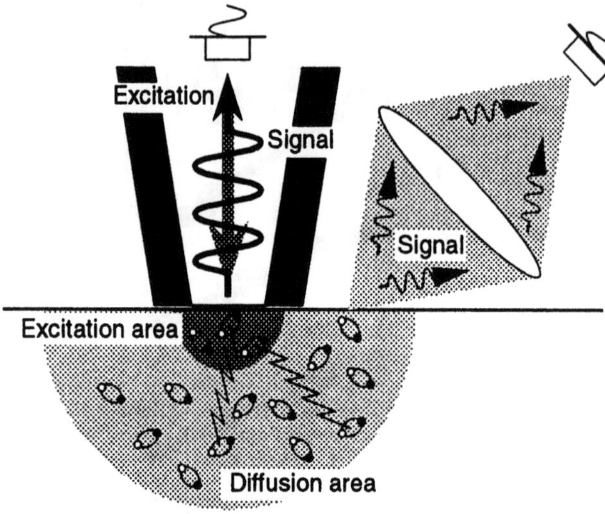

Fig. 9.4. Illustration for a comparison between the illumination mode with far-field collection and the illumination–collection hybrid mode

9.2 Multidiagnostics of Lateral p–n Junctions

9.2.1 Sample and Experimental Set-up

A lateral p–n junction is a novel device which is expected to be applied to the carrier confinement structure of a surface-emitting diode, the active regions in lasers, and so on [47–50]. Moreover, the one-step formation of such structures on a patterned substrate is a promising process which is damage-free compared with the etching and overgrowth procedures. This section is devoted to the multidiagnostics of GaAs lateral p–n junctions for both the analysis of new devices and to demonstrate the ability of the near-field spectroscopy in semiconductors.

The conduction type of GaAs layers with Si dopant depends on the growth conditions and the orientation of the substrate. By using the amphoteric nature of Si, both n- and p-type regions can be grown simultaneously on a patterned substrate, and lateral p–n junctions are formed at the boundary of the two regions. Figure 9.5 shows a schematic representation of such a structure, defining upper and lower junctions. The luminescence peak wavelength strongly depends on the conduction type and the carrier concentration of GaAs layers. By measuring the PL spectra with high spatial resolution, we can precisely examine the carrier distribution in the transition region of a p–n junction.

The fabrication procedure of the sample is described in detail in Ref. [47], and we show it briefly here. A semi-insulating GaAs (111)A substrate is etched with a photolithography technique to obtain a triangular (111)A surface surrounded by three (311)A slopes. After thermal cleaning, a Si-doped

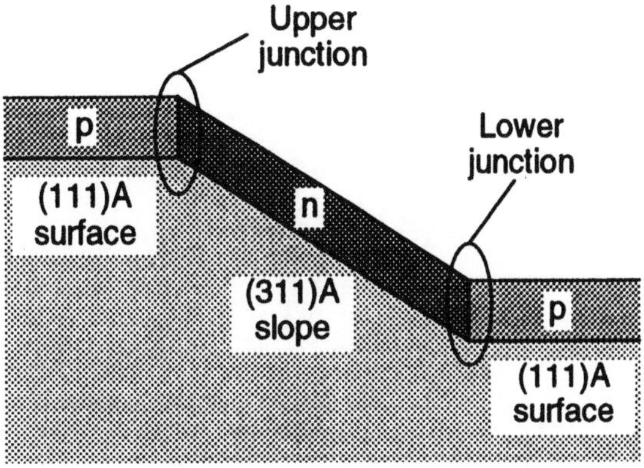

Fig. 9.5. Cross-sectional profile of lateral p–n junctions grown on a patterned GaAs (111)A substrate

GaAs layer with a thickness of 1 μm is grown on the patterned substrate at 600°C. The Si concentration is estimated as 1×10^{18} cm^{-2}.

Schematic diagrams of the experimental set-up for the multidiagnostics of p–n junctions are shown in Fig. 9.6 [41]. PL spectra, integrated PL intensity, electroluminescence (EL) intensity, and PC spectroscopy can be obtained by employing the configurations in Fig. 9.6a–d, respectively. Simply by changing the connections, all the measurements can be performed consecutively on the same area of the sample. Details of each measurement are described in the following subsections. We use a double-tapered probe metallized with gold 200 nm thick [42]. Its typical d_f is around 200 nm. The throughput is estimated as 1.0×10^{-3} by collecting the far-field output light from the probe with a 0.4 NA objective lens. For the regulation of tip–sample distance, the shear-force feedback technique is employed. To monitor the amplitude of the vibration, we use a 1.55-μm laser diode, whose photon energy is far below the absorption edge of GaAs. By positioning the tip <10 nm above the sample, near-field excitation and collection can be realized.

Figure 9.7 shows a shear-force topographic image of the sample in the vicinity of the slope. The width and height of the slope are approximately 10 μm and 6 μm, respectively. From the top surface to the slope, some bumps, which may be produced in the etching process of the substrates, appears.

9.2.2 Spatially Resolved Photoluminescence Spectroscopy

In order to determine the position and local optical properties of the transition regions of p–n junctions, spatially resolved measurements of PL spectra are investigated in illumination-mode operation [11]. A schematic diagram of the experimental set-up for PL spectroscopy is shown in Fig. 9.6a. As an excitation light source, 0.5 mW of He-Ne laser (λ=632.8 nm) is coupled into the fiber probe. By controlling the probe in close proximity to the sample,

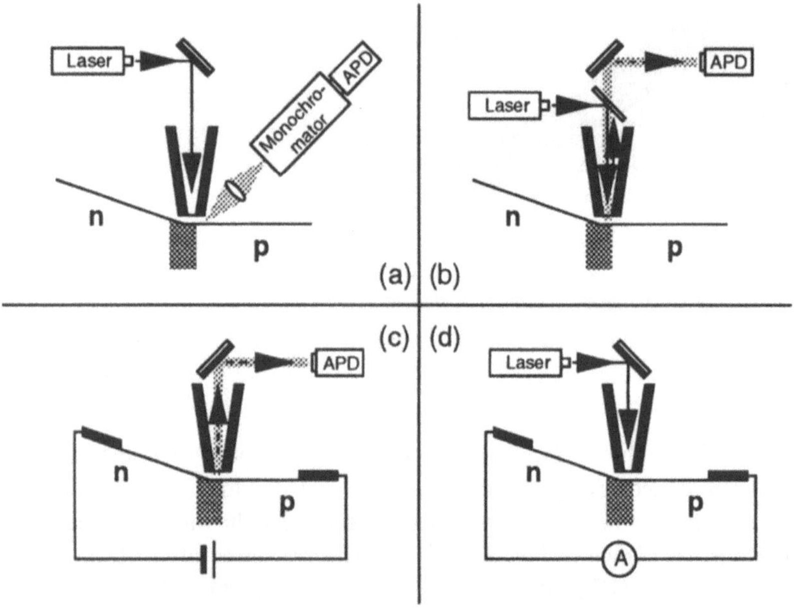

Fig. 9.6. Block diagrams of the measurements for **a** photoluminescence spectra, **b** integrated photoluminescence intensity, **c** electroluminescence intensity, and **d** photocurrent spectroscopy

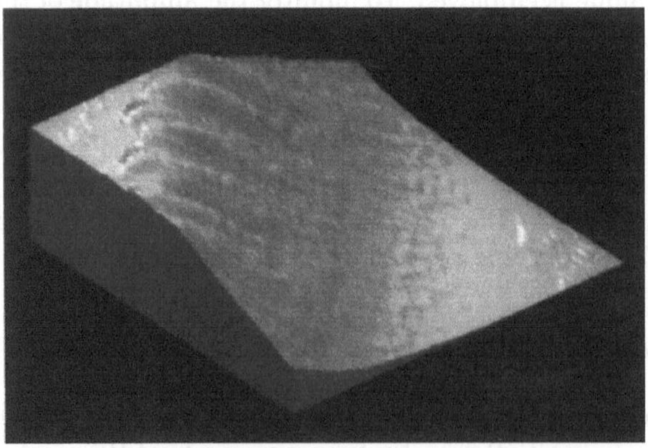

Fig. 9.7. Perspective view of a topographic image in the vicinity of the slope including upper and lower junctions. The image size is 15 μm$\times$15 μm

the excitation region is restricted to ~100 nm×100 nm in area and 200 nm in depth, which are determined by d_f and the penetration depth of GaAs. PL from the sample was collected on the same side of the sample with a 0.4 NA objective lens, and transported to a 20-cm monochromator with an avalanche photodiode for photon counting detection.

Normalized PL spectra at some characteristic points on the slope are shown in Fig. 9.8. The peak wavelength at the top surface (position A in the insert in Fig. 9.8) and at the bottom surface (E) shows the same value of 870 nm. This peak wavelength corresponds to that of a flat (111)A surface with the same Si concentration. Its conduction type (p-type) is confirmed by the measurement of capacitance–voltage (C–V) characteristics. On the (311)A slope (C), the emission peak shifts to the higher energy side. Its value of 855 nm is also in agreement with that of the n-type flat (100)A surface investigated previously. In the transition region (B and D), where the conduction type and carrier concentration gradually change, the emission peaks lie at intermediate values. To examine the transition region more precisely in relation to the sample structure, in Fig. 9.9, we plot the emission peak and the total intensity along the slope.

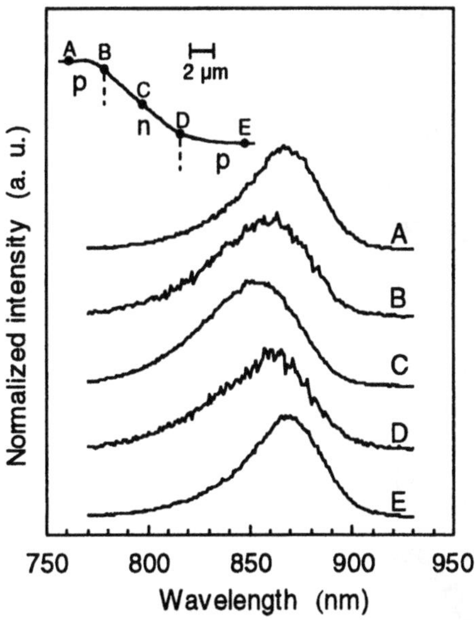

Fig. 9.8. Normalized photoluminescence spectra at some points in the vicinity of the slope. The inset shows the cross-sectional view of the p–n junctions indicating the measuring points (A–E)

A variety of features are observed in the transition region of p–n junctions. At the lower junction, the emission peak shows graded changes with a transition width of 5 μm. Since the conduction type and carrier concentration vary with the tilt angle of the substrate from the (111)A surface, the transition width is closely relevant to the structure of the junction. The gradual change in the tilt angle at lower the junction, as shown in Fig. 9.9a, causes the wide transition width of the carrier concentration. At the upper junction, on the other hand, the top surface and the slope make a clear ridge. This abrupt change results in the narrower transition width of 1 μm. From the fast rise of the emission peak at the upper junction, it is confirmed that we attain a spatial resolution of <400 nm. Such a resolution is determined rather by the diffusion length of the photoexcited carriers than by d_f. Carefully comparing the spectral change with the topographic image, it can be found that the transition region is formed on the slope, not at the ridge. So far, to explain the results of C–V measurements [49], it has been inferred that the modulation of carrier concentration occurs on the sloping side of the intersection. This has been supported by cathodoluminescence measurements with the resolution of around 1 μm. We have determined the position of the boundary with higher resolution through the accurate correspondence between the surface structure and the optical response.

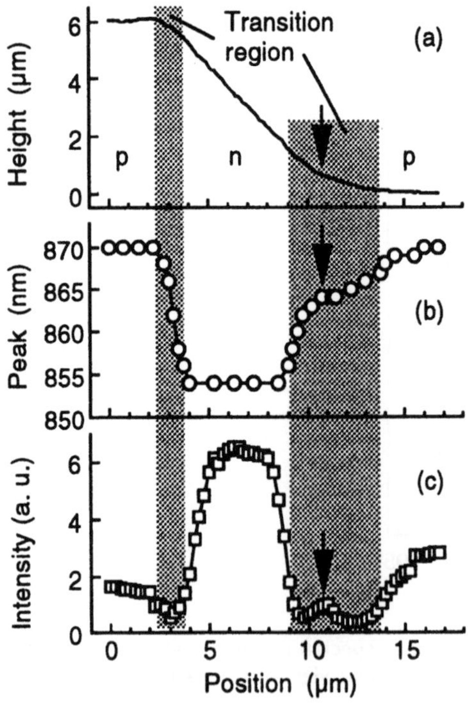

Fig. 9.9. **a** Cross-sectional view of the slope and plots of **b** emission peak, and **c** emission intensity

9.2.3 Two-Dimensional Mapping of Photoluminescence Intensity

As is clearly shown in Fig. 9.9, the total emission intensity in the transition region is much lower than in the other regions. This is mainly due to the separation and drift of photoexcited electrons and holes by the internal electric field in the transition regions. From the slow rises of emission intensity on the p- and n-side, the width of depletion layers can be estimated quantitatively. In the lower junction, some anomalous optical structures, such as a plateau in the changes in the emission peak and an increase of emission intensity, are observed, as indicated by the arrows in Fig. 9.9. It is important to investigate these local optical properties in the vicinity of the junctions in detail, since the distribution of defects and strains, and the nonuniformity of dopants in this active region, will affect the emission efficiency as a light emitting device.

To measure the two-dimensional distribution of the PL intensity, an operation is performed in illumination–collection hybrid mode [11]. In this mode, as the probe locally collects the emission in addition to being a local excitation source, we can achieve higher resolution (<200 nm) which is not affected by the carrier diffusion. Figure 9.10 shows the PL intensity image in the same scanning area as the corresponding topographic image in Fig. 9.7. In this measurement, the maximum and minimum counting rates are 2×10^4 and 1×10^4 counts/s, respectively. In the transition region, as shown previously, the emission intensity decreases considerably, compared with the other regions. Some bright areas, where the emission intensity increases locally, appear in the lower junction. These signals indicate essentially optical properties, since no correlation is found in the topographic image. A nonuniform distribution of Si dopants and of the resultant internal electric field, or that of defects and strains, will affect the local optical responses. Although, in this examination, we cannot specify the origin of these signal behaviors, further experiments will provide valuable insights into the dynamics of induced carriers.

9.2.4 Collection-Mode Imaging of Electroluminescence

In the operation of the lateral p–n junction as the surface emitting diode, the quality of EL is the most important feature. Moreover, a precise determination of the light emission section is necessary for the design of the active region in a transverse junction stripe laser. The collection-mode imaging of EL is used to study the local properties of active regions and to compare to them with those of the transition regions investigated in Sect. 9.2.2 [18].

A schematic diagram of collection-mode detection of EL from the lower junction is shown in Fig. 9.6c. Electrodes are deposited on the sample surface for the injection of carriers. The emitted light is collected through the probe in the near-field region and transported to the avalanche photodiode.

Figure 9.11a shows the EL image of the lower junction together with its topographic image. The light emission is observed at the foot of the slope in

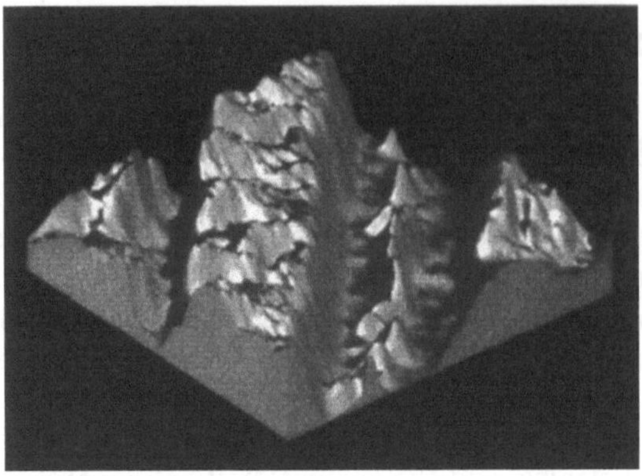

Fig. 9.10. Perspective view of a photoluminescence intensity image in illumination–collection hybrid mode. The scanning section is the same as that of the topographic image in Fig. 9.7

a straight line. The cross-sectional profiles of the emission signal are shown in Fig 9.11b for both the lower junction and the upper one, together with the height of the surface. The shaded areas indicate the transition region of lower and upper junctions as determined by the spatially resolved spectroscopy detailed in Sect. 9.2.2. The peak position of the EL from each junction is located on the p-type side of the corresponding transition region.

The difference in the incorporation behavior of Si in GaAs on different surfaces can be understood in terms of surface bonds. On the (111)A surface, there are only single dangling-bond sites, while the (311)A surface consists of alternate single and double dangling-bond sites, which are (111)A-like and (100)-like bond sites, respectively. The conduction type is governed by the competition of the Si incorporation mechanisms between the (111)A and (100) type, in which Si occupies the Ga site or the As site, respectively. The result that the lower junction is formed on the almost (111)A surface shows that the incorporation of Si in the Ga site is much more dominant on the non-(111)A surface, although almost all the Si is incorporated in As sites once the (111)A surface is attained. This observation of switching was realized for the first time as a result of the high spatial resolution of the detection system.

Moreover, the width of the EL region for both junctions is 1.1 μm. This result implies that the competition between the two incorporation mechanisms takes place at both junctions in the same manner, in spite of the large difference in the widths of the transition regions obtained from the PL measurements.

9.2.5 Multiwavelength Photocurrent Spectroscopy

So far, the near-field technique has been employed for the observation of micro/nano structures on the surface or thin materials. In the study of bulk de-

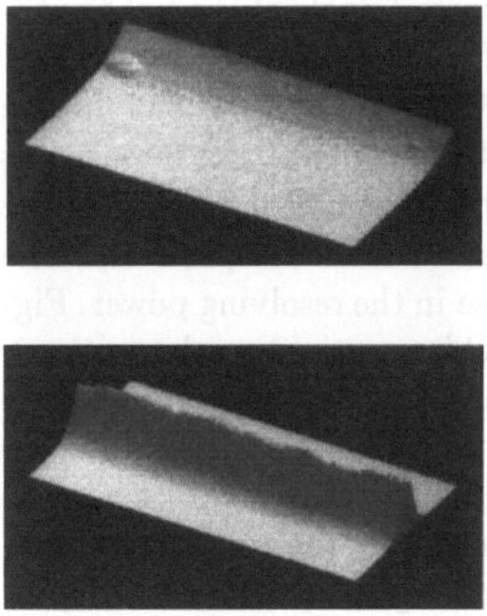

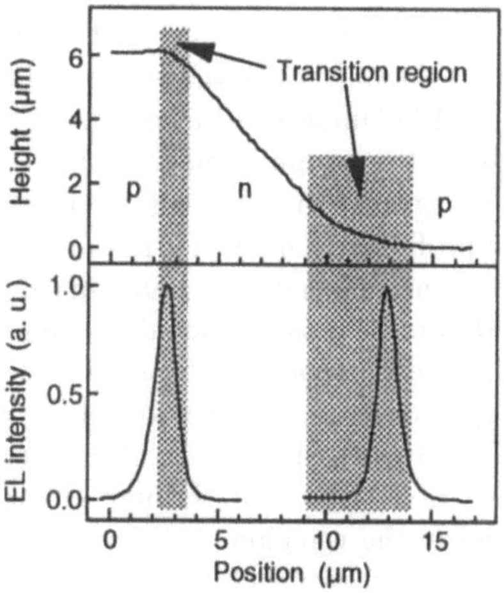

(a) (b)

Fig. 9.11. a Topographic image of the lower junction (top) and its corresponding two-dimensional distribution of electroluminescence intensity (bottom). The image size is 15 μm×8 μm. b Cross section of the slope and electroluminescence line scans at the upper and lower junctions. The shaded areas indicate the transition regions determined in Fig. 9.9

vices, however, their internal optical and transport properties should also be examined in detail. For this purpose, near-field PC measurement with propagating modes into the sample is expected to make an important contribution. Although propagating modes do not have any resolving power, tomographic information of the material investigated can be obtained by systematically varying the optical penetration depth over a wide range [38].

A block diagram of near-field PC measurement is shown in Fig. 9.6d. As multiwavelength light sources, an Ar^+ laser (λ=488 nm), a He–Ne laser (λ=633 nm) and a Ti:sapphire laser (λ=780 and 830 nm) are coupled into the probe. By using these lights, the optical penetration depths in GaAs can be tuned in a wide range from 80 nm to 1.0 μm. The PC induced by the light through the probe is collected at electrodes and amplified with a current injection preamplifier. The PC signal is synchronously detected with a lock-in amplifier.

As already mentioned in Sect. 9.1.2, the interaction between the evanescent modes and the large refractive index semiconductor plays an important role in the generation of propagating modes into the sample. In free space, if $d_f<\lambda/2$, there is a dominant occupation of the evanescent modes in the region $k_{||}>k_0$. When the probe is positioned close to the optically dense material (refractive index n), part of the evanescent modes present in the region $k_{||}<nk_0$ are coupled into propagation modes in the material. In the case of d_f=200 nm and n=3.5 (refractive index of GaAs), almost all the evanescent modes are coupled into propagating ones for the entire wavelength range

(λ=488–830 nm), where $\pi/d_f < nk_0$. The beam spread angle of the light propagating into GaAs (shown as f in Fig. 9.14) is finally determined by d_f.

Near-field PC images at the excitation wavelengths of 488 and 830 nm are shown with a topographic image in Fig. 9.12. Uniformity of PC intensity is observed along the p–n active region. The full widths at half-maximum of the PC signal profile are 0.6 and 1.7 μm at λ=488 and 830 nm, respectively. The increase in penetration depth results in a decrease in the resolving power. Figure 9.13 shows the cross-sectional profiles of PC intensities on a logarithmic scale. At the excitation wavelength of 488 nm, owing to the shallow penetration depth of 80 nm, the resolution is determined by d_f and the diffusion length of photoexcited carriers. The different diffusion lengths of electrons and holes can be observed through the slower rise of the signal in the p-region than in the n-region. Moreover, on increasing the penetration depth, the decay length becomes longer and the asymmetric behavior reverses. We believe that the longer decay length in the n-region than in the p-region at larger penetration depth can be explained by the slant of the p–n interface (shown as θ in Fig. 9.14).

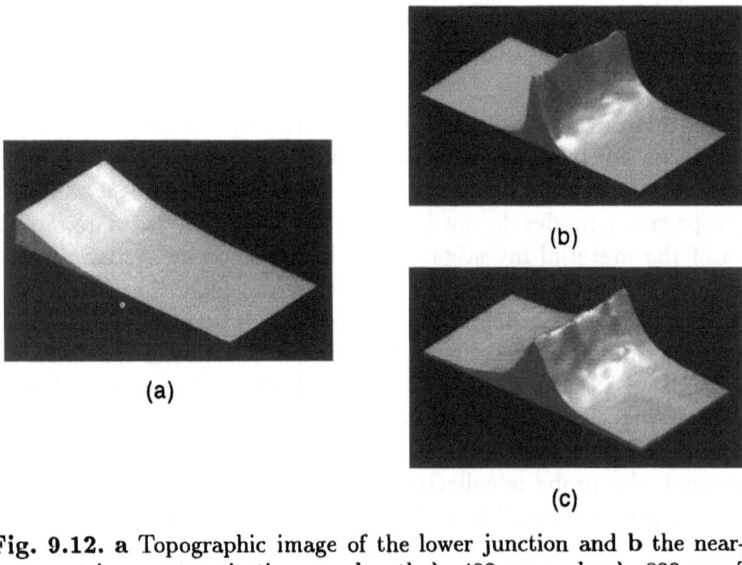

Fig. 9.12. **a** Topographic image of the lower junction and **b** the near-field photocurrent image at excitation wavelength λ=488 nm and **c** λ=830 nm. The image size is 5 μm×10 μm

Here, we make an analysis of the asymmetric signal behavior using a one-dimensional model, as shown in Fig. 9.14. The fitting parameters are the slant angle of the p–n interface θ and the beam spread angle ϕ. Since the detailed procedure of this analysis is described in Ref. [38], we only show the results here. As plotted in Fig. 9.14, the experimental curves are perfectly fitted by the numerical analysis. From this calculation, we obtain the slant angle of

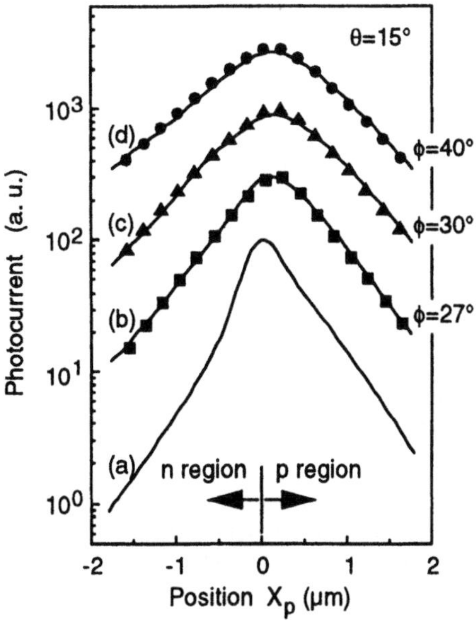

Fig. 9.13. Cross-sectional profiles of the near-field photocurrent at excitation wavelengths (a) λ=488 nm, (b) λ=633 nm, (c) λ=780 nm, and (d) λ=830 nm. The symbols correspond to the results of calculations with the fitting parameters being θ =15°, ϕ=27° (b), ϕ=30° (c), ϕ=40° (d)

the p–n interface, θ=15 ± 8°, and the beam spread, ϕ, for each excitation wavelength.

A total slant of the p–n interface of 30±8° to the p-side, which is the sum of the observed slant angle θ of 15±8° and the intended tilt angle of 15° in the experimental set-up, can be explained by the crystal orientation dependence of the growth nature. The most significant origin is that the growth rate of GaAs on (311)A is faster than that of (111)A, which causes a shift of the n-type region to the p-side during growth. Other factors are also examined, taking into account the experimental results on PL and EL measurements.

The wavelength-dependent feature of the beam spread angle ϕ also gives us important information on the mode conversion of the evanescent light. Figure 9.15 shows a plot of ϕ as a function of the wavelength of propagating light. The tangential wavevector of the propagating light, which is determined by the spread angle ϕ, should be equal to the cutoff wavevector of the diffraction spectrum of the aperture; $nk_0 \sin\phi = k_c$. The fitted curve in Fig. 9.15 is obtained by assuming that k_c lies at π/d_f. This estimation of k_c is consistent with the result obtained in Sect. 9.1.2 where k_c is determined by the critical value of d_f needed to achieve near-field resolution.

Finally, as an application of near-field PC measurement, we show an analysis of the Si p–n structure incorporated into an integrated circuit device chip. Due to the small absorption coefficient of Si compared with GaAs, PC detec-

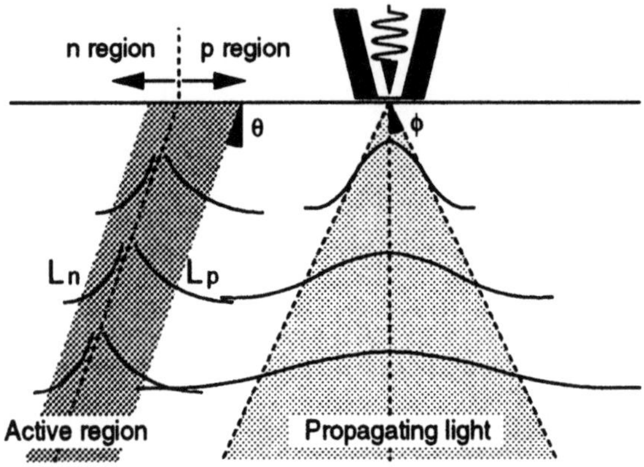

Fig. 9.14. Schematic diagram of the slanting p–n interface and light propagating into the GaAs. L_p and L_n are the diffusion lengths of electrons in the p region and holes in the n region, respectively. The slant angle θ and the beam spread angle ϕ are the fitting parameters in the calculations

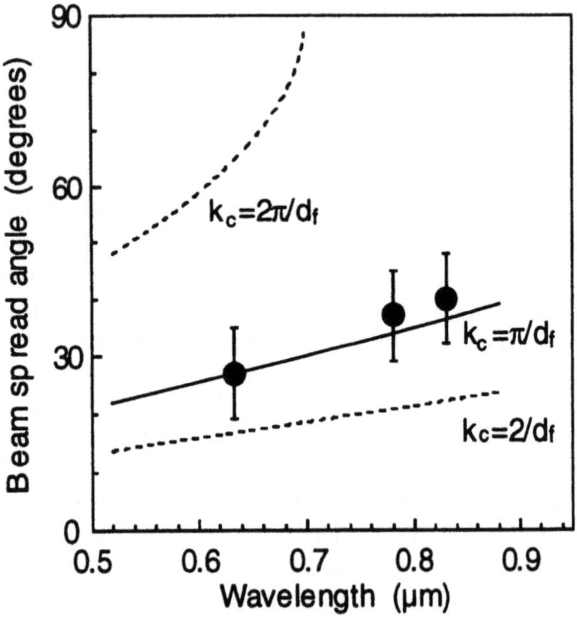

Fig. 9.15. Plots of the beam spread angle as a function of wavelength. The curve is obtained by assuming that the cutoff wavevector lies at π/d_f. d_f=200 nm and n=3.5

tion of the Si p–n junction seems to be rather difficult. Fukuda et al. obtained near-field PC images with different Ar^+ laser lines, as shown in Fig. 9.16. The penetration depths corresponding to 457.9 nm and 514.5 nm excitation are 280 nm and 690 nm, respectively. Reflecting these values, the PC signal profiles are largely different. The asymmetry behavior of the profile originates from the difference in diffusion lengths of electron and holes, which are estimated as 830 nm and 530 nm, respectively, from the analysis of the profile.

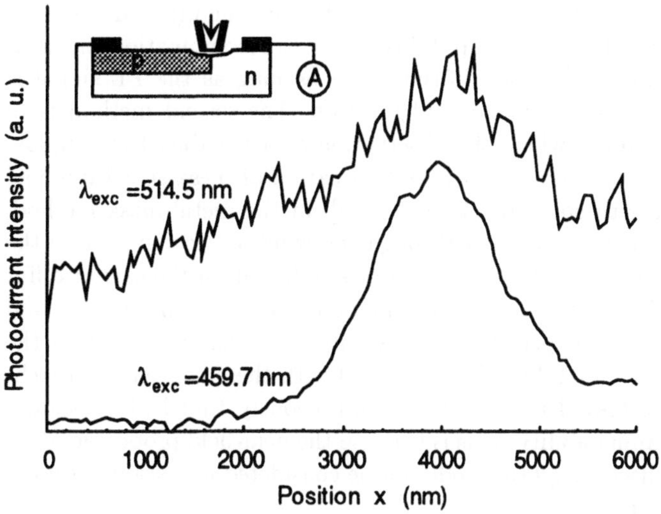

Fig. 9.16. Cross-sectional profile of the near-field photocurrent at excitation wavelengths λ=457.9 nm and λ=514.5 nm. The inset shows a schematic diagram of the sample investigated

9.3 Low-Temperature Single Quantum Dot Spectroscopy

9.3.1 Near-field single quantum dot spectroscopy

With the progress of fabrication techniques for self-assembled quantum dots (QDs), unique features of semiconductor nanostructures have received increasing attention in the last few years. One of the most attractive properties of a zero-dimensional confined structure is a narrow, strong optical transition due to its atomic-like discrete density of state, which is favorable for optoelectronic devices such as lasers with low threshold current density [51]. PL spectroscopy is the most common way to gain information on the electronic structures and relaxation processes of photoexcited carriers in QDs. Single

QD spectroscopy is essential for the precise evaluation of the intrinsic capability of QDs since the conventional far-field studies measure ensembles of dots with size and shape fluctuations, which result in inhomogeneously broadened spectral features. In order to realize single QD spectroscopy, therefore, we should restrict the observation area by applying novel microscopic techniques. As a powerful tool for such single particle observation, in this section, we describe low-temperature near-field spectroscopy of a single QD.

Here we survey several techniques for single QD PL spectroscopy. The first method is to use local electron injection methods such as cathodoluminescence [52, 53] and tunneling luminescence [54, 55]. As the injected electrons or generated electron–hole pairs can diffuse prior to recombination, the spatial resolution is limited by the diffusion area as long as the PL signal is collected by a lens in the far-field configuration. The second method is to use micro-PL techniques, where the observation area is reduced by employing microfabrication techniques, such as fabrication of mesa structures [56] and a metal mask with small windows [57–59]. In the metal-mask method, since the PL signal is also collected through the window, we can achieve the spatial resolution determined by the window size free from the carrier diffusion effect. Recently, the metal-mask technique has been the most successful way to realize single QD spectroscopy. Its only disadvantage is lack of the scanning ability of the metal mask. Compared with these conventional techniques, NOM has achieved both merits simultaneously, that is, high spatial resolution and scanning ability. This is because the near-field probe operating in illumination–collection hybrid mode can be considered as a scanning mask with a small aperture.

9.3.2 Low-Temperature NOM

The schematics of low-temperature NOM [32–35] are shown in Fig. 9.17. The head is composed of a fiber probe with a dither piezo, its coarse approach mechanism, and a sample scanner. Since we successfully utilize the illumination–collection hybrid mode operation, the structure of the head is very simple, and has no far-field illumination or collection lenses. It is designed to fit in a continuous He gas flow optical cryostat. For the coarse approach of the probe to the sample surface, a stable translation stage is employed in combination with a high-resolution stepping motor. A sample is mounted on a piezotube of dimensions 70 mm length and 10 mm diameter, which allows for a scan range of 4 μm×4 μm at 5 K. For the shear force measurement, both the illumination of the laser diode (λ=1.55 μm) and the detection of transmitted (or reflected) light by the photodiode are performed through the two optical windows of the cryostat.

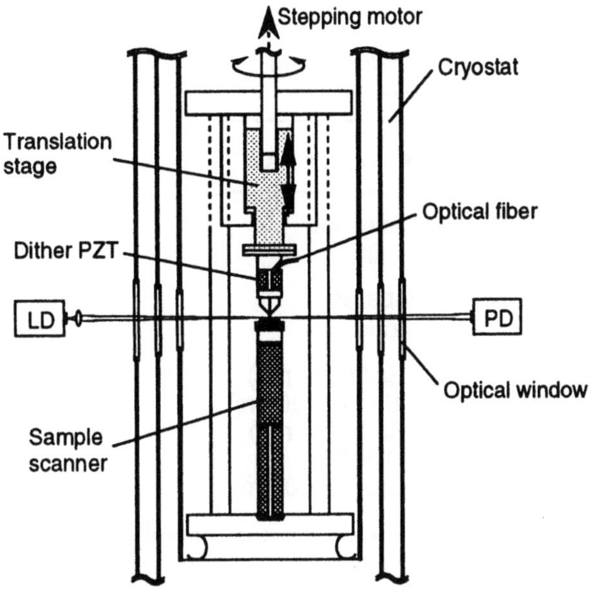

Fig. 9.17. Diagram of the mechanical set-up of low temperature NOM in a cryostat

9.3.3 Sample and Experimental Set-up

Figure 9.18 shows a schematic representation of the QD sample structure and the experimental configuration. $In_{0.5}Ga_{0.5}As$ self-assembled QDs are grown on (100) GaAs substrate by gas-source molecular beam epitaxy with a density of 5×10^9–1×10^{10} dots/cm^2 [60]. Typical dot diameters of around 30 nm and heights of 15 nm are observed with an atomic force microscope. These QDs are covered with cap layers composed of GaAs and AlGaAs with total thickness of 80 nm. The wide gap layers of AlGaAs are introduced in order to prevent carrier leakage from the GaAs layer to the substrate or surface.

We employ a chemically etched fiber probe with a flat aperture of d_f=500 nm, which is around a half of the PL wavelength of QDs. The shape of the tapered part is optimized to attain high sensitivity, as mentioned in Sect. 4.3. The QD sample on the scanning piezotube is illuminated with a He–Ne laser light (λ=633 nm) through the probe in close proximity to the surface. Carriers are generated not only in the InGaAs QDs, but also in the barrier layers of GaAs and AlGaAs. Most of the photoexcited carriers migrate in the barrier layers and are captured by the confined states of QDs. In order to achieve a high spatial resolution free from carrier diffusion effects, the resultant PL signal is collected by the same probe. In the case of signal collection through the probe, the resolution is determined by the various factors of the probe rather than d_f, as is detailed in the following subsection. After rejecting the excitation light with a long pass filter, the signal is focused into a 50-cm monochromator and is detected by a cooled photomultiplier tube

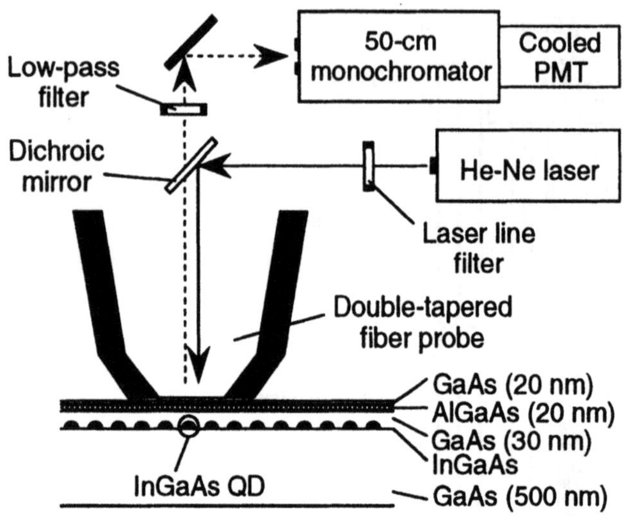

Fig. 9.18. Schematic representation of a sample structure, optical near-field configuration, and block diagram of the measurement

(Hamamatsu R5509-41, <10 dark counts/s, 0.2–1% quantum efficiency at the detection wavelength of 0.92–1.0 μm) using the photon counting technique. All the measurements are performed at 5–10 K in the cryostat.

9.3.4 Fundamental Performance of the System

Figure 9.19 shows a near-field PL spectrum with weak excitation power density of 0.2 W/cm^2. By restricting the observation area with near-field excitation and collection, we obtain a sharp emission line from a single QD which has been buried in the inhomogeneously broadened far-field spectrum. The linewidth of 0.5 meV is determined by the spectral resolution of the monochromator. Within an excitation area of $\pi(0.25\ \mu\text{m})^2$, the excitation light of 0.4 nW through the probe generates 1×10^9 electron–hole (e–h) pairs per second. From the value of the QD density, the injection rate of e–h pairs into individual QDs is estimated to be $<1 \times 10^8$ e–h/s. In this estimation, we do not take into account the effect of carrier diffusion. Since the injection rate of e–h pairs is much smaller than $1/\tau$, where τ is a ground state emission lifetime of about 1 ns, we conclude that the observed emission line originates from the recombination of ground-state exciton.

We roughly estimate the collection efficiency of the PL signal of the probe. When the injection rate of excited carriers per single QD is 1×10^8 e–h/s, the resultant photon counting rate of the PL signal is around 50 counts/s. By using the apparatus parameters of the detection system, such as the filters, the monochromator, and the photomultiplier tube (0.5% quantum efficiency at 970 nm), a product of quantum efficiency of dots (ϕ_{qe}) and the collection efficiency of the probe (ψ_{ce}) can be evaluated as $\phi_{\text{qe}}\psi_{\text{ce}} \sim 1 \times 10^{-3}$. If ϕ_{qe} is

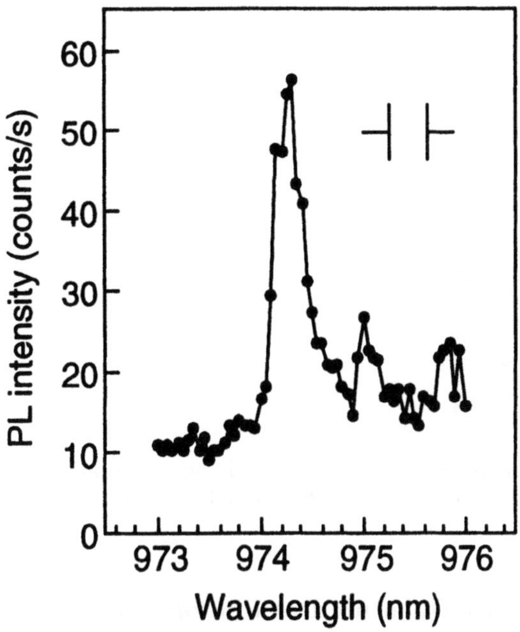

Fig. 9.19. Photoluminescence spectrum of an InGaAs single quantum dot detected in near-field configuration at 5 K. The linewidth of 0.5 meV is limited by the spectral resolution of the monochromator

equal to 10%, ψ_{ce} reaches 1%, which is as high as the collection efficiency in the high NA far-field configuration.

Figure 9.20 shows a monochromatic PL image constructed by fixing the detection wavelength at 965 nm and scanning the probe in a 0.8 μm$\times$0.8 μm area. Each circular spot corresponds to a single QD having an emission wavelength of around 965 nm. The minimum and average sizes of the spot images at full width at half-maximum are 160 nm and 220 nm, respectively. A significantly high spatial resolution of $\lambda/6$–$\lambda/5$ was achieved. It should be noted that the spot size obtained is much smaller than that with d_f of 500 nm ($\lambda/2$). Since the carriers generated in the barrier layers of GaAs diffuse into the extended region, the spot size is determined only by the collection area of the PL signal through the probe. For d_f=500 nm to 1 μm, it has been demonstrated that the resolution is not strongly dependent on d_f even when a fairly large aperture of d_f=1 μm is utilized, a similar resolution of around 200 nm can be obtained. The mechanism to achieve such a high resolution has not been fully understood. We speculate that the close proximity of the glass apex to the sample, the large cone angle of the tip, and the modal occupation of the tapered region contribute to the realization of something like a high NA lens. Such a lens is also expected to ensure the high collection efficiency through the probe.

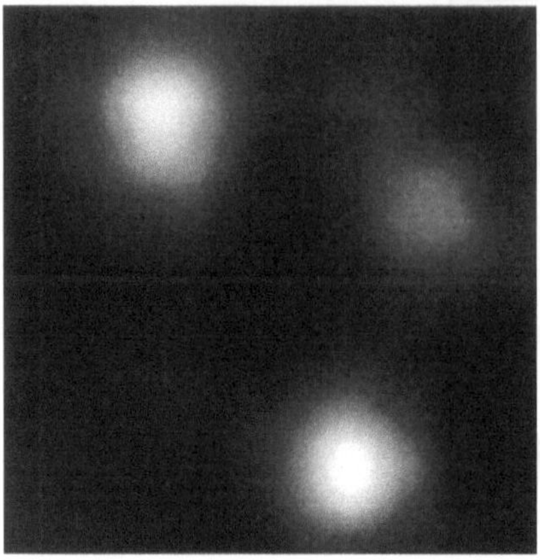

Fig. 9.20. Monochromatic photoluminescence image of single quantum dots with an emission wavelength of 965 nm and spectral resolution of 1 nm. The scanning area is 0.8 μm×0.8 μm

9.3.5 Physical Insight of Single Quantum Dot Photoluminescence

The evolution of the single QD PL spectrum with excitation intensity gives us information on the intrinsic physical properties of a zero-dimensionally confined system, such as the quantization energy of electrons and the binding energy of electron–hole pairs. Figure 9.21 shows an excitation power dependence of PL spectra of a single QD. The peak intensity of each spectrum is normalized by the excitation power density. For an excitation power lower than 5 W/cm^2, the PL spectra consists of a single line. As mentioned in the previous section, in such a power density, the single PL line originates from the recombination of single excitons from the ground state. With increasing excitation power, other elementary features are observed in the PL spectra. First, after the saturation of the ground-state emission, a new PL line appears 30 meV above the ground-state transition. This line corresponds to the emission from the first excited state, which arises from the state filling of the ground state. Furthermore, Fig. 9.21 shows another feature 2.2 meV below the single-exciton emission. The emission intensity grows superlinearly with the excitation intensity. From the energetic position and the excitation power dependence, this emission line can be associated with the recombination of biexcitons, the rate of which increases quadratically with the generation rate of the single excitons in a single QD [61, 62].

Figure 9.22 shows the monochromatic PL images in the same scanning area of 3 μm×3 μm for various excitation power densities. The detection wavelength is fixed at 975 nm with a bandwidth of 1 nm. The number of emission spots from individual dots drastically changes with increasing excitation intensity, while the spot sizes are unchanged. Such a power-dependent

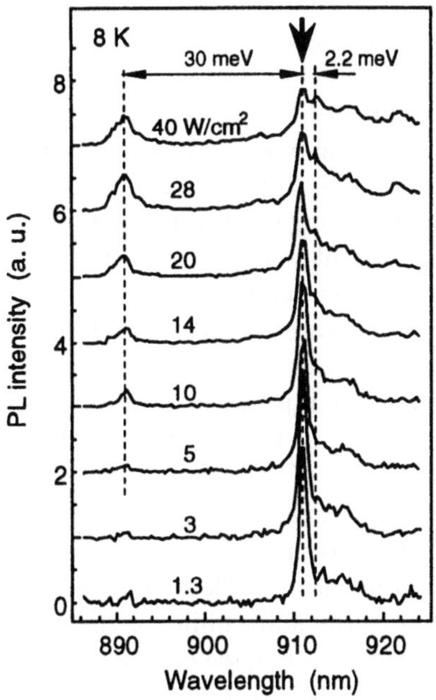

Fig. 9.21. Single quantum dot photoluminescence spectra observed at 8 K with various excitation densities. The emission wavelength of a ground-state single exciton is indicated by the arrow

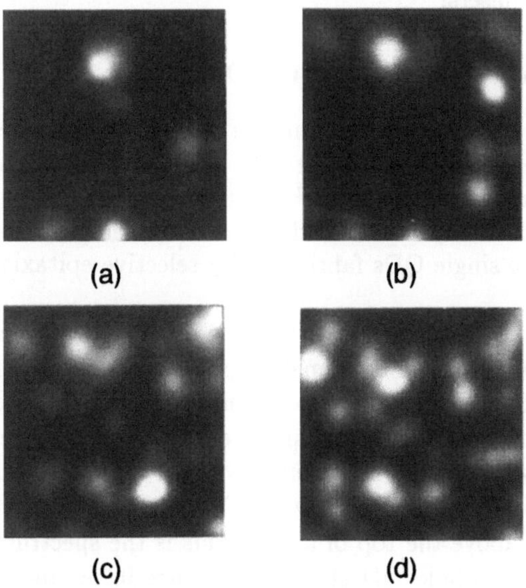

Fig. 9.22. Monochromatic photoluminescence images obtained at 8 K with various excitation densities. **a** 3.9 W/cm^2; **b** 30 W/cm^2; **c** 120 W/cm^2; **d** 700 W/cm^2. The detection wavelength is 975 nm with a spectral bandwidth of 1 nm, and the scanning area is 3 $\mu m \times 3$ μm

behavior shows the appearance of emissions from excited states and biexciton states, which are also observed in the PL spectra in Fig. 9.21.

In order to determine the origins of the individual emissions, Fig. 9.23 shows the plots of the spot intensities as a function of the excitation power density for each spot observed in Fig. 9.22. Specific power-dependent behaviors can be classified into three types. The plots in Fig. 9.23a show a linear increase in the weak excitation region. The beginning of saturation at an excitation intensity of around 10 W/cm^2 is consistent with the result obtained in Fig. 9.21. Such evolutions of emission intensities confirm that these emissions originate from the recombination of ground-state excitons. The decrease in the emission intensity for higher excitation is due to the increase in the generation rate of biexcitons and the change in the injection rate of e–h pairs into the individual dots. Another behavior shown in Fig. 9.23b is quadratic dependence on the excitation density, which is one of the most characteristic features of biexciton recombination. The other emissions in Fig. 9.23c are classified into four groups in terms of the threshold intensities in the appearance of individual spots and their saturation intensities. These plots display the appearance of higher excited states due to the state filling of the lower excited and ground states. Based on these consideration, we can identify the origins of individual emission spots in the image of 500 W/cm^2, which is summarized in Fig. 9.24. At such a high excitation region, spots of excited-state emission are dominant in the spatial distribution. The two-dimensional specification of the emission feature is important, especially in the investigation of the lasing mechanism of QD lasers.

9.3.6 Observation of Other Types of Quantum Dots

As well as self-assembled QDs, there are various kinds of fabrication methods for the QD structure. Selective etching epitaxial growth is a useful candidate because of its damage-free formation and the high regularity of the QD arrangement. This subsection is devoted to the introduction of low-temperature near-field spectroscopy of GaAs single QDs fabricated by selective epitaxial growth [24].

Figure 9.25a shows the cross-sectional structure of a GaAs QD grown on SiO_2-patterned GaAs (100) substrates with metalorganic chemical vapor deposition. The size of the QD pattern (190 nm×160 nm×12 nm) and the separation of QDs by 2 μm are observed by scanning electron microscope. A single-tapered fiber probe with d_f=100 nm is employed for the near-field illumination of the sample. The PL spectrum shown in Fig. 9.25b is obtained by positioning the tip 200 nm above the top of a QD. This is the spectrum from the carriers excited in the whole QD structure. Figure 9.25c and d show the PL spectra with the tip position less than 10 nm above the region of the QD and that of the SiO_2 mask, respectively. Figure 9.26a–c show monochromatic PL images corresponding to the energy regions indicated by the arrows labeled A, B, and C, which originated from the GaAs bulk, the

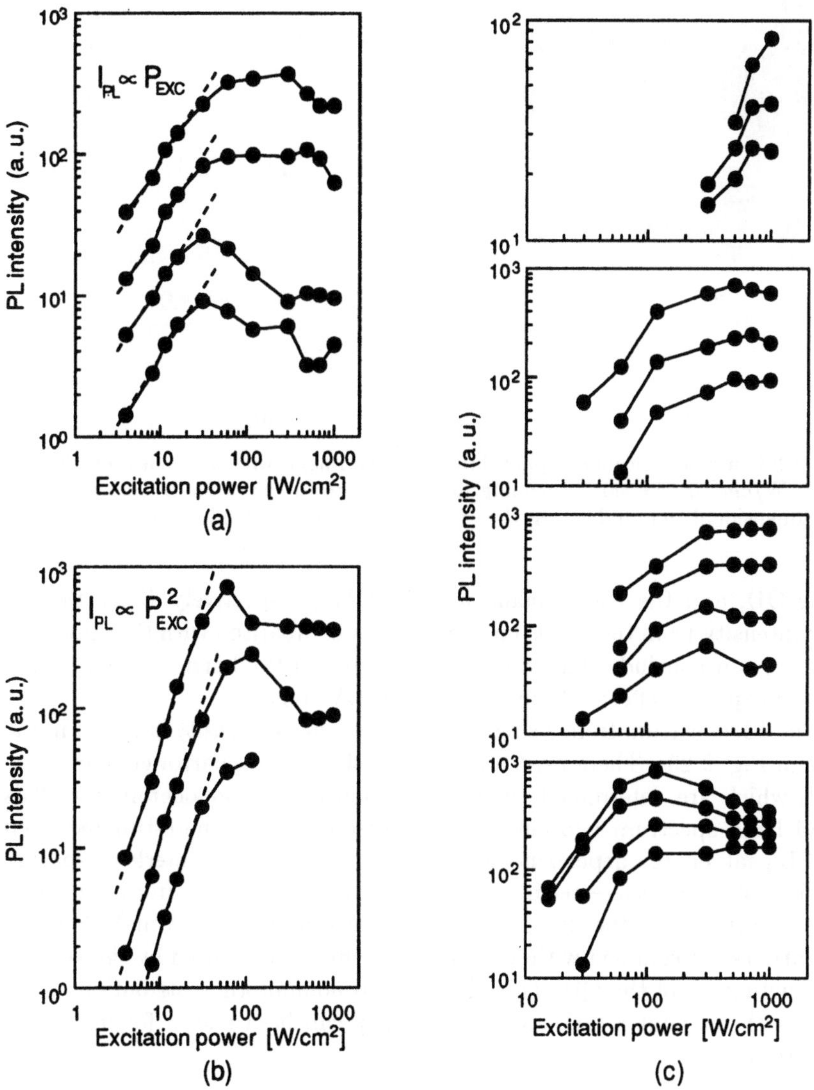

Fig. 9.23. Plots of emission spot intensities as a function of excitation densities for the spots observed in Fig. 9.22. **a** Linear increase, **b** quadratic increase in a weak excitation region, and **c** with a threshold in the appearance of emission. Plots in **c** are classified into four groups in terms of the threshold intensity and saturation intensity

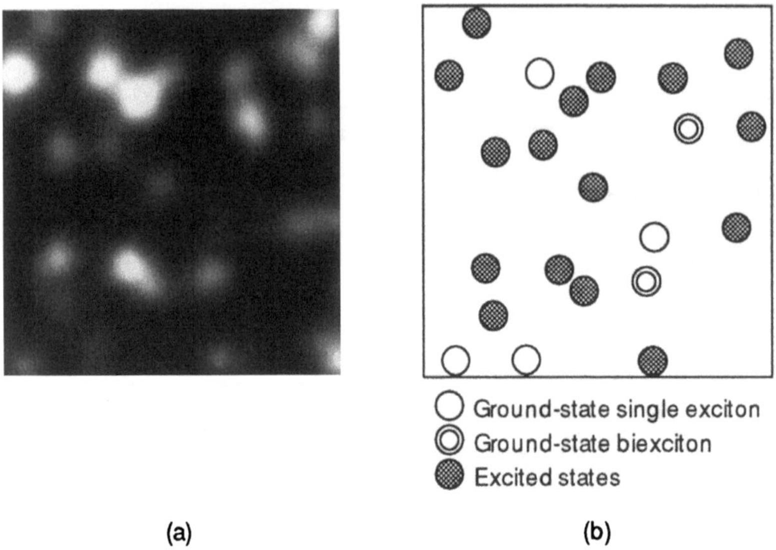

Fig. 9.24. a Monochromatic photoluminescence image with an excitation density of 500 W/cm^2 in the same scanning area as shown in Fig. 9.22. b Corresponding two-dimensional map indicating the emission origins of individual spots in a

GaAs QD, and the GaAs quantum well (QW), respectively. Since the PL peak intensity from the QD is large enough in comparison with that from the bulk, we can conclude that the carriers excited in the barrier region diffuse and are captured effectively in the QD and QW regions.

The PL spectrum from a single QD at lower power density is shown in the inset in Fig. 9.25c. There exist a number of sharp lines, narrower than a few meV, which are determined by the resolution of the monochromator. This broad band spectrum is occurs for two reasons. As can be estimated from the PL peak energy dependence on the magnetic field, the interlevel spacing is 7.5 meV. First, there seem to be a number of transitions from the upper excited states since the interlevel spacing is estimated as 7.5 meV. Second, the carriers are confined within the valleys, which are induced by the vertical size fluctuation of the QD. This gives rise to nonuniform distribution of the carriers throughout a QD, and consequently can be a reason for multipeak generation.

9.4 Ultraviolet Spectroscopy of Polysilane Molecules

9.4.1 Polysilanes

Polysilanes (PSs) are s-conjugated polymers which can be regarded as ultimate quantum wires made of Si [63]. PSs are expected to be applied to the ultraviolet (UV) light-emitting device owing to their wide electronic band gap

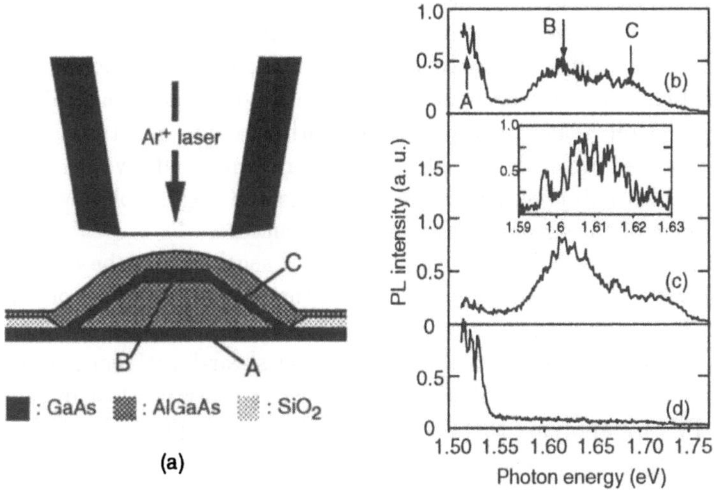

Fig. 9.25. **a** Schematic representation of the cross-sectional profile of a single GaAs quantum dot. The sections indicated by A–C correspond to the structure of the GaAs quantum dot, the GaAs quantum well, and the GaAs bulk, respectively. **b–d** Spatially resolved photoluminescence (PL) spectra at 18 K: **b** maintaining the tip 200 nm above the quantum dot (QD); **c** less than 10 nm above QD; **d** less than 10 nm above the SiO_2 mask. The inset shows a PL spectrum with low excitation power in the same configuration as **c**

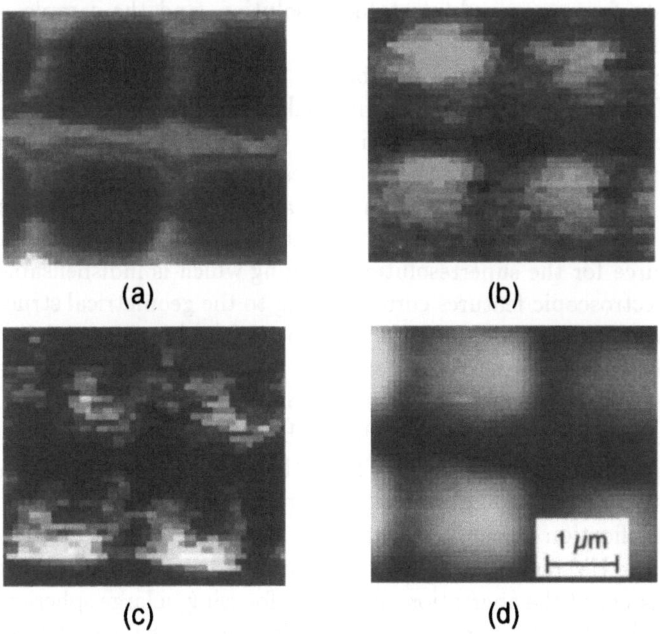

Fig. 9.26. Monochromatic photoluminescence images at 18 K with the energy regions labeled **a** A, **b** B, and **c** C in Fig. 9. 25. **d** Simultaneously obtained topographic image

and high quantum efficiency. The electronic structures and optical properties of PSs can be controlled by changing the conformation of the Si backbones, i.e., the shape of molecules. Recently, by using an atomic force microscope, Ebihara et al. [64] have succeeded in the observation of a single helical PS molecule prepared on a hydrophobic ultrasmooth sapphire substrate. The length of the observed molecule was 2 μm. Such a measurement gives us hope that we will be able to observe the optical properties of a single molecule with an accurate image of the actual conformational structure.

NOM has achieved a sufficiently high degree of sensitivity to be considered a very powerful tool in the investigation of the optical responses of single molecules [65–70] and aggregates [71, 72]. Furthermore, we obtain their geometrical structures simultaneously through the shear-force topographic image. In this section, we described the near-field PL study of aggregate structures of PSs by applying the UV spectroscopy technique and an optimized fiber probe.

9.4.2 Near-Field Ultraviolet Spectroscopy

The inset in Fig. 9.27 shows a schematic diagram of the trans-planar-type conformation of polydihexylsilane (PDHS). At room temperature, dried PDHS has absorption and PL peaks at wavelengths of 370 nm and 380 nm, respectively. The quantum efficiency is evaluated as high as 5%. PDHS is cast on the silica substrate from a very dilute toluene solution, and the sample is dried in a vacuum.

First, we describe the fiber probe suitable for high-resolution imaging in the UV region. The core of the fiber is composed of pure silica to suppress the strong scattering loss in the propagation of UV light. As described in Sect. 4.3, by shortening the length of the optical loss region due to evanescent propagation and metal-cladding absorption, we have successfully obtained a high throughput. Moreover, the protrusion part at the apex functions as the localized light source for the superresolution imaging which is indispensable to obtain local spectroscopic features corresponding to the geometrical structure. The fabrication process for this type of probe is found in Sect. 3.6.2.

Figure 9.27 shows a diagram of the near-field UV measurement. For the optical excitation, a 351-nm line of Ar^+ laser is coupled into the fiber probe. The PDHS molecules dispersed on the silica substrate are illuminated by the near-field light through the probe. PL from the excited PS molecules is collected by a low-magnification objective lens or pure silica spherical lens. Here, a high-magnification (large NA) lens is not available owing to its low transmittance in the UV region. If the detection area of the photodetector is wide enough to disregard the aberrations in the PL focusing, a large spherical lens is more suitable for the highly efficient collection of the signal than a small NA lens. In two-dimensional optical imaging, the collected signal (transmission of the excitation light or PL) is detected by photomultiplier

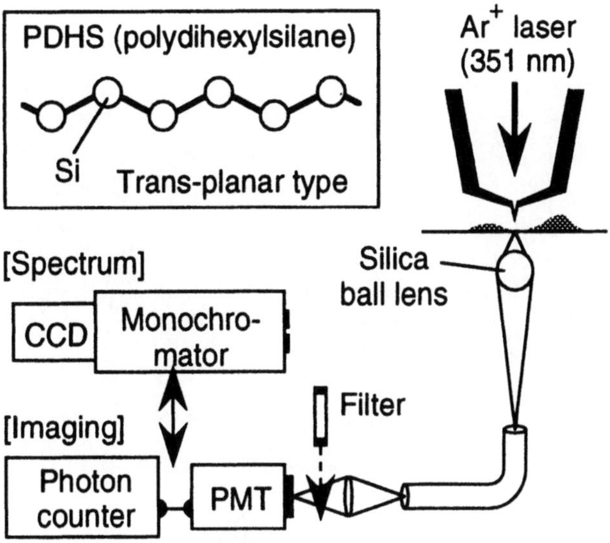

Fig. 9.27. Schematic diagram of the experimental set-up for near-field ultraviolet spectroscopy. The inset shows the transplanar type conformation of polydihexylsilane

tube through the notch and bandpass filters. A monochromator with a cooled CCD is employed for the spectroscopic analysis of the PL signal.

9.4.3 Imaging and Spectroscopy of Polysilane Aggregates

Figure 9.28a and b show a transmission image of PDHS aggregates and the corresponding topographic image, respectively, in a 2 μm×2 μm scanning area. The maximum height of the sample is as low as 40 nm. Since each structure in the transmission image is the reverse of that in the topographic image, the dominant mechanism of contrast generation is the absorption of illuminated UV light by PDHS molecules. The strength of the contrast can be estimated from the absorbance of PDHS thin film at a wavelength of 351 nm as $10^{(-40\mathrm{nm}/400\mathrm{nm})} \sim 0.8$. This value is in good agreement with the contrast obtained in the transmission image of 0.75.

As shown in Fig. 9.29, a PL image of a thin aggregate structure can be obtained under an excitation condition as low as 40 pW illumination through the probe. At such a low excitation, fatigue of the PL signal can be prevented. The aggregate structures are thinner than 3 nm. The background signal in the dark region is almost zero, in contrast to the maximum counting rate of the PL signal of 50 counts per 100 ms. This implies that the fluorescence from the fiber or substrate due to the UV excitation does not affect the measurements. The PL signal intensity obtained agrees well with a numerical estimation using the parameters of absorption coefficient, quantum efficiency and so on.

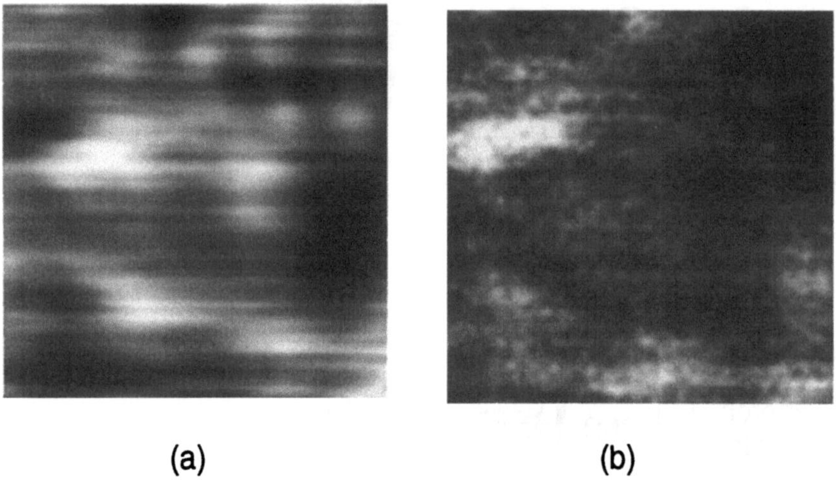

Fig. 9.28. **a** Topographic image of polydihexylsilane aggregates, and **b** the corresponding mapping of the transmission intensity of ultraviolet light. The scanning area is 2 μm×2 μm

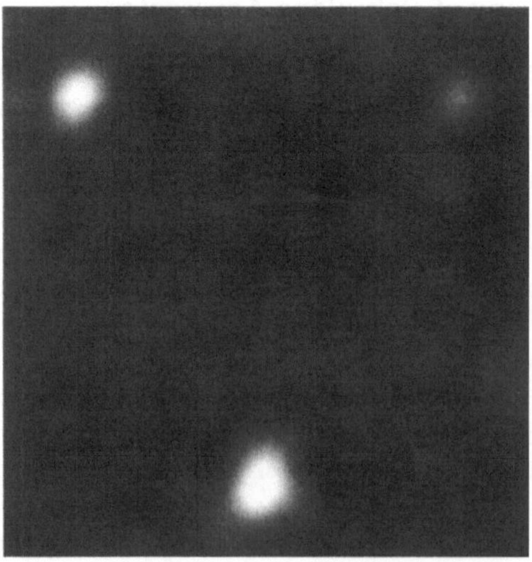

Fig. 9.29. Near-field photoluminescence image of thin polydihexylsilane aggregate structures. The scanning area is 5 μm×5 μm

PL spectra from a thick aggregate structure are acquired every 3 s by a charge-coupled device (CCD) camera using a rather large d_f with 100 nW excitation. Several spectra are presented in Fig. 9.30. A peak wavelength of 380 nm and spectral width of 9 nm are obtained. These values are not very different from the results of the far-field spectroscopy of PS thin film. Under such a high-power excitation at room temperature in the atmosphere, the spectrum is rapidly bleached in about 60 s, which is a serious problem when observing a small structure with low signal intensity. As in the PL imaging, to avoid this fatigue phenomena, the excitation power should be as low as possible.

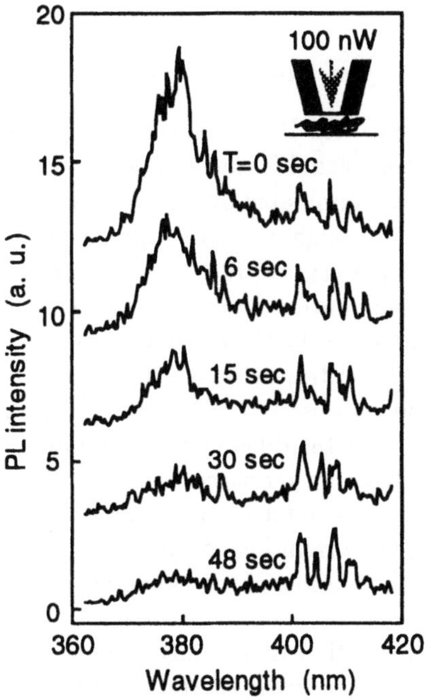

Fig. 9.30. Time-dependent features of photoluminescence spectra obtained from thick polydihexylsilane aggregate structures with high excitation intensity

9.5 Raman Spectroscopy of Semiconductors

9.5.1 Near-Field Raman Spectroscopy

Raman spectroscopy has been widely applied to study the material phase, the distribution of stress, the identification of chemical bonding, and so on. The combination of Raman spectroscopy and near-field techniques is expected to provide us further investigation with high spatial resolution and topographic information from the surface [73-77]. In such advanced measurements, low sensitivity due to the weak scattering effect has frequently been reported. However, this problem is expected to be solved by employing a high-throughput probe, as described in Sect. 4.3 or one with a functional structure such as a local plasmon.

The fundamental configuration of the experimental set-up is similar to the PL spectroscopy of opaque materials. At present, however, the illumination–collection hybrid operation is difficult to apply to Raman signal detection; there is strong Raman scattering from the optical fiber, which obstructs the extraction of weak signals from the sample. The experiments described in this section were performed in illumination mode with the collection of a Raman signal in a far-field scheme with an objective lens (NA 0.8).

9.5.2 Raman Imaging and Spectroscopy of Polydiacetylene and Si

As a demonstration of the feasibility of near-field Raman spectroscopy, we show the two-dimensional mapping of Raman signal features of polydiacetylene (PDA). A tabular PDA single crystal is illuminated by 10 nW of second harmonics of Nd:YAG laser light (532 nm) through a double-tapered probe with d_f=100 nm. In a 1 μm×1 μm area of PDA crystal, 10 × 10 points of Raman spectra were measured, with 600 s accumulation for each point. Raman spectra of PDA have been well investigated due to their strong scattering efficiency and resonance effect. In near-field spectra, as shown in Fig. 9.31a, two specific peaks relating to the C=C bond are observed at 1457 cm^{-1} (peak L) and 1520 cm^{-1} (peak S). These peaks reflect the difference in the number of successive *cis*-bonds; peak L originates from longer successive bonds and peak S from shorter ones. Figure 9.31b shows the intensity ratio distribution of peak S to peak L. Since, in this image, no correlation to the topographic image is observed, the spatial difference in the number of successive *cis*-bonds is clearly obtained.

Raman spectroscopy of Si with high resolution is certainly required for the analysis of electronic devices. Due to its rather weak scattering efficiency and small wavenumber, the desired signal competes with the background signal of Raman scattering from the optical fiber. In this measurement, an optical fiber with a pure silica core, which is also used in UV spectroscopy, was employed. By using this fiber, the Raman background signal from the fiber can be reduced to less than 10% compared with a fiber with a GeO_2-doped (25 mol%) core. Figure 9.32 shows an example of a Raman spectrum of a Si crystal using a triple-tapered probe with d_f=100 nm (see Sect. 3.6.2) and short propagation of excitation light (wavelength 532 nm) in the fiber. A sharp peak from Si is clearly seen at 520 cm^{-1}. The broad signal around 400 cm^{-1} is assigned to the Raman scattering from the SiO_2 composing the fiber. The successful acquisition of a weak Raman signal from Si is mainly attributed to a reduction of the background Raman scattering through the careful selection of the composition of the fiber.

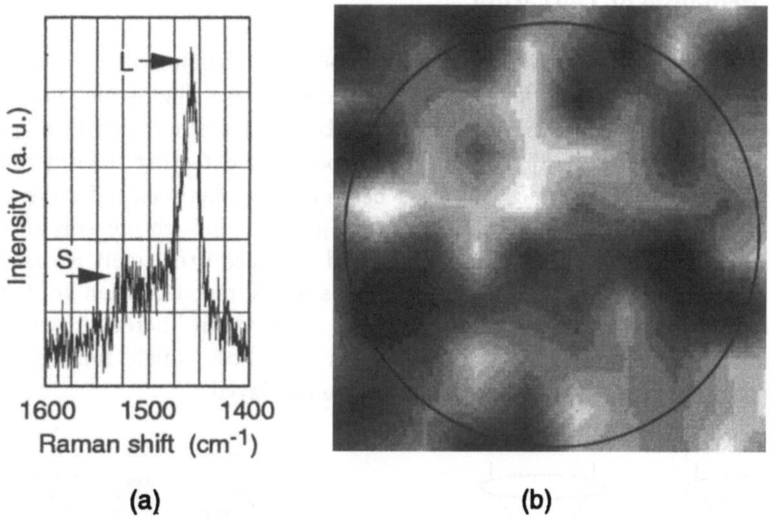

Fig. 9.31. a Near-field Raman spectrum of polydiacetylene. b Two-dimensional mapping of the intensity ratio of the Raman signal at two specific peaks, reflecting the difference in the number of successive C=C *cis*-bonds

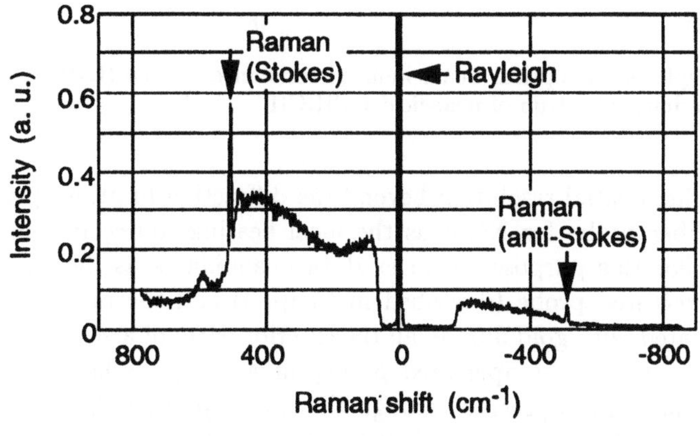

Fig. 9.32. Near-field Raman spectrum of an Si crystal obtained by a probe with d_f=100 nm

9.6 Diagnostics of Al Stripes in an Integrated Circuit

9.6.1 Principle of Detection

The detection of voids in Al stripes is an important technique in failure analysis of integrated circuit devices. As a nondestructive diagnostic tool, the optical beam induced resistance change (OBIRCH) method has been proposed by Nikawa et al. [78]. This method is based on detecting the difference in resistance change, produced by laser beam heating, between defective and defect-free areas. Figure 9.33a illustrates the principle of such measurements. When a laser beam scans, the heat generated is transmitted freely across areas that are free of voids, while such transmission is impeded when voids are encountered. This creates differences in the temperature increases between irradiation points which are not near voids and those which are, and the resulting differences in resistance changes are displayed.

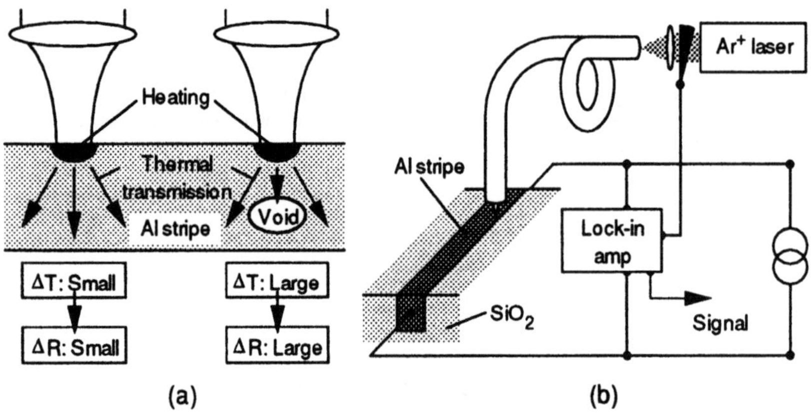

Fig. 9.33. **a** Principle of the optical beam induced resistance change (OBIRCH) method and **b** experimental set-up of near-field OBIRCH

To achieve higher spatial resolution beyond the diffraction limit, we propose to use the fiber probe for NOM as the local heating source in place of a laser beam. For this purpose, there exist two candidates as the heating source, the apertured probe (described in Chap. 4) and the metallized probe, which is coated with gold film on all the exterior parts of a sharpened fiber probe. In the case of the apertured probe, an Al stripe is heated by the absorption of laser light passing through the small aperture. Since the penetration (skin) depth of Al is very shallow, the resolution is determined by d_f. On the other hand, the heating mechanism using the metallized probe is the conduction of heat through the contact between the Al stripe and the apex of the probe, which is heated by the strong absorption of propagating light in the probe [79–82]. In principle, since the resolution of this method is determined by the contact area, a similar resolution to that of topographic imaging is expected to be achieved. In this section, we employ both types of probe and compare them from the aspects of resolution and sensitivity.

Figure 9.33b shows the experimental configuration for the detection of resistance change through near-field heating. Ar^+ laser light coupled to the fiber illuminates the stripe through the aperture in the near-field region. In the case of a metallized probe, the coupled laser light heats the apex of the probe due to absorption by the coating metal. Using the shear-force feedback, the probe apex is kept in near-contact interaction with the sample surface. The Al stripe is connected to the constant current source, and the voltage change due to local heating by the probe is detected with a lock-in amplifier.

9.6.2 Heating with a Metallized Probe

For the metallized probe, we show the dependence of the signal intensity on the sample–probe separation in Fig. 9.34a. The shear-force curve obtained simultaneously is also shown. The signal rises abruptly in close proximity to the sample, where the shear-force signal decreases due to the near-contact interaction. This result demonstrates that the dominant mechanism heating the stripe is conduction by the contact between the probe and the sample. The signal remaining even in the noncontact region seems to originate from conduction by air.

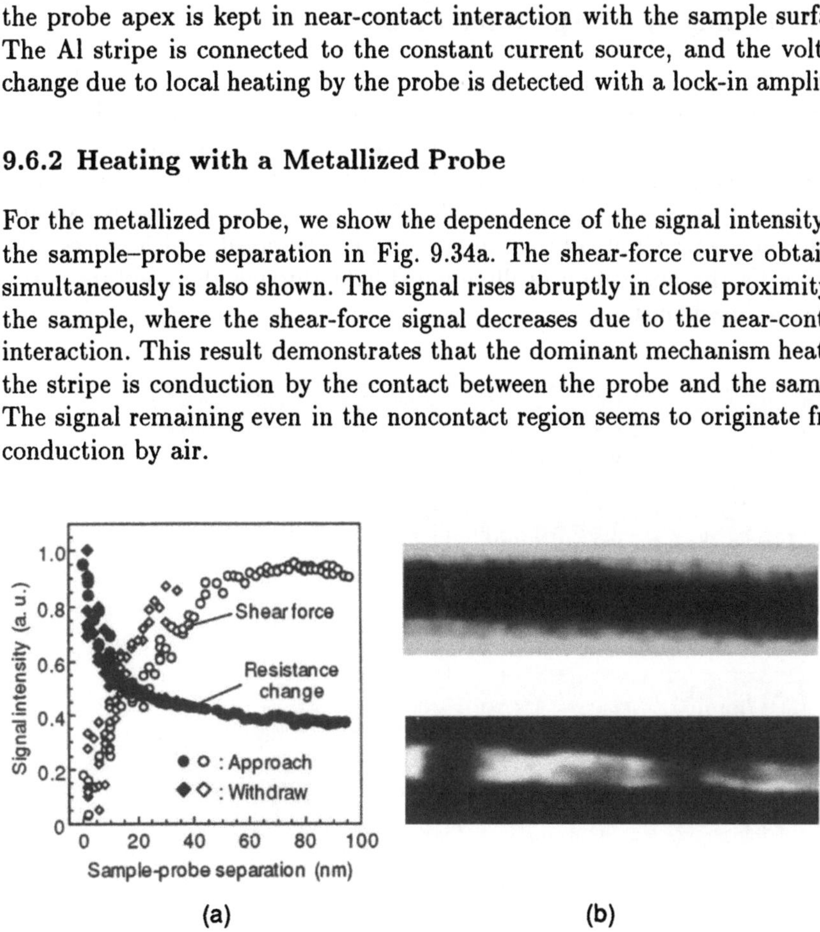

Fig. 9.34. a Plots of the signal intensity of resistance change as a function of sample–probe separation. The shear-force curve is also shown. b Topographic image of an Al stripe (top) and a two-dimensional map of the signal intensity (bottom). The bright area corresponds to the section where the resistance change is large. The image size is 2 μm$\times$0.5 μm

Figure 9.34b shows an image of the resistance change in the Al stripe with corresponding shear-force topographic image. The bright part corresponds to the area where the resistance change is large. The width and the depth of the stripe are 0.4 μm and 0.5 μm, respectively. Apart from the large structure to the left, there is no correlation between the shear-force topographic image and the resistance change image. This implies that the clear contrast reflects the existence of defects beneath the surface of the stripe. We also confirm the achievement of high spatial resolution as small as 50 nm, which is related to the contact area between the probe apex and the sample surface.

9.6.3 Heating by an Apertured Probe

The ability of an apertured probe to be a local heating source is also demonstrated. In order to obtain an adequate heating effect, we use a rather large aperture with a diameter of 300 nm. The throughput of this probe is around 4×10^{-2}. The dependence of the signal intensity on the sample–probe separation is very gradual compared with that for the metallized probe. No abrupt change in the proximity of the sample surface is observed, which proves that heating by the absorption of illuminated light through the aperture is the dominant mechanism.

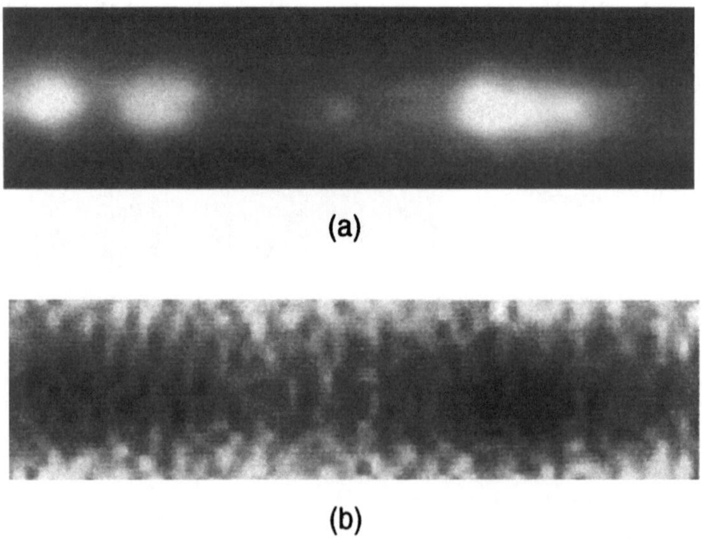

Fig. 9.35. Resistance change images obtained by a near-field OBIRCH, and b conventional OBIRCH methods. The observation sections (4 μm$\times$1 μm) are the same in both images

Figure 9.35a shows a resistance change image obtained by a 300 nm aperture probe with 40 μW irradiation. The stripe is a trench structure 0.3 μm wide and 0.5 μm deep. Figure 9.35b shows the resistance change image by the OBIRCH method in the same scanning area. A good correspondence of signal structure in both images was obtained. Furthermore, it should be noted that in the near-field image, the fine structures to the left are clearly resolved, while in the OBIRCH image they are not. This result demonstrates the high resolving power of near-field illumination through the probe. The existence of voids corresponding to the near-field observations has been verified by scanning ion microscope.

References

1. S. K. Buratto, J. W. P. Hsu, E. Betzig, J. K. Trautman, R. B. Bylsma, C. C. Bahr, M. J. Cardillo, Appl. Phys. Lett. **65**, 2654 (1994)
2. S. K. Buratto, J. W. P. Hsu, J. K. Trautman, E. Betzig, R B. Bylsma, C. C. Bahr, M. J. Cardillo, J. Appl. Phys. **76**, 7720 (1994)
3. B. B. Goldberg, M. S. Ünlü, W. D. Herzog, H. F. Ghaemi, E. Towe, IEEE J. Selected Topics in Quantum Electron. **1**, 1073 (1995)
4. U. Ben-Ami, N. Tessler, N. Ben-Ami, R. Nagar, G. Fish, K. Lieberman, G. Eisenstein, A. Lewis, J. M. Nielsen, A. Møeller-Larsen, Appl. Phys. Lett. **68**, 2337 (1996)
5. Ch. Lienau, A. Richter, A. Klehr, T. Elsaesser, Appl. Phys. Lett. **69**, 2471 (1996)
6. A. Richter, J. W. Tomm, Ch. Lienau, J. Luft, Appl. Phys. Lett. **69**, 3981 (1996)
7. W. D. Herzog, M. S. Ünlü, B. B. Goldberg, G. H. Rhodes, C. Harder, Appl. Phys. Lett. **70**, 688 (1997)
8. J. W. P. Hsu, E. A. Fizgerald, Y. H. Xie, P. J. Silverman, Appl. Phys. Lett. **65**, 344 (1994)
9. T. J. Silva, S. Schultz, D. Weller, Appl. Phys. Lett. **65**, 658 (1994)
10. J. K. Rogers, F. Seiferth, M. Vaez-Iravani, Appl. Phys. Lett. **66**, 3260 (1995)
11. T. Saiki, S. Mononobe, M. Ohtsu, N. Saito, J. Kusano, Appl. Phys. Lett. **67**, 2191 (1995)
12. M. J. Gregor, P. G. Blome, R. G. Ulbrich, P. Grossmann, S. Grosse, J. Feldmann, W. Stolz, E. O. Göbel, D. J. Arent, M. Bode, K. A. Bertness, J. M. Olson, Appl. Phys. Lett. **67**, 3572 (1995)
13. J. Liu, T. F. Kuech, Appl. Phys. Lett. **69**, 662 (1996)
14. J. Almeida, T. dell'Orto, C. Coluzza, G. Margaritondo, O. Bergossi, M. Spajer, D. Courjon, Appl. Phys. Lett. **69**, 2361 (1996)
15. J. Liu, N. R. Perkins, M. N. Horton, J. M. Redwing, M. A. Tischler, T. F. Kuech, Appl. Phys. Lett. **69**, 3519 (1996)
16. C. Durkan, I. V. Shvets, J. C. Lodder, Appl. Phys. Lett. **70**, 1323 (1997)
17. A. A. McDaniel, J. W. P Hsu, A. M. Gabor, Appl. Phys. Lett. **70**, 3555 (1997)
18. N. Saito, F. Sato, K. Takizawa, J. Kusano, H. Okumura, T. Aida, T. Saiki, M. Ohtsu, Jpn. J. Appl. Phys. **36**, L896 (1997)

19. H. F. Hess, E. Betzig, T. D. Harris, L. N. Pfeiffer, K. W. West, Science **264**, 1740 (1994)
20. R. D. Grober, T. D. Harris, J. K. Trautman, E. Betzig, W. Wegscheider, L. Pfeiffer, K. West, Appl. Phys. Lett. **64**, 1421 (1994)
21. T. D. Harris, D. Gershoni, R. D. Grober, L. Pfeiffer, K. West, N. Chand, Appl. Phys. Lett. **68**, 988 (1996)
22. A. Richter, G. Behme, M. Süptitz, Ch. Lienau, T. Elsaesser, M. Ramsteiner, R. Nötzel, K. H. Ploog, Phys. Rev. Lett. **79**, 2145 (1997)
23. H. F. Ghaemi, B. B. Goldberg, C. Cates, P. D. Wang, C. M. Sotomayor Torres, M. Fritze, A. Nurmikko, Superlattices and Microstructures **17**, 15 (1995)
24. Y. Toda, M. Kourogi, M. Ohtsu, Y. Nagamune, Y. Arakawa, Appl. Phys. Lett. **69**, 827 (1996)
25. F. Flack, N. Samarth, V. Nikitin, P. A. Crowell, J. Shi, J. Levy, D. D. Awschalom, Phys. Rev. B **54**, R17312 (1996)
26. L. -Q. Li, L. M. Davis, Rev. Sci. Instrum. **64**, 1524 (1993)
27. J. B. Stark. U. Mohideen, R. E. Slusher, Tech. Dig. Ser. Conf. Ed. **16**, 82 (1995)
28. A. Lewis, U. Ben-Ami, N. Kuck, G. Fish, D. Diamant, L. Lubovsky, K. Lieberman, S. Katz, A. Saar, M. Roth, Scanning **17**, 3 (1995)
29. J. Levy, V. Nikitin, J. M. Kikkawa, A. Cohen, N. Samarth, R. Garcia, D. D. Awschalom, Phys. Rev. Lett. **76**, 1948 (1996)
30. A. Vertikov, M. Kuball, A. V. Nurmikko, H. J. Maris, Appl. Phys. Lett. **69**, 2465 (1996)
31. A. H. La Rosa, B. I. Yakobson, H. D. Hallen, Appl. Phys. Lett. **70**, 1656 (1997)
32. R. D. Grober, T. D. Harris, J. K. Trautman, E. Betzig, Rev. Sci. Instrum. **65**, 626 (1994)
33. H. Ghaemi, C. Cates, B. B. Goldberg, Ultramicroscopy **57**, 165 (1995)
34. J. Levy, V. Nikitin, J. M. Kikkawa, D. D. Awschalom, N. Samarth, J. Appl. Phys. **79**, 6095 (1996)
35. W. Göhde, J. Tittel, Th. Basché, C. Bräuchle, U. C. Fischer, H. Fuchs, Rev. Sci. Instrum. **68**, 2466 (1997)
36. R. D. Grober, T. Rutherford, T. D. Harris, Appl. Opt. **35**, 3488 (1996)
37. M. S. Ünlü, B. B. Goldberg, W. D. Herzog, D. Sun, E. Towe, Appl. Phys. Lett. **67**, 1862 (1995)
38. T. Saiki, N. Saito, J. Kusano, M. Ohtsu, Appl. Phys. Lett. **69**, 644 (1996)
39. G. Kolb, K. Karraï, G. Abstreiter, Appl. Phys. Lett. **65**, 3090 (1994)
40. C. Obermüller, K. Karraï, G. Kolb, G. Abstreiter, Ultramicroscopy **61**, 171 (1995)
41. T. Saiki, N. Saito, M. Ohtsu, Mat. Sci. and Eng. B **48**, 162 (1997)
42. T. Saiki, S. Mononobe, M. Ohtsu, N. Saito, J. Kusano, Appl. Phys. Lett. **68**, 2612 (1996)
43. T. Yatsui, M. Kourogi, M. Ohtsu, Appl. Phys. Lett. **71**, 1756 (1997)

44. L. Novotny, D. W. Pohl, B. Hecht, Ultramicroscopy **61**, 1 (1995)
45. V. S. Gurevich, M. Libenson, Ultramicroscopy **57**, 277 (1995)
46. J. Takahara, S. Yamagishi, H. Taki, A. Morimoto, T. Kobayashi, Opt. Lett. **22**, 475 (1997)
47. N. Saito, M. Yamaga, F. Sato, I. Fujimoto, M. Inai, T. Yamamoto, T. Watanabe, Inst. Phys. Conf. Ser. **136**, 601 (1993)
48. M. Fujii, T. Yamamoto, M. Shigeta, T. Takebe, K. Kobayashi, S. Hiyamizu, I. Fujimoto, Surf. Sci. **267**, 26 (1992)
49. M. Inai, T. Yamamoto, M. Fujii, T. Takebe, K. Kobayashi, Jpn. J. Appl. Phys. **32**, 523 (1993)
50. T. Nishinaga, K. Mochizuki, H. Yoshinaga, C. Sasaoka, M. Washiyama, J. Cryst. Growth **98**, 98 (1989)
51. Y. Arakawa, H. Sakaki, Appl. Phys. Lett. **40**, 939 (1982)
52. R. Leon, P. M. Petroff, D. Leonard, S. Fafard, Science **267**, 1966 (1995)
53. M. Grundmann, J. Christen, N. N. Ledentsov, J. Böhrer, D. Bimberg, S. S. Ruvimov, P. Werner, U. Richter, U. Gösele, J. Heydenreich, V. M. Ustinov, A. Yu. Egorov, A. E. Zhukov, P. S. Kop'ev, Zh. I. Alferov, Phys. Rev. Lett. **74**, 4043 (1995)
54. L. Samuelson, A. Gustafsson, D. Hessman, J. Lindahl, L. Montelius, A. Petersson, M. -E. Pistol, Phys. Stat. Sol. (a) **152**, 269 (1995)
55. M. E. Rubin, G. Medeiros-Ribeiro, J. J. O'shea, M. A. Chin, E. Y. Lee, P. M. Petroff, V. Narayanamurti, Phys. Rev. Lett. **77**, 5268 (1996)
56. J.-Y. Marzin, J.-M. Gérard, A. Izraël, D. Barrier, G. Bastard, Phys. Rev. Lett. **73**, 716 (1994)
57. D. Hessman, P. Castrillo, M.-E. Pistol, C. Pryor, L. Samuelson, Appl. Phys. Lett. **69**, 749 (1996)
58. D. Gammon, E. S. Snow, B. V. Shanabrook, D. S. Katzer, D. Park, Science **273**, 87 (1996)
59. H. Kamada, J. Temmyo, M. Notomi, T. Furuta, T. Tamamura, Jpn. J. Appl. Phys. **36**, 4194 (1997)
60. K. Nishi, R. Mirin, D. Leonard, G. M.-Ribeiro, P. M. Petroff, A. C. Gossard, J. Appl. Phys. **80**, 3466 (1996)
61. K. Brunner, G. Abstreiter, G. Böhm, G. Tränkle, G. Weimann, Phys. Rev. Lett. **73**, 1138 (1994)
62. M. Michel, A. Forchel, F. Faller, Appl. Phys. Lett. **70**, 393 (1997)
63. H. Tachibana, M. Matsumoto, Y. Tokura, Y. Moritomo, A. Yamaguchi, S. Koshihara, R. D. Miller, S. Abe, Phys. Rev. B **47**, 4363 (1993)
64. K. Ebihara, S. Koshihara, M. Yoshimoto, T. Maeda, T. Ohnishi, H. Koinuma, M. Fujiki, Jpn. J. Appl. Phys. **36**, L1211 (1997)
65. E. Betzig, R. J. Chichester, Sciemce **262**, 1422 (1993)
66. W. P. Ambrose, P. M. Goodwin, J. C. Martin, R. A. Keller, Phys. Lev. Lett. **72**, 160 (1994)
67. J. K. Trautman, J. J. Macklin, L. E. Brus, E. Betzig, Nature **369**, 40 (1994)

68. X. S. Xie, R. C. Dunn, Science **265**, 361 (1994)
69. W. E. Moerner, T. Plakhotnik, T. Irngartinger, U. P. Wild, Phys. Rev. Lett. **73**, 2764 (1994)
70. T. Ha, Th. Enderle, D. S. Chemla, S. Weiss, IEEE J. Selected Topics in Quantum Electron. **2**, 1115 (1996)
71. D. A. Higgins, P. J. Reid, P. F. Barbara, J. Phys. Chem. **100**, 1174 (1996)
72. D. A. V. Bout, J. Kerimo, D. A. Higgins, P. F. Barbara, Acc. Chem. Res. **30**, 204 (1997)
73. D. P. Tsai, A. Othonos, M. Moskovits, D. Uttamchandani, Appl. Phys. Lett. **64**, 1768 (1994)
74. C. L. Jahncke, M. A. Paesler, H. D. Hallen, Appl. Phys. Lett. **67**, 2483 (1995)
75. D. A. Smith, S. Webster, M. Ayad, S. D. Evans, D. Fogherty, D. Batchelder, Ultramicroscopy **61**, 247 (1995)
76. J. Grausem, B. Humbert, A. Burneau, J. Oswalt, Appl. Phys. Lett. **70**, 1671 (1997)
77. S. R. Emory, S. Nie, Anal. Chem. **69**, 2631 (1997)
78. K. Nikawa, C. Matsumoto, S. Inoue, Jpn. J. Appl. Phys. **34**, 2260 (1995)
79. D. I. Kavaldjiev, R. Toledo-Crow, M. Vaez-Iravani, Appl. Phys. Lett. **67**, 2771 (1995)
80. B. I. Yakobson, A. LaRosa, H. D. Hallen, M. A. Paesler, Ultramicroscopy **61**, 179 (1995)
81. V. Kurpas, M. Libenson, G. Martsinovsky, Ultramicroscopy **61**, 187 (1995)
82. M. Stähelin, M. A. Bopp, G. Tarrach, A. J. Meixner, I. Zschokke-Gränacher, Appl. Phys. Lett. **68**, 2603 (1996)

Chapter 10

Toward Nano-Photonic Devices

10.1 Introduction

The strength of the local electromagnetic interaction between a specimen and a probe increases with increases in the optical near-field intensity. This increase allows modifications to the specimen's surface and also nanometric processing, which leads to the creation of novel nanometric materials. Further, novel functional photonic devices are expected from the use of a resonant interaction between the optical near field and the matter, as well as using the local optical nonlinear effect.

In view of the development of nano-photonic devices, this chapter reviews the surface plasmon and its applications in Sect. 10.2, and the application of the optical near field to high density optical storage in Sect. 10.3.

10.2 Use of Surface Plasmons

10.2.1 Principles of Surface Plasmons

Surface plasmons (SPs) are quanta of collective oscillations of charges on conductive surfaces, and have excellent spatial coherence due to long-range Coulomb interactions [1]. These plasmons propagate along the surface, coupling with electromagnetic waves around the conductive (1)/dielectric (2) interface, as depicted in Fig. 10.1. The normal component of this field is given by $E_z = E_0 \exp[\mathrm{i}(k_x x \pm k_z z - \omega t)]$. The characteristics of these waves can be described by Maxwell's theory. The dispersion relation of SPs is given by

$$\frac{k_{z1}}{\varepsilon_1} + \frac{k_{z2}}{\varepsilon_2} \qquad (10.1)$$

Further, the relation between the two components of the wave number is given by

$$k_x^2 + k_{zi}^2 = \varepsilon_i \left(\frac{\omega}{c}\right)^2, \; i = 1, 2 \qquad (10.2)$$

These equations involve the most important properties of SPs, i.e., if k_x is larger than the wave number of the propagating light $k_i^{\mathrm{light}} = \sqrt{\varepsilon_i}(\omega/c)$ in

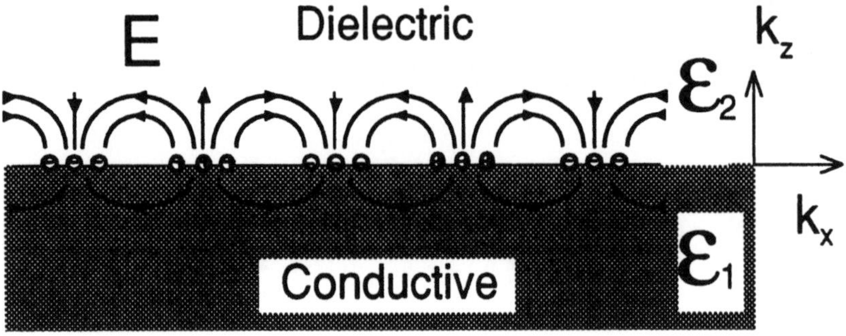

Fig. 10.1. Schematic model of a propagating surface plasmon coupling with an electromagnetic wave localized around the conductive (1)/dielectric (2) interface

the general system of a metal/air interface (in which $\varepsilon_1 < -1$ and $\varepsilon_2 = 1$ for the visible light range), k_{zi} becomes imaginary. Then E_z exponentially decays away from the interface, and the electromagnetic field is concentrated near the interface. This property reveals the applicability of SPs for photonic devices [2], especially owing to their nanometer-scale sensitivity to chemical and physical changes on the surface [3, 4].

For the excitation of SPs, optical near fields with wave vectors larger than those of the propagating light are required. In particular, those generated by a total internal reflection (TIR) of coherent light can resonantly produce a given mode of SPs with high efficiencies. This is well known as the attenuated total reflection (ATR) method. The excited mode strongly depends on the optical conditions of the interfaces. One of the typical examples of application is spectroscopic analysis of adsobates or ultra-thin films on metal surfaces using ATR devices [3, 5]. This method works as a powerful tool for longitudinal nanometer-scale characterization of surfaces owing to its experimental simplicity. It has also been successfully applied to develop surface plasmon microscopy in order to obtain two-dimensional (2D) mapping of adsobates [3, 6].

Direct imaging of the propagating SPs has been tried by NOM [7]. It was tried in an attempt to confirm the scattering phenomena, or the interaction with the local changes in the chemical and physical properties of the surface. Local field enhancement is considered to occur according to the multiple-scattering or interference phenomena of propagating SPs. Weak [8, 9] and strong (Anderson) [9, 10] localization of fields have been reported on rough surfaces. Enhancement of the optical near field on the surface has sufficient possibilities to encourage the development of novel nonlinear optical devices. Second harmonic generations, for example, have been realized on those surfaces [11]. We will discuss localization techniques like those, and more gentle ones, in Sect. 10.2.3, but first we discuss observation techniques for scattering or interference phenomena using the high spatial coherency of SPs.

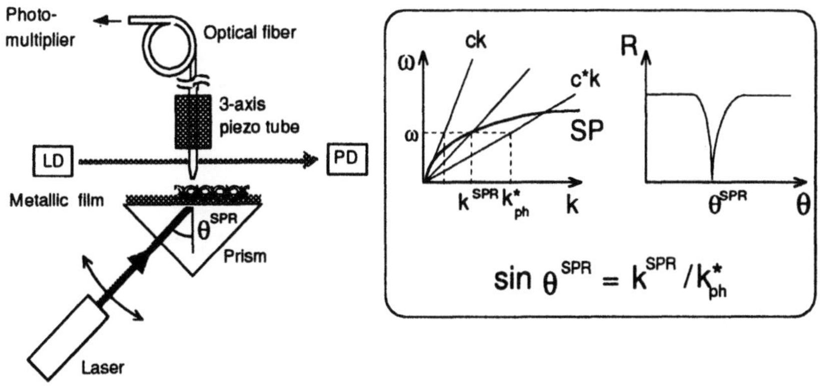

Fig. 10.2. Experimental set-up for imaging excited SPs. Surface plasmons can be excited by prism coupling. The inset shows the dispersion curve for SPs and photons in air, and the dependence of the reflectivity R on the incidence angle θ

10.2.2 Observation of Surface Plasmons

The enhanced optical near field due to the excitation of SPs has been directly observed using a c-mode NOM combined with the Kretshmann geometry of an ATR device, as shown in Fig. 10.2 [7, 9–13]. In this experimental set-up, a clear near-field image can be obtained even by an optical fiber probe with no aperture if the sample-probe separation is maintained at a few nanometers only. This is because the electromagnetic field is almost nonradiative on the planar metallic surface, and the apex of the scanning probe works as a scattering center. However, on a thermally deposited metal surface, the radiation damping of the propagating SPs makes the image acquisition complicated. That is, if the amplitude of the surface roughness is large, or localized defects like adsobates exist on the surface, they will generate intense radiation. Then it is difficult to divide clearly the near-field and the far-field components because the far-field components couple easily to the tapered portion of the fiber probe. One way to extract only the near-field components is by vertically vibrating the probe, which can be done with a cantilever or metallic probe [14]. A simpler and clearer technique to obtain near-field images is to use a fiber probe with small aperture. We have obtained well-pronounced 2D periodic images of the near field caused by the mutual interference of excited SPs, i.e., we have tried to observe directly the interference between two counterpropagating SPs on a planar gold thin film deposited on a glass plate by means of the incidence of two counterpropagating near-field components.

Figure 10.3 shows the experimental set-up, where an apertured fiber probe with a 200-nm-thick gold coating and a 50 nm diameter aperture [15] was kept at a constant height about 5 nm away from the surface by shear-force feedback control. In Kretschmann geometry, the two counterpropagating SPs were excited on a 48-nm-thick gold film by two incoming p-polarized He–Ne laser lights ($\sim$ 1 mW, 633 nm) from opposite directions.

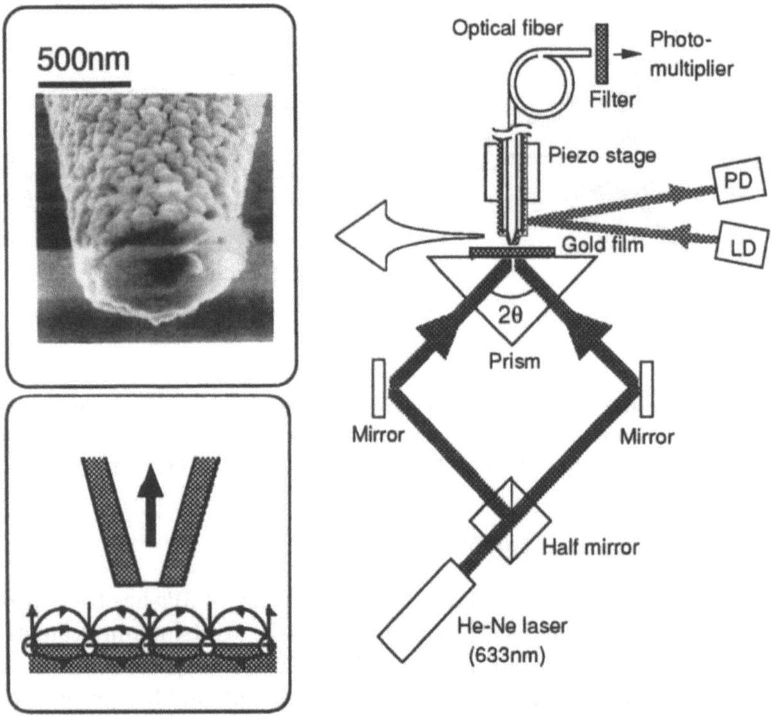

Fig. 10.3. Experimental set-up for imaging a *standing-wave-like* near field caused by the interference between the two counterpropagating SPs. The *upper* inset shows a SEM micrograph of the aperture of the probe. The aperture diameter was about 50 nm. The *lower* inset shows a cross-sectional view of the sample–probe system

The obtained image shown in Fig. 10.4a shows the generation of standing-wave-like near-field enhancement with 2D periodicity. In the single propagating mode, the asymmetry of the optical near field with a given wave vector may affect the image acquisition process using the probe as a scattering center. On the other hand, the contrast of the image can be clearly understood as a result of the differences in interactions of SPs with the probe apex in the two polarization directions, as claimed in Ref. [16]. Real parts of the complex wave vectors of excited SPs have already been obtained by power spectral analysis of Fourier transform (FT) of the near-field image of Fig. 10.4a. Both elastic and inelastic scattering phenomena of SPs have been analyzed by the combination of near-field and their FT images. Moreover, the local field enhanced phenomena caused by localized small defects such as nanometer-size contaminants have been observed in the near-field images. This mechanism is now under study, and may be seen to be the key effect in controlling the localization or confinement of a nonradiative optical near field.

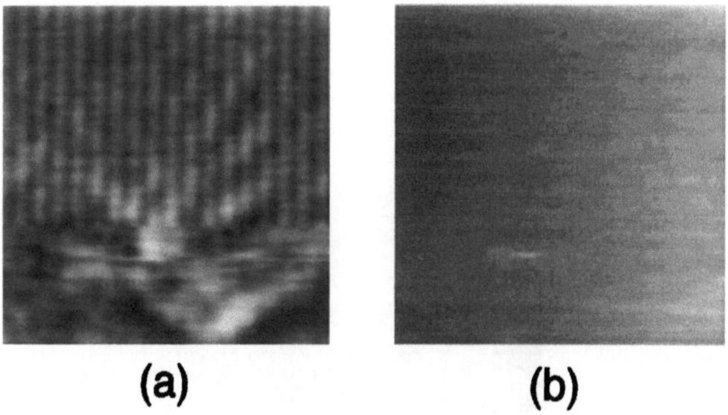

Fig. 10.4. a Near-field image obtained by the experimental set-up in Fig. 10.3. Vertical stripes indicate the interference fringe between two counterpropagating SPs. b Topographical image simultaneously obtained by the shear-force technique. Both image frames are 5 μm$\times$5 μm

10.2.3 Toward Two-Dimensional Devices

The high coherency and monochromaticity of SPs are capable of use in the development of many novel functional devices. For example, a wavelength selector based on the resonance of SPs has recently been proposed using Kretshmann geometry by virtue of the monochromaticity [17]. Field enhancement is considered to be caused by the scattering or interference of SPs. Weak or strong localization of the enhanced field by SPs on random and rough surfaces is important in understanding these localization phenomena. However, the competition between strong localization and radiation damping leads to a problem that is yet to be solved [18].

In this subsection, we review the scattering phenomena at the sharpened edges of metal films in order to control the scattering of 2D propagating SPs by suppressing radiation damping. Although the incidence of propagating SPs at the sloping edge has been imaged using a NOM [7], no clear scattering phenomena have yet been observed. For this observation, we fabricated much sharper edges of gold and silver films by a combination of electron beam lithography and the lift-off technique.

Figure 10.5 shows SEM micrographs of the fabricated edges of silver film about 50 nm thick deposited on a glass plate. The edge angle was estimated to be larger than 80° by high-resolution SEM micrographs. The near-field image obtained using an aperture-type probe is shown in Fig. 10.6a. The topographical image simultaneously obtained by shear-force feedback control is shown in Fig. 10.6b. The experimental set-up is similar to that in Fig. 10.3, where a He–Ne laser light beam ($\lambda = 633$ nm) was incident to the silver film near the fabricated edge at the resonance angle. The dithering direction of the probe for shear-force feedback control was almost parallel to the edge. Well-pronounced interference fringes were observed due to the interference

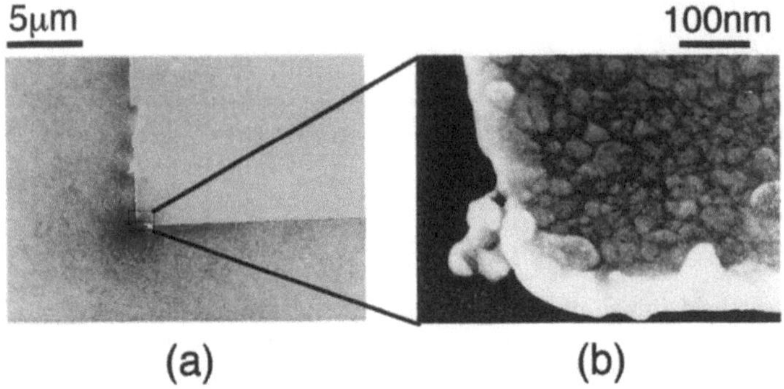

Fig. 10.5. a A SEM micrograph of a sharpened thin film edge fabricated by a combination of electron beam lithography and the lift-off technique. b Magnified image of a corner

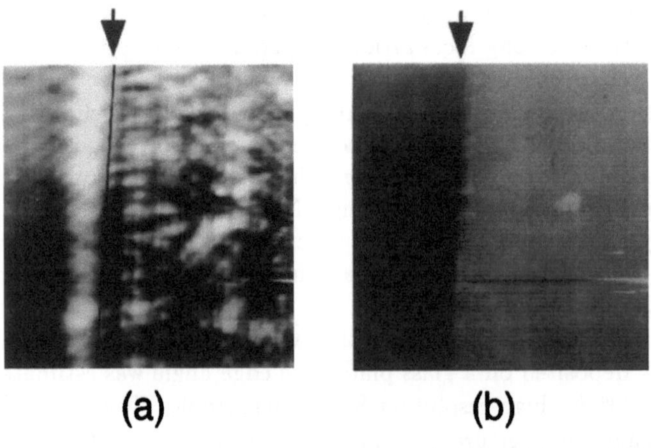

Fig. 10.6. a Near-field image obtained with the incidence of SPs onto a film with a sharpened edge. b Topographical image simultaneously obtained by the shear-force technique. Both image frames are 15.2 μm×15.2 μm. The lines indicated by *arrows* correspond to the edges

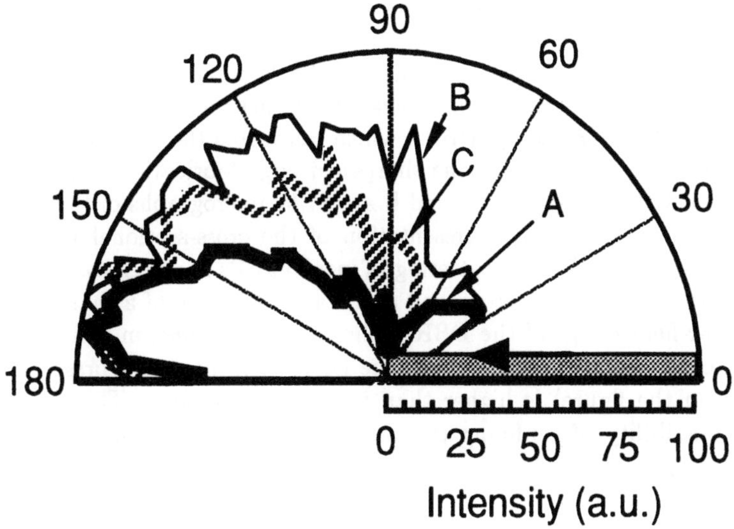

Fig. 10.7. Angular distributions of the detected intensities radiating from the edge site. The *large arrow* represents the direction of the excited SPs. Curve A, surface plasmon resonance (SPR) condition ($\theta = \theta^{SPR}$); curve B, incidence of a p-polarized laser beam with the same photon energy as curve A, but out of SPR condition ($\theta = \theta^{SPR} \pm 0.5°$); curve C, incidence of an s-polarized laser beam with the same photon energy and incidence angle as curve A

between the excited SPs and the elastically scattered SPs at the fabricated edge. The contrast of the fringes was about one-tenth as strong as those of counter interferences in Fig. 10.4a. Reflectivity at the edge can be estimated by those differences in contrast. However, this estimation is not applicable to the results obtained by different types of probes because the contrast of each image strongly depends on the short-range electromagnetic interaction between the probe and the edged surface by multiple scattering of the optical near field. The detected intensity was very high just outside the edge, as shown in Fig. 10.6a, for which two explanations are possible. One is that the probe detected radiations from the side of the edge. The other is that a novel localized mode of SPs was generated at the side of the edge.

Angular distributions of radiations from the edge were evaluated by measuring the intensities. Figure 10.7 shows the measured results for a sharpened edge similar to the one used for the near-field observation. Curve A shows the angular distribution of the resonant excitation of SPs. Curve B is the off-resonant condition ($\theta = \theta^{SPR} \pm 0.5°$). Curve C is for the incident angle θ^{SPR} as curve A, but by s-polarized incident light. Comparison of these curves indicates the limitation of the radiation directions of SPs at the fabricated edges. Although the coupling with the prism has not been fully evaluated in this experiment, it shows the possibility of controlling the radiation direction by varying the shape of the film edge. This technique can be applied to realize a highly efficient edge coupler.

Sharpened edges cannot work as a reflector of SPs because of their strong radiation damping. One of the concepts available to get high reflectivity is to use a distributed Bragg reflector (DBR), which can forbid the propagation of polariton mode. We have applied the sharpened edge fabrication technique to make such a DBR. Figure 10.8a and b are SEM micrographs of the fabricated DBR and a schematic representation of the cross-sectional profile, respectively. For this DBR, the scattering of incident SPs has been evaluated using the same experimental set-up and probe as in the case of a sharpened edge. The near-field image of the DBR is shown in Fig. 10.8c, in which the near-field intensity near the boundary of the DBR (i.e., the area represented by an arrow in the figure) is higher than that for other areas. This result reveals the possibility of confining the propagating SPs while suppressing radiation damping.

10.2.4 Toward Three-Dimensional Devices

If the reflectors completely surround an area as small as a given wavelength on a metal (silver or gold) surface, localized field enhancement with a specified wavelength will be realized without any radiation damping. Concentration of the electromagnetic field due to this localized field enhancement will change the optical properties of the localized area, and will give rise to optical nonlinear effects. The localized mode of SPs can exist on a small metallic particle. Scattering phenomena on such small conductive particles have been studied for a long time [1, 2], and the localized field enhancement around small particles is considered to be one of the main reasons for surface-enhanced Raman scattering [19, 20]. In the case of a spherical particle, eigenmodes in a nonretarded case are expressed as [1]

$$\varepsilon_1 = -\varepsilon_2 \frac{l+1}{l},\ l = 1, 2, 3, \cdots \tag{10.3}$$

On the other hand, the corresponding condition on the planar surface is given by $\varepsilon_1 = -\varepsilon_2$ from Eqs. 10.1 and 10.2. Equation 10.3 gives the localized mode in small particles. Generally, this localized mode is radiative and can be coupled with propagating light. Assuming that the particle size is much smaller than the wavelength of the excited field, radiation damping can be suppressed [21].

In this subsection we discuss the possibility of realizing three-dimensional devices for which polarization-dependence is also taken into account. In spherical particles there is no dependence on the polarization direction of an excitation field. On the other hand, in ellipsoidal particles, the local field can be enhanced when the field component of excitation light is parallel to the longer principal axis of the ellipsoid. The maximum degree of enhancement

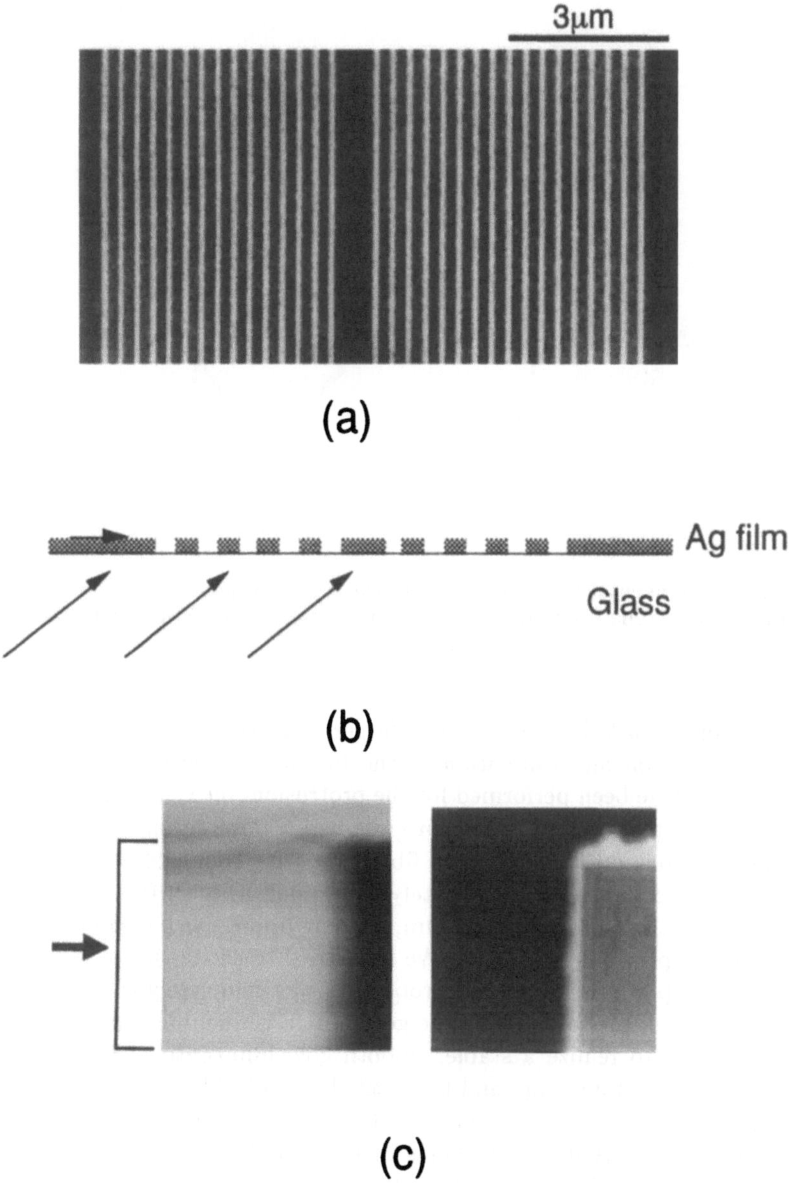

Fig. 10.8. **a** A SEM micrograph of a distributed Bragg reflector fabricated by a combination of electron beam lithography and the lift-off technique. **b** Cross sectional profile. **c** *Left.* Near-field image obtained during the excitation of SPs in the area sandwiched between the two constructions. The *arrow* represents the area near the boundary of the DBR. *Right.* Topographical image simultaneously obtained by the shear-force technique. Both image frames are 15.2 μm$\times$15.2 μm

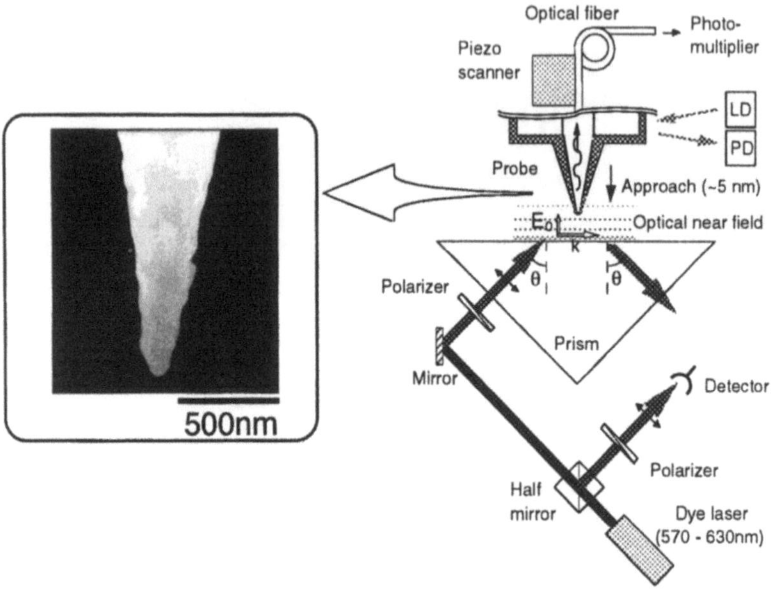

Fig. 10.9. Experimental set-up for exciting the localized mode of SPs on a fabricated probe apex. The inset shows a SEM micrograph of a fabricated metallized probe apex

in an ellipsoidal particle can be much larger than that in spherical particles [1]. Realization and evaluation of the localized mode excitation on a hemi-ellipsoid have been performed for the protrusions on a plane surface using lithographic techniques [21]. We propose here a novel technique to excite the localized mode on the apex of a fiber probe for NOM [22]. The shape of the probe apex can be approximately ellipsoidal. If its surface is covered by a very smooth conductive thin film, the conditions can be same as the ellipsoid in the previous study [21]. We have tried to excite such a localized mode on the top of our metallized probe that was completely covered with a thin gold film. Prior to applying the gold film, a germanium thin film was applied in order to realize a stable, smooth gold film coating. Figure 10.9 shows an experimental set-up, and the inset shows a SEM micrograph of the fabricated metallized probe. The cross-sectional profile of the probe depicted in this figure shows that the thickness of the gold film increased from the apex to the foot side, i.e., it was about 25 nm at the apex and about 80 nm at the foot.

In order to confirm the local mode generation, the excitation photon energy was varied by tuning a R6G dye laser (570–630 nm). The incident angle of the p-polarized laser beam was set to 45° for the TIR condition, where the critical angle is always less than 42° because the refractive index of the prism is 1.53 for 570–630 nm wavelength range. The distance between the probe and

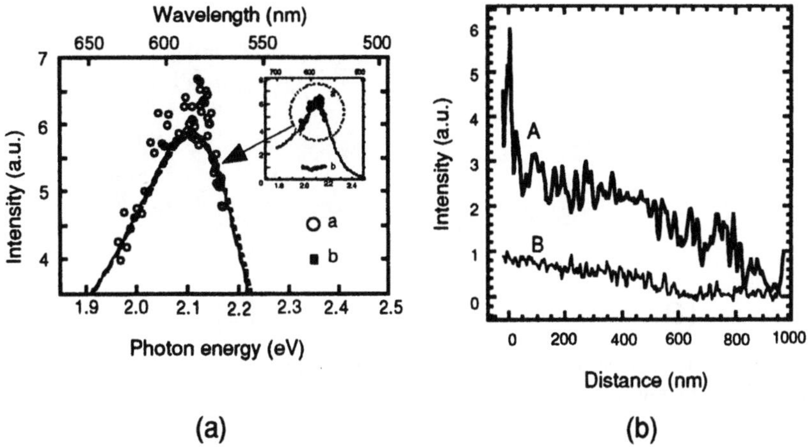

Fig. 10.10. a Dependence of the detected intensity on the excitation photon energy. *Open circles*, the metallized probe. A theoretical curve fitted to the open circles is also drawn. *Closed squares*, the uncoated probe. b Dependence of the detected intensity on the distance between the probe and the prism surface. Curve A, the metallized probe; curve B, the uncoated probe

the prism surface was kept to be 5 nm by shear-force feedback control. Dependence of the detected intensity on the excitation photon energy is shown in Fig. 10.10a. In this figure, open circles represent experimental values for the metallized probe, and a theoretical curve was fitted to them based on a local field theory involving a radiation damping factor [21, 23]. For reference, closed squares represent the values for an uncoated probe. Comparison between the open circles and the closed squares confirms the realization of the local field enhancement in the metallized probe. The resonant photon energy in the metallized probe was about 2.11 eV, corresponding to 588 nm in wavelength. The detected intensity at the resonant point is about six times as high as that of the uncoated probe. Theoretical enhancement of the near field, however, is estimated to be about 150 times the incident intensity. This difference is due to the experimental fact that many of the unwanted low spatial Fourier frequency components and far-field components of the light are included in the detected light intensity for the uncoated probe.

At the resonant condition, the environment(i.e., the substrate surface and the gap) of the probe was not unity but about 1.5, which reveals that the close proximity of the probe to the substrate will change the optical property of the probe. This can be called the proximity effect (or substrate effect). To examine this effect, we have also evaluated the dependence of intensity on the distance between the probe and the prism surface. The results are shown in Fig. 10.10b. Curves A and B are the results for the metallized and uncoated probes, respectively. The excitation photon energy was fixed at the resonance condition (2.11 eV energy). The detected intensity of curve A increases with decreasing distance, and a dramatic increase can be seen at about 50 nm.

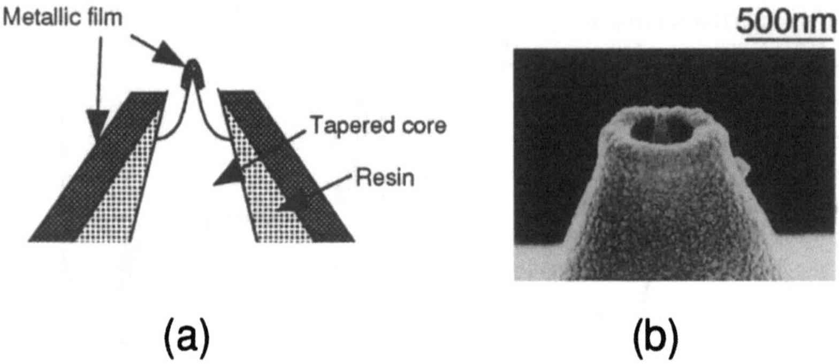

Fig. 10.11. a Cross-sectional structure of a protruded metallized probe with an aperture. b SEM micrograph of the fabricated probe

Further, the slope of this curve for a distance less than 50 nm is much larger than that of the exponential functional curve. Such a large increase and steep slope are considered to be because of the localized fields around the probe apex. These fields are enhanced only when the effective dielectric constant of the environment of the probe is shifted to a value in close proximity to those of the probe and the substrate. This discrimination sensitivity of the intensity with respect to the distance can be higher than that of the tunneling effect of electrons. Moreover, this drastic increase could not be obtained in the case of s-polarized photons.

Due to high discrimination sensitivity, this technique for enhancing the localized field has many applications such as operating nanometric three-dimensional functional devices, mapping the local dielectric constant of a sample surface, near-field imaging with much higher vertical resolution and sensitivity, distribution of near-field polarization, and the sensing of single molecules in liquids. Further, since our metallized probe can be useful in scanning tunneling microscopy, it can be used to study nanometer-sized local interactions between a photon and an electron.

10.2.5 A Protruded Metallized Probe with an Aperture

We have developed a protruded metallized probe with an aperture in order to improve the performance of the metallized probe discussed in the previous subsection. Figure 10.11a shows its cross-sectional structure, and the two metallic films which are coated on separately at the apex and the foot of the tapered core. The metallic film on the apex is to increase the scattering efficiency, and to enhance the optical near-field intensity for the c-mode and i-mode NOMs, respectively, by which an improvement in measurement sensitivity is expected. The metallic film on the foot forms an aperture to suppress the detection and generation of light with low spatial Fourier components, by which increase in the contrast of the image is expected. The fabrication

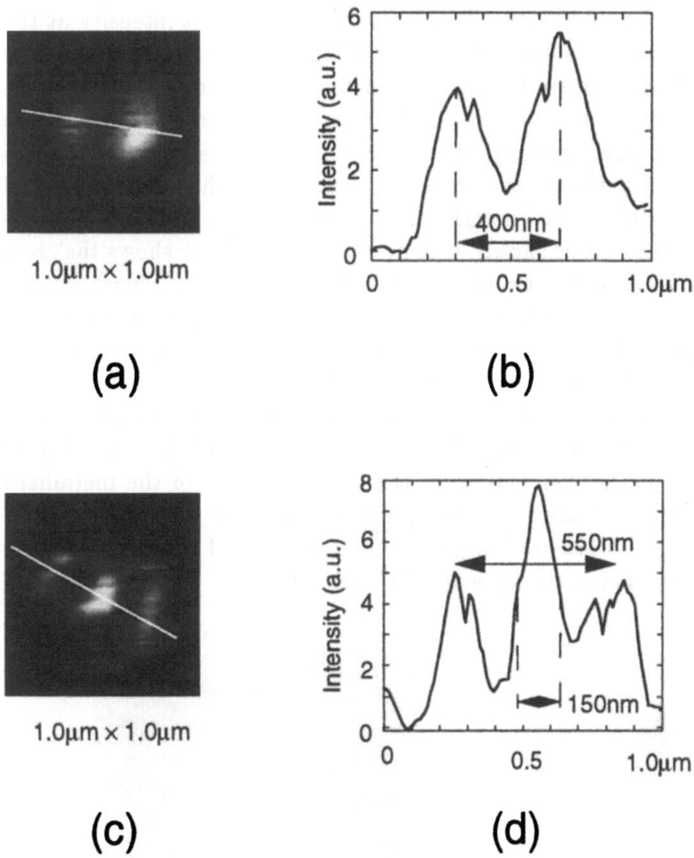

Fig. 10.12. a Measured spatial distribution of the output light intensity of the probe with an uncoated apex. Except for the apex, the structures are the same as for the fabricated probe in Fig. 10.11. **b** The cross-sectional distribution along the white line in Fig. 10.12a. **c** Measured spatial distribution of the output light intensity measured for the probe in Fig. 10.11b. **d** Cross-sectional distribution along the white line in Fig. 10.12c

process for this probe is as follows. (A) The fiber core is tapered and the cladding diameter is decreased by chemical etching (cf. Sect. 3.3.1.1). (B) The tapered core, except for the absolute apex, is coated with a resin film. (C) The uncoated part of the apex of the core is chemically etched to further reduce its size. (D) Gold film is coated onto the foot of the tapered core to form the aperture, and the fiber axis is tilted in order to avoid coating the apex. (E) The apex of the tapered core is coated with a gold film.

Figure 10.11b shows a SEM micrograph of the fabricated probe, where the gold-coated apex diameter, d, is 35 nm (the uncoated apex diameter is smaller than 5 nm). The thickness of the gold film at the foot is 100 nm, and the aperture diameter (i.e., the foot diameter), d_f, is 400 nm.

We measured the spatial distribution of the output light intensity at the apex of the probe by coupling a 680-nm-wavelength light into this fiber. The experimental set-up is equivalent to that in Fig. 4.6, where a flat-top apertured probe with an aperture diameter of 100 nm was scanned. Figure 10.12a shows the measured distribution for a probe with an uncoated apex as a reference. Note that the structure except for the apex, is the same as that of the probe in Fig. 10.11. The butterfly-shaped distribution in this figure corresponds to the HE_{11} mode shown in Sect. 4.2. Figure 10.12b shows the cross sectional distribution along the white line in Fig. 10.12a. The separation between the two peaks is 400 nm, which corresponds to the aperture diameter. Figure 10.12c shows the measured distribution for the probe in Fig. 10.11b. The high intensity at the center of the butterfly represents the effect of the metal coating on the apex. The cross-sectional distribution along the white line in this figure is given in Fig. 10.12d. The half-width of the central peak is 150 nm, which is the convolution of the apex diameter of the metallized core and the aperture diameter of the probe scanned for the measurement. Comparison of Fig. 10.12a–d clearly confirms the effect of the metal coating on the apex in enhancing the optical field intensity.

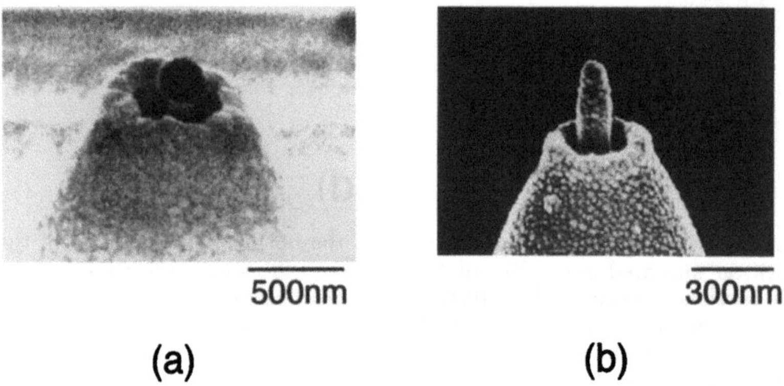

Fig. 10.13. SEM micrographs of a protruded metallized probe with an aperture. **a** Spherical and **b** ellipsoidal metallized apex

Figure 10.13a and b show the SEM micrographs of probes with a spherical and an ellipsoidal metallized apex, respectively, which are fabricated by slightly modifying step E of the process just described. These figures demonstrate the versatility of the process in realizing a variety of apex shapes, by which an optimized shape can be obtained to obtain, for example, a plasmon resonance for further enhancement of the field intensity. Further, apart from the gold film, this process can coat a variety of materials such as silver and semiconductors, from which novel functional probes are expected.

10.3 Application to High-Density Optical Memory

An optical memory is an example of one of the promising applications of optical near fields to nano-photonics. Although conventional optical memory has realized storage densities as high as 2 Gb/inch^2 for DVD-RAM by using a propagating far-field light of 650 nm wavelength, its highest density is limited by the diffraction of the light. For higher densities, the use of a shorter-wavelength light source (e.g., a blue-violet semiconductor laser) or volume holography, etc., have been proposed. Use of a solid immersion lens has also been proposed in order to reduce the diffraction-limited focused spot size of the laser beam, which has realized a storage/read-out of a 350 nm diameter pit on a TbFeCo film (a magneto-optical storage medium) [24, 25].

The use of an optical near field is an attempt to go beyond the diffraction limit and to realize a storage density as high as 1 Tb/inch^2, which is limited by the apex size of the probe. Preliminary experiments have been carried out for Pt/Co film (a magneto-optical storage medium) using an i-mode NOM to obtain 170 Gb/inch^2 storage density [26]. This can be called *thermal-mode* storage because it uses the local heating of the storage medium by the optical near field. An experiment in *photon-mode* storage/read-out/erasing, using a photochemical reaction, has also been carried out for a Langmuir-Blodgett film of a photochromic material to realize a storage density as high as the one achieved by thermal-mode storage [27]. The advantages of photon-mode storage are an inherently fast response time of the electronic transition of the photochromic material, free from the thermal damage of the probe, and so on.

Since this pioneering work, thermal-mode storage using $Ge_2Sb_2Te_5$, a phase-change medium, has been carried out at a lower storage temperature than that for the magneto-optical storage medium [28]. Highly sensitive photochromic materials have also been developed for photon-mode storage [29]. Since these studies were only carried out to demonstrate the feasibility of near-field optics for high-density optical storage/read-out by scanning a conventional fiber probe, they do not clearly represent future directions for practical/commercial optical memory. The next section is devoted to a review of the expected directions of future progress.

10.3.1 Problems to Be Solved

The problems involved in realizing a low-cost and reliable optical memory system are related not only to the hardware, but also to the software of the system, and further, they are correlated with each other. The problems can be covered under four main headings.

1. Software.
   a) Establishing an application software to be supplied to the customer, e.g., data bases on medical care for individuals, library data bases, information on weather forecasting, or providing motion pictures.

   b) The form and architecture of the memory, e.g., read-only, erasable, etc., must be decided depending on the type of software, and this also governs the strategy of developing types of hardware.
   c) Fail-safe software for read-out, e.g., to skip nanometric surface singularities in the medium.
2. System.
   a) The form of the storage medium, e.g., a linear tape or a circular disk, which depends on the kind of information to be stored.
   b) Scanning the head for storage and read-out (SR head).
   c) Controlling the nanometric separation between the storage medium surface and the SR head.
   d) Tracking feedback of the SR head.
   e) Removing the memory from the system to realize a portable memory.
   f) Fast detection of a very weak light intensity for read-out.
3. Device.
   a) A high-throughput SR head for high-speed scanning.
   b) An interface connecting a light source, an SR head, and a photodetector.
   c) An actuator for scanning the SR head.
4. Storage medium.
   a) Exploring novel phenomena of sensitive and local interactions between the optical near field and the nanometric matter, and materials sensitive to these interactions.
   b) Developing materials with nanometric grain and domain.
   c) Re-investigating physical and thermodynamic concepts and quantities for nanometric matter, such as heat dissipation, temperature, and diffusion.
   d) Developing a protection layer of nanometric thickness.

10.3.2 Approaches to Solving the Problems

We now review our attempts to solve problems 2 (b–d) and 3 in the lists in the previous subsection.

10.3.2.1 Structure of the Read-Out Head

The solid line in Fig. 10.14 shows the calculated relation between the throughput of the probe and the data transmission rate. The assumptions for this calculation are: (1) the read-out signal is the light scattered from the surface of the medium; (2) use of a pulse code modulation (PCM); (3) the bit error rate is 1×10^{-12}, which is due to the shot noise in the detected light of 600 nm wavelength; (4) a fiber probe is used as a read-out head; (5) the light power coupled from the light source to the fiber is 1 mW, which is below the threshold of thermal damage to the probe; (6) the contrast of the scattered light intensity(i.e., the ratio of the light intensities scattered at the stored pit and at the unstored medium surface) is 0.1. This value is estimated from the

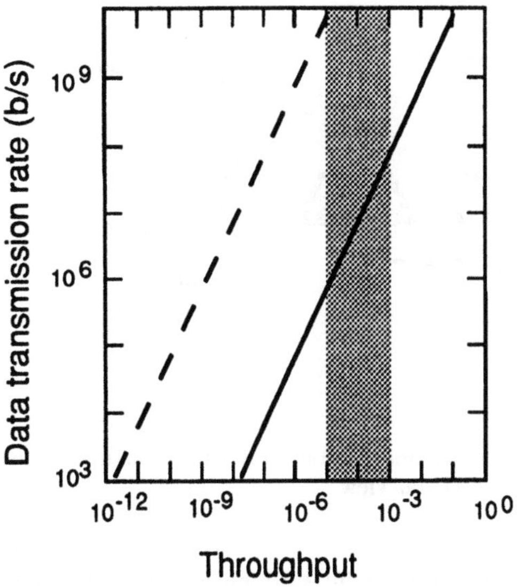

Fig. 10.14. Relation between the throughput of the probe and the PCM read-out data transmission rate needed to maintain a bit error rate as low as 1×10^{-12}. *Solid* and *broken lines* indicates the use of a single fiber probe and a probe array, respectively. *The meshed area* represents the value of the throughput of our fiber probes (cf. Fig. 4.14)

experimental results of our photon-mode storage/read-out [27], and can also be applicable to the heat-mode storage/read-out.

The meshed area in this figure represents the value of the throughput of our fiber probes, which is roughly in the range 1×10^{-5} to 1×10^{-3} (cf. Fig. 4.14). From a comparison between the meshed area and the solid line, the expected data transmission rate is only about 50 Mb/s. This means that a novel read-out head is required because the data transmission rate should be increased to as high as 1 Gb/inch^2 in order to be compatible with the 1 Tb/inch^2 density storage which is expected in the future. One way of realizing such a high-speed read-out head is to develop a two-dimensional planar probe array, as shown in Fig. 10.15 [30]. The scanning speed v of this array is given by $v = sr/n$, where s is the separation between the adjacent stored pits, r is the data transmission rate, and n is the number of apertures in the array. By substituting typical values for 1 Tb/inch^2 density (i.e., $s =$ 25 nm, $r =$ 1 Gb/s, and $n = 100 \times 100$) into this relation, we obtain $v =$ 2.5 mm/s, which is a reasonable value that could be realized even by using a conventional scanning actuator. The broken line in Fig. 10.14 represents the data transmission rate using this array. In the case of a single fiber probe ($n = 1$), note that the value of v is 25 m/s, which is too large to be realized by a conventional scanning actuator.

The two-dimensional planar probe array in Fig. 10.15 has further advantages, not only a reduction in the scanning speed. These are: (1) the

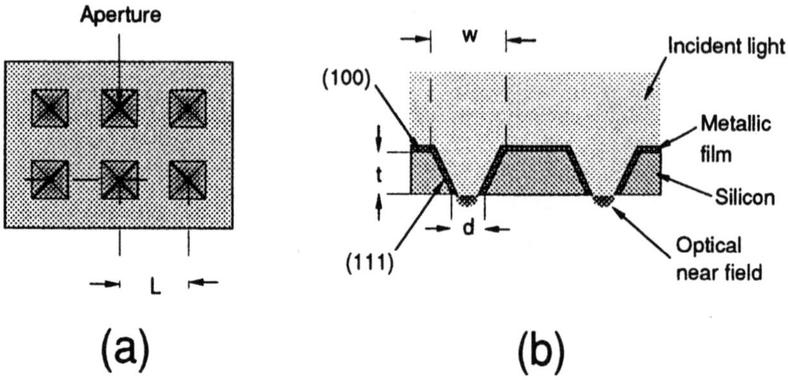

Fig. 10.15. Profile of the two-dimensional planar probe array with apertures on a silicon substrate. **a** Top view. **b** Cross-sectional view

separation between the array and the storage medium surface can be kept constant due to the flat bottom surface of the array; (2) the array can be smoothly scanned on the storage medium while maintaining a narrow gap, e.g., for the use of a lubricant oil film of nanometric thickness to coat on the storage medium, as is the case with a contact-type hard disk; (3) track-less scanning is possible, as will be described in Sect. 10.3.2.3; (4) integration with interfacing devices such as a slab waveguide or a photodetector array is possible.

10.3.2.2 Storage Probe Array

Figure 10.16a shows schematically how to store and read out the memory. The array is scanned over the storage medium surface by sliding on a coated lubricant oil film of about 10 nm thickness. The light intensity incident into each aperture of the array is modulated by a spatial modulator, depending on the information to be stored. Thus, the optical near-field intensity on the bottom surface of each aperture is modulated to be used for storage. Free-space propagation of the light from the light source to the upper surface of the array allows the array to be independent of the other far-field optical components, and thus maintain the total weight of the array at a very low level. This is advantageous for scanning the array fast. The separation, L, between the adjacent apertures has to be larger than the wavelength of the incident light in order to avoid interference between the modulated lights incident into the adjacent apertures. This large separation is also advantageous because no special nanometric fabrication processes are required to prepare a photomask for fabricating the array.

10.3.2.3 Track-less Read-out

Figure 10.16b shows that the array is scanned along the direction represented by the arrow while the axis of the array is tilted with respect to that of the two-dimensionally stored pits. The tilt angle is θ. The broken lines in this

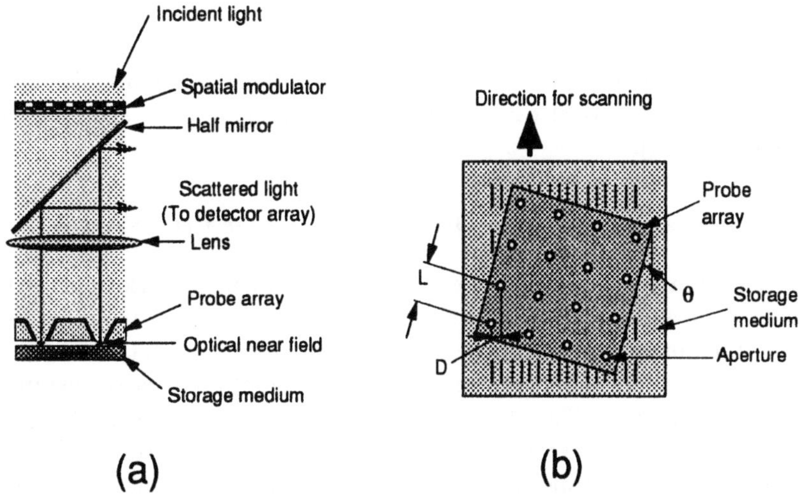

Fig. 10.16. Schematic representation of the storage and read-out. **a** Set-up for storage by using a spatially modulated propagating light incident into the probe array, and for read-out by detecting the scattered light. **b** Read-out by scanning the array with a tilt angle θ. Broken lines represent the scanning trajectories of the apertures in the array. L, separation between the adjacent apertures in the array; D, separation between the adjacent broken lines

figure represent the scanning trajectories of the apertures. The separation D between adjacent short lines is given by $L \sin\theta$. By using a conventional focusing lens, scattered lights from the storage medium coming through the apertures are separately detected by the photodetector array because the separation L is larger than the wavelength of the incident light. By storing the temporal sequence of the output electric signals from the detector array into a memory circuit, the relation between the light intensity and the position of a scatterer (i.e., the stored pit) on the storage medium can be found because the scanning speed is known. Thus, the stored pits can be read-out even though the separation between the pits is shorter than the aperture separation L.

Since the aperture diameter d is smaller than the stored pit diameter d_M, all the spatial Fourier frequency components of the scattered light lower than $1/d_M$ can be picked up by the array, which enables the user to retrieve the spatial distribution of the scattered light intensity along the scanning direction. For a lateral scanning direction, however, the sampling theorem ensures complete regeneration if $D \leq d_M/2$. In the case where $s = 25$ nm and $n = 100 \times 100$, as in Sect. 10.3.2.1, the size of the array is maintained as low as 250 μm$\times$250 μm. Further, the inaccuracy in scanning trajectory control can be as large as several 10 μm. Therefore, by storing the positions of the pits aligned lateral to the scanning direction in an auxiliary memory circuit, the positions of the stored pits can be found and accessed by the SR head. This means that a tracking servo-control of the array is not essential.

A more reliable read-out can be expected by adding error-correcting codes along and lateral to the scanning direction, as well as using the interleave technique. A track-less read-out similar to that in Fig. 10.16b has also been proposed using a vertical cavity surface-emitting diode laser array as the read-out head [31].

10.3.3 Fabrication of a Two-Dimensional Planar Probe Array

We have fabricated a two-dimensional planar probe array by anisotropic etching of a silicon substrate [30]. The concave pyramids in Fig. 10.15 are formed by using the difference in etching rates along the (100) and (111) axes of the crystal. The nanometric hole formed at the bottom of the concave pyramid is used as an aperture. The slope of the pyramid (i.e., 70.5°) is governed by the direction of the (111) axis. The aperture size, d, is determined by the thickness t of the substrate and the size of the upper surface of the concave pyramid, w, which is given by $d = w - \sqrt{2}t$. The accuracy of w depends on that of the lithography. In order to maintain the homogeneity and the accuracy of t at a sufficiently high level, we used a SOI (silicon on insulator) substrate in which a SiO_2 layer is buried. That is, we used an insulator substrate as a blocking layer to inhibit further etching. After forming the apertures, the slopes are coated with a gold film. The aperture size d is also determined by the thickness of the gold film coating.

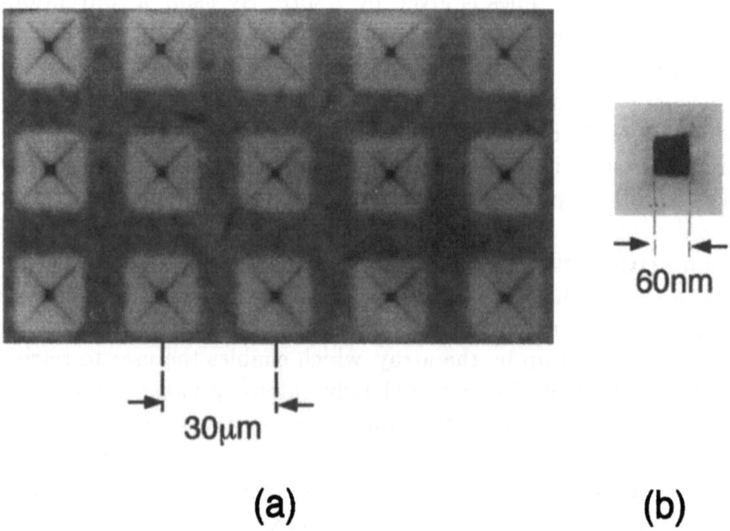

Fig. 10.17. SEM micrograph of a fabricated array. **a** Top view. **b** Bottom view of the smallest aperture

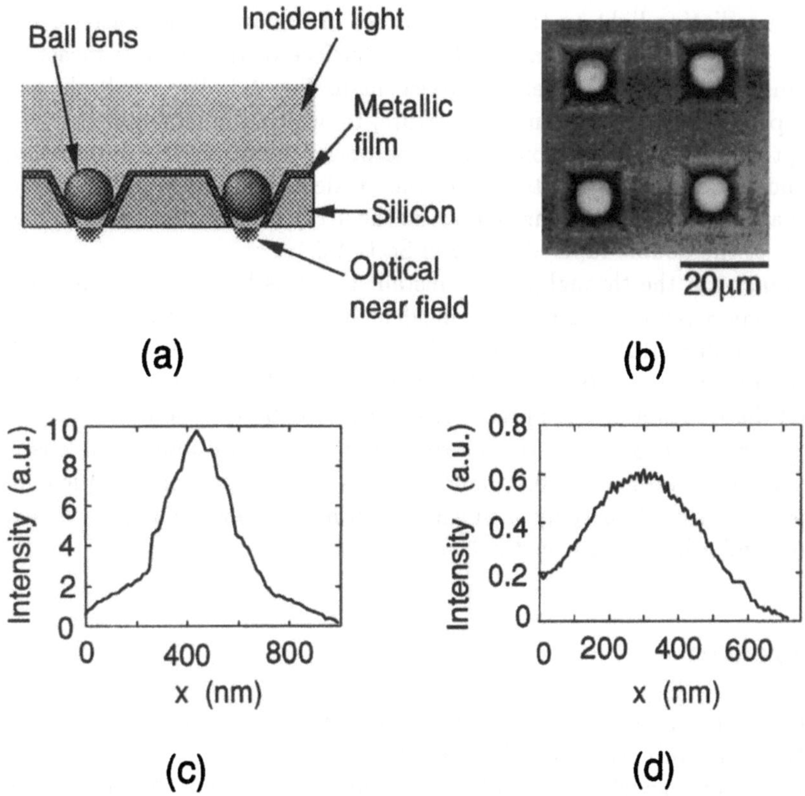

Fig. 10.18. **a, b** Cross sectional structure and SEM micrograph, respectively, of the array with a glass ball lens. **c, d** Measured spatial distribution of the output light from the aperture in arrays with and without a ball lens, respectively

The advantage of this process is that the aperture array is fabricated effectively and homogeneously by using only the established technique of lithography for the silicon planar process, even though the size of the aperture is smaller than the resolution limit of lithography. A conventional photomask technique can maintain the accuracy of the separation L sufficiently for trackless read-out because the value of L is larger than the optical wavelength. Fluctuations of the aperture size d may induce fluctuations of the scattered light intensity from each aperture, which may decrease the signal-to-noise ratio for signal detection. However, this decrease can be avoided by appropriately adjusting the threshold level to detect the temporal signal sequence.

The quadratic apertures were fabricated reproducibly. Figure 10.17a shows a SEM micrograph of a fabricated array, where the value of L was fixed to be as large as 30 μm as a preliminary experiment. The aperture size d is 200 nm after a coating of a 50 nm-thick gold film. It should be noted that an array with an aperture size d as small as 60 nm is also fabricated (see Fig. 10.17b).

By irradiating light on the upper surface of the array, the spatial distribution of the output light at the bottom surface of the array is measured by using an experimental set-up equivalent to Fig. 4.6. The result shows a single peaked distribution on the aperture, by which a reliable generation of the optical near field was confirmed. The power density of the output light is found to be nearly equal to that of the incident light at the upper surface of the array, by which the throughput of the array is confirmed to be as high as that of the double-tapered probe in Sect. 4.3.1.

To increase the throughput, we installed a glass ball lens into the slope of the convex pyramid so that the incident light was focused onto the aperture [30]. Figure 10.18a and b show the cross-sectional structure and a SEM micrograph, respectively, of the fabricated array. Figure 10.18c shows the spatial distribution of the output light measured by the method described above. Figure 10.18d shows the distribution measured without a ball lens. A comparison between Fig. 10.18c and d shows that the use of a ball lens increases the light intensity by as much as 16 times, from which highly efficient storage and a high-speed read-out are expected.

References

1. H. Raether, *Surface Plasmons on Smooth and Rough Surface and on Grating* (Springer-Verlag, Berlin, 1988)
2. V. M. Agranovich, D. L. Mills (eds.), *Surface Polaritons* (North-Holland, Amsterdam, 1982)
3. H. Knobloch, G. von S. -Borryszkowski, S. Woigk, A. Helms, L. Brehmer, Appl. Phys. Lett. **69**, 2336 (1996)
4. I. I. Smolyaninov, D. L. Mazzoni, C. C. Davis, Phys. Rev. Lett. **77**, 3877 (1996)
5. G. B. Sigal, M. Mrksich, G. M. Whitesides, Langmuir **13**, 2749 (1997)
6. M. Specht, J. D. Pedarnig, W. M. Heckl, T. W. Hänsch, Phys. Rev. Lett. **68**, 476 (1992)
7. P. Dawson, F. de Fornel, J-P. Goudonnet, Phys. Rev. Lett. **72**, 2927 (1994)
8. O. A. Aktsipetrov, V. N. Golovkina, O. I. Kapusta, T. A. Leskova, N. N. Novikova, Phys. Lett. A **170**, 231 (1992)
9. S. I. Bozhevolnyi, Phys. Rev. B **54**, 8177 (1996)
10. S. I. Bozhevolnyi, I. I. Smolyaninov, A. V. Zayats, Phys. Rev. B **51**, 17916 (1996)
11. S. I. Bozhevolnyi, K. Pedersen, Surf. Sci. **377/379**, 384 (1997)
12. S. I. Bozhevolnyi, D. L. Mazzoni, J. Mait, C. C. Davis, Phys. Rev. B **56**, 1601 (1997)
13. J. R. Krenn, R. Wolf, A. Leitner, F. R. Aussenegg, Opt. Commun. **137**, 46 (1997)

14. F. Zenhausern, Y. Martin, H. K. Wickramasinghe, Science **269**, 1083 (1995)
15. S. Mononobe, M. Naya, T. Saiki, M. Ohtsu, Appl. Opt. **36**, 1496 (1997)
16. W. Jhe, K. Jang, Ultramicroscopy **61**, 81 (1995)
17. P. J. Kajenski, Opt. Eng. **36**, 263 (1997)
18. J. A. Sánchez-Gill, A. A. Maradudin, Phy. Rev. B **56**, 1103 (1997)
19. R. K. Chang, T. E. Furtak (eds.), *Surface Enhanced Raman Scattering* (Plenum, New York, 1982)
20. I. Baltog, N. Primeau, R. Reinisch, J. Opt. Soc. Am. B **13**, 656 (1996)
21. A. Wokaun, J. P. Gordon, P. F. Liao, Phys. Rev. Lett. **48**, 957 (1982)
22. M. Ashino, M. Ohtsu, Appl. Phys. Lett. **72**, 1299 (1998)
23. G. T. Boyd, Th. Raising, J. R. R. Leite, Y. R. Shen, Phys. Rev. B **30**, 519 (1984)
24. S. M. Mansfield, G. S. Kino, Appl. Phys. Lett. **57**, 2615 (1990)
25. B. D. Terris, H. J. Mamin, D. Rugar, W. R. Studenmund, G. S. Kino, Appl. Phys. Lett. **65**, 388 (1994)
26. E. Betzig, J. K. Trautman, R. Wolfe, E. M. Gyorgy, P. L. Finn, M. H. Kryder, C.-H. Chang, Appl. Phys. Lett. **61**, 142 (1992)
27. S. Jiang, J. Ichihashi, H. Monobe, M. Fujihira, M. Ohtsu, Opt. Commun. **106**, 173 (1994)
28. S. Hosaka, T. Shintani, M. Miyamoto, A. Kikukawa, A. Hirotsune, M. Terao, M. Yoshida, K. Fujita, S. Krammer, J. Appl. Phys. **79**, 8082 (1996)
29. M. Hamano, M. Irie, Jpn. J. Appl. Phys. **35**, 1764 (1996)
30. M. Ohtsu, in Technical digest, Joint international symposium on magneto-optical recording and optical memory, Yamagata, Japan, October 1997, pp. 180–181
31. K. Goto, in Technical digest, Joint international symposium on magneto-optical recording and optical memory, Yamagata, Japan, October 1997, pp. 184–185

Chapter 11

Near-Field Optical Atom Manipulation: Toward Atom Photonics

11.1 Introduction

Precise control of atomic motion is an up-to-date and important subject not only for atomic physics, including high-resolution laser spectroscopy, but also for the sophisticated technology of crystal growth. Since an optical near field can be localized in a nanometric region, it is very suitable for atom manipulation with high spatial accuracy. This chapter covers several methods of guiding and manipulating atoms by means of an optical near field. First, before discussing details, we will give an outline of the control of atomic motion based on the near-resonant mechanical interaction between an atom and an optical near field.

11.1.1 Control of Gaseous Atoms: From Far Field to Near Field

Each individual atom of a gas is moving at random at a high speed: heavy atoms of alkali–metal vapor such as rubidium (Rb) or cesium (Cs) are ballistically flying with a mean velocity of several hundred meters per second, even at room temperature. As is well known, the only way of controlling thermal atoms without charges is to use laser light because electric and magnetic fields exert no powerful force on neutral atoms. In the last two decades, immense progress has been made in the control of atomic motion with near-resonant laser light, as in the development of frequency-tunable lasers, including diode lasers.

The mechanical action of light on atoms consists of two parts: the dissipative part (the radiation pressure), and the reactive part (the dipole force) [1]. Although the major subject of this chapter concerns the dipole force of the optical near field on atoms, we will also refer briefly to the force of the propagating light on atoms for comparison. The effectiveness of the radiation pressure, which is sometimes called the spontaneous force, has been proved in the laser cooling of neutral atoms [2]. Let us consider the case where an atom is moving at a velocity v against a laser light beam with a wavenumber k. When the laser frequency ω_L is taken to be kv lower than the resonant frequency ω_0 of an atomic dipole transition in proportion to the atomic Doppler shift kv, the atom absorbs a photon of the laser light and then spontaneously emits a photon in a random direction. Then, if the absorption–spontaneous emission cycle is repeated many times, the atom takes a net force statistically, i.e., the radiation pressure, so that it loses its momentum in the direction of light propagation. This leads us to the fact that if we adjust the frequency detuning of the laser light in accordance with the change of the atomic Doppler shift, we can continuously reduce the atomic velocity. This is called Doppler cooling.

In addition, based on Doppler cooling, we can make a magneto-optical trap (MOT) composed of three pairs of two counterpropagating circularly polarized light beams orthogonalized to one another and anti-Helmholtz coils [2, 3]. Such an MOT produces a highly dense atomic ensemble with a mean temperature below 100 μK and a diameter of about 1 mm in the center of the trap. Recently, an atomic gas has been cooled to below 100 nK in an elaborate magnetic trap, which follows the MOT, with evaporative cooling. As a result, the Bose-Einstein condensation of an alkali-metal atomic gas has been realized for the first time [4]. In recognition of their pioneering work on laser cooling, S. Chu of Stanford University, C. Cohen-Tannoudji of the Collège de France, and W. D. Phillips of the National Institute of Standards and Technology were awarded the 1997 Nobel prize for physics.

The dipole force of propagating light has also been employed for the collimation of an atomic beam. In particular, a standing wave composed of two counterpropagating light beams works as a lens, and as a diffraction grating or a beam splitter for an atomic beam. This research area is called atom optics [5, 6]. It should be noted that the dipole force acts in the direction that the light intensity changes, while the radiation pressure acts in the direction that the light propagates. The strength of the dipole force is in proportion to the gradient of the light intensity. We will describe the characteristics of the dipole force in the next subsection.

Now, let us return to the subject of this chapter. Although the MOT is very useful for collecting ultracold atoms in a momentum-position phase space, it has no function to control atomic position with a precision below a wavelength. The traditional atom-optical methods with propagating light do not work for the precise control of atomic motion either. This stems from the fact that the use of propagating light is limited to the spatial accuracy of atom manipulation, which is determined by the wavelength used, owing to diffraction. Conversely, an optical near field has the advantage of being free from the diffraction effect. Therefore, we can use the optical near field to control atoms with high spatial accuracy beyond the diffraction limit of light waves.

Here, recall that the strength of the dipole force is in proportion to the gradient of the light intensity. Since the optical near field exponentially decays in a region below a wavelength, in which the decay length depends on the size of the substances (cf. Sect. 2.1), it exerts a strong dipole force on atoms. Accordingly, studies in the control of atomic motion using the optical near field deal with the near-resonant dipole interaction. In the subsections that follow a short review of the dipole force, we will see the developing process of atom manipulation by means of the optical near field, in which the spatial accuracy is increased from the wavelength level to the nanometer level.

11.1.2 Dipole Force

According to the quantum mechanics of atoms [7], an atom possesses a fine structure made up of discrete energy levels, owing to the coupling $\boldsymbol{J} = \boldsymbol{L} + \boldsymbol{S}$ of the orbital angular momentum $\boldsymbol{L}$ with the spin angular momentum $\boldsymbol{S}$ of electrons around a nucleus: the fine energy levels are classified with the quantum number $J = |L + S|, |L + S - 1|, \cdots, |L - S|$. Moreover, each fine energy level branches into some hyperfine levels due to the coupling $\boldsymbol{F} = \boldsymbol{J} + \boldsymbol{I}$ of the electron angular momentum $\boldsymbol{J}$ with the nuclear spin momentum $\boldsymbol{I}$: the hyperfine levels are labeled with the quantum number $F = |J + I|, |J + I - 1|, \cdots, |J - I|$. After Sect. 11.3, we use the sign $n\mathrm{X}_J$ to represent the atomic fine energy levels with a principal quantum number n, where X is written as S for $L = 0$, P for $L = 1$, D for $L = 2$, etc. For example, the ground state of a Rb atom with a mass number 85 (^{85}Rb) is expressed as $5\mathrm{S}_{1/2}$.

As is well known, an atom can absorb a photon of light with a frequency equivalent to an energy difference between the two hyperfine energy levels. On this occasion, the atom is polarized so that an electric dipole moment is induced. The dipole force originates from the fact that the electric field of the light operates on the induced electric dipole: if the light intensity is spatially inhomogeneous, there is a net force on the atom. Since the net force is in proportion to the intensity gradient of the light field, it is sometimes called the gradient force.

The direction of the dipole force $\boldsymbol{F}_{\mathrm{d}}$ depends on the detuning $\delta = \omega_{\mathrm{L}} - \omega_0$ of the light frequency ω_{L} from an atomic resonant frequency ω_0. Here, for simplicity, let us consider a two-level atom. The two-level atom picture is a good approximation of the alkali–metal atoms such as Rb employed in the following experiments. In the case where a dipole moment with a matrix element μ is induced by light with an electric field amplitude E, the dipole force on the two-level atom is written as [1, 8]

$$\boldsymbol{F}_{\mathrm{d}} = -\frac{\hbar\delta\nabla\Omega^2}{4\delta^2 + \Gamma^2 + 2\Omega^2} \tag{11.1}$$

where Γ is a natural linewidth of the atomic dipole transition, and $\Omega = \mu E/\hbar$ is an atomic Rabi frequency. From Eq. 11.1, we can see that the dipole force works in the direction where the light intensity becomes weak if $\delta > 0$, while the direction is opposite if $\delta < 0$: the dipole force is repulsive under a blue-detuning condition of $\delta > 0$, but attractive under a red-detuning condition of $\delta < 0$. This frequency dispersion properties of the dipole force are shown in Fig. 11.1a.

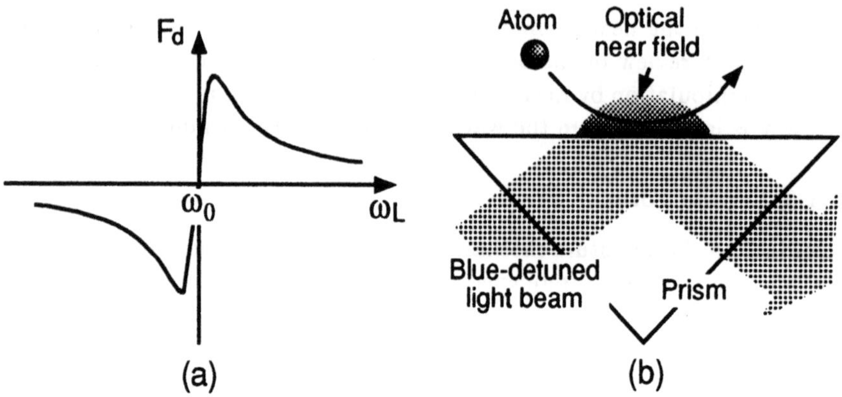

Fig. 11.1. **a** Dipole force F_d as a function of the light frequency ω_L, in which ω_0 is an atomic resonant frequency. The dipole force is repulsive in the blue-detuning region but attractive in the red-detuning region. **b** Atomic quantum sheet: reflection of an atom by a planar optical near field induced over the surface of a prism via total internal reflection of a blue-detuned light beam

11.1.3 Atomic Quantum Sheets: Atom Reflection Using a Planar Optical Near Field

By using the dipole force produced on atoms by an optical near field, we can make an *atomic quantum sheet* (AQS) as shown in Fig. 11.1b. The AQS is induced over the surface of a prism via total internal reflection of a blue-detuned light beam at the glass–vacuum interface. Such a planar optical near field leaked from the surface is often called an evanescent wave. If the light-field intensity is sufficiently strong and a proper blue detuning is chosen, an atom approaching the prism from the vacuum side is reflected without hitting the surface. It should be noted that when an alkali–metal atom such as Rb collides with a dielectric surface, it will stick there with a high probability. In this way, the planar optical near field acts as an atomic mirror, so that it is used for two-dimensional manipulation of atoms.

Let us consider the case where a light beam with an incident power P_0 and a wavelength λ shines on the surface of a prism with a refractive index n at an incident angle θ lager than the critical angle of total internal reflection. Then, as a function of the distance r from the surface, the intensity of the optical near field is given by $P(r) = P_0 \exp(-2r/\Lambda)$, where $\Lambda = (\lambda/2\pi)/\sqrt{n^2 \sin^2\theta - 1}$ is the decay length [8, 9]. In this case, from Eq. 11.1, the dipole force is written as [8, 10]

$$F_\mathrm{d}(r) = \frac{2\hbar\Delta\Omega(r)^2/\Lambda}{4\Delta^2 + \Gamma^2 + 2\Omega(r)^2} \tag{11.2}$$

where $\Delta = \omega_\mathrm{L} - \omega_0 - k_\mathrm{t} v_\mathrm{z}$ is the frequency detuning including the Doppler shift $k_\mathrm{t} v_\mathrm{z}$ due to the atomic velocity v_z along the surface, and $k_\mathrm{t} = (\omega_\mathrm{L}/c) n \sin\theta$

is the propagation constant. The position dependent Rabi frequency $\Omega(r)$ can be also given by $\Omega(r) = \sqrt{P(r)/2P_{\text{sat}}}\Gamma$, where $P_{\text{sat}} = \hbar\Gamma\omega_L^3/12\pi c^2$ is the saturation intensity of the atomic dipole transition. Equation 11.2 indicates that the optical near field produced over a dielectric plane fades away exponentially with the decay length Λ.

The first demonstration of an AQS was made with a thermal atomic beam of sodium (Na) [10], and then multiple bouncing on the AQS was performed with laser-cooled Cs atoms released from an MOT [11–13]. Recently, the AQS has been applied to observation of the van der Waals interaction between a ground-state Cs atom and a dielectric plane [14]. Incidentally, the specular reflection of atoms can be enhanced with a metal coating on the prism surface due to surface-plasmon excitation. An experiment with a surface-plasmon AQS has been also demonstrated with a thermal Rb beam [15]. These experiments in atom optics including the AQS are usually carried out at a vacuum pressure below 10^{-6} Pa.

11.1.4 Atomic Quantum Wires: Atom Guidance Using a Cylindrical Optical Near Field

For enhancement of the manipulation accuracy as compared with an AQS, an optical near field induced in a narrow space is required. To this end, hollow optical fibers are available. In fact, the guiding of neutral atoms has been demonstrated by means of a hollow optical fiber. Figure 11.2 shows a way of sending atoms down a hollow fiber escorted by a cylindrical optical near field. When a blue-detuned light beam is coupled to a hollow fiber, an optical near field leaks to the hollow region so that an optical tunnel to reflect atoms is induced around the inner wall. Then atoms entering the hollow region are guided through the hollow fiber without much loss, bouncing off the optical tunnel many times. This scheme enables one to one-dimensionally manipulate atoms: in other words, the cylindrical optical near field works as an *atomic quantum wire* (AQW).

Following some theoretical work [16–18], experiments with an AQW have been demonstrated with a 20-μm hollow fiber having no cladding by Renn et al. [19], and independently with a micron-sized hollow optical fiber having an annular structure composed of a thin core and a cladding by Ito et al. [20]. In the latter experiment, the laser spectroscopy of the guided atoms was made by means of two-step laser-photoionization. The photoionization experiments will be described in Sects. 11.3 and 11.4. Another scheme for an AQW with an optical near field has been proposed by Jhe et al. [21], in which atoms are guided through a narrow vacant region surrounded by four excentric-core optical fibers. This experiment has not yet been carried out.

There is another type of AQW with a far-field light whose frequency is red-detuned [8, 22]. This scheme is based on the fact that a hollow fiber can support grazing incident electromagnetic modes propagating in the hollow region, although they are very leaky. The transverse profile of the lowest

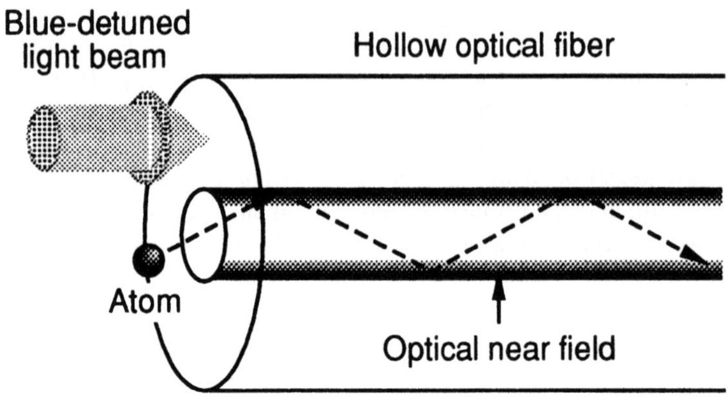

Fig. 11.2. Atomic quantum wire: atom guidance with a hollow optical fiber. The coupling of a blue-detuned light beam to a hollow fiber gives rise to excitation of a cylindrical near field working as an optical tunnel to guide atoms

order, the EH_{11} mode, of such propagating modes with a wavenumber k and a propagation constant β includes the zeroth order Bessel function of the first kind $J_0(u\rho)$, so that the light intensity is maximum along the hollow axis, where $u = \sqrt{n^2k^2 - \beta^2}$ and ρ are the transverse characteristic constant and the radial coordinate, respectively. Therefore, atoms are drawn to the axis by the attractive dipole force under a red-detuning condition. The AQW with a far field has been also demonstrated by Renn et al. [23].

In spite of the successful guiding of atoms, the AQW with a far field has significant drawbacks. Atoms are always exposed to the strong light field so that they undergo heating by multiple spontaneous emission events. In addition, the propagating modes exponentially decay away as they propagate, because of loss by total internal reflection and the loose and grazing incident confinement. This indicates that we cannot make the hollow diameter of the far-field AQW smaller, compared with the near-field AQW. Therefore, the scheme of the far-field AQW is not suitable for the precise control of atomic motion.

11.1.5 Atomic Quantum Dots: Atom Manipulation Using a Localized Optical Near Field

The AQW has greatly improved the spatial accuracy of atomic control. However, the optical near field induced over the inner-wall surface of a hollow fiber is not necessarily localized in a region far below a wavelength. As shown in the next section, the transverse profile of the lowest order, the LP_{01} mode, of the optical near field has the zeroth order modified Bessel function $I_0(v\rho)$ of the first kind, where $v = \sqrt{\beta^2 - k^2}$ is the transverse decay constant. From this, assuming a wavelength $\lambda = 780$ nm for guiding Rb atoms, we can see that the field intensity inside the hollow region is almost homogeneous in a region below about 100 nm, and as a result the effective optical potential to reflect atoms cannot be produced in the case of a hollow fiber with a hollow diameter below 100 nm.

Let us now consider the use of a three-dimensionally localized optical near field for the purpose of atom manipulation with a nanometric spatial accuracy. Important to this idea is the fact that an optical near field can be produced in an extremely narrow space below a wavelength without a diffraction limit. For example, the optical near field localized in a nanometer region can be induced over the tip of a sharpened fiber with a small apex. Such a nanometric sharpened fiber supporting the localized optical near field has the potential for trapping an atom in a narrow region, which leads to single-atom manipulation. Therefore, we can call the nanometric near-field optical device an *atomic quantum dot* (AQD).

The original idea of trapping an atom near the hemispherical tip of a sharpened fiber was proposed by Ohtsu and Hori and their collaborators [24]. In that scheme with a red-detuned optical near field to make a stable atom trap, we must balance three kinds of forces: the attractive dipole force in the radial direction of a spherical coordinate system with respect to the hemispherical tip, the radiation pressure in the polar-angle direction, and the centrifugal force around the polar axis. Although this atom trap works well if the radius of curvature is chosen so that the balance condition is satisfied, there is still a significant problem: that is, we must also take into account the influence of the cavity quantum electrodynamic (QED) effects, including the van der Waals and Casimir–Polder forces, because they can be greatly enhanced in the vicinity of the nanometric tip. Such being the case, we will now introduce new static atom traps composed of sharpened optical fibers together with a scheme of atomic deflection. These schemes include the dipole force and the van der Waals force as one of the cavity QED effects, in which the frequency detuning is large enough to ensure that the spontaneous emission event seldom occurs.

Figure 11.3 illustrates atom manipulation by means of an AQD with a sharpened optical fiber: a is the atomic deflection, and b is the catch and release of an atom. In Fig. 11.3a, the repulsive dipole force produced by a blue-detuned optical near field changes the path of the atomic ballistic motion. On the other hand, an atom trap, as shown in Fig. 11.3b, is based on the balance between the repulsive dipole force and the attractive van der Waals force, in which an atom is trapped around the point where the potential is minimum. In this atom trap, by changing the frequency detuning and the light intensity, we can also release the trapped atom. This technique will be very useful for creating atomic-scale matters. The advanced techniques of atom manipulation with sharpened optical fibers will be described in Sect. 11.6.

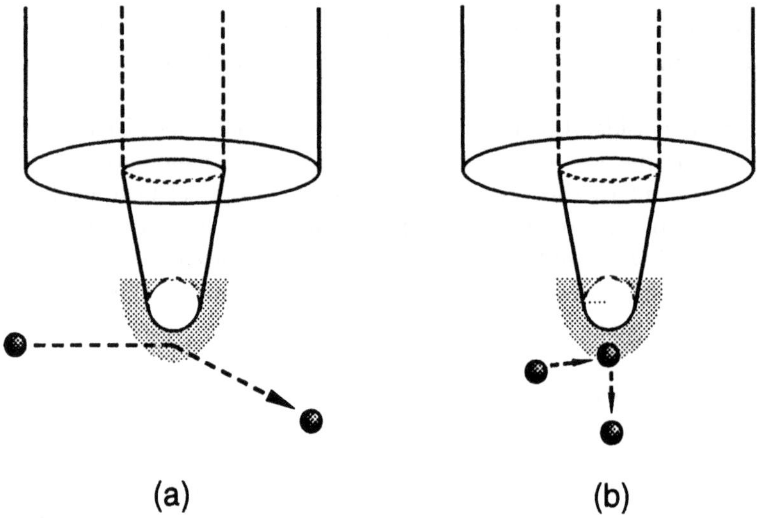

Fig. 11.3. Atomic quantum dot: atom manipulation with a sharpened optical fiber: a atomic deflection by a blue-detuned optical near field, and b atom trap with a function to catch and release an atom. The potential produced near the nanometric tip consists of the repulsive dipole force and the attractive van der Waals force

11.2 Cylindrical Optical Near Field for Atomic Quantum Wires

As we have pointed out, the scheme of AQW with a blue-detuned optical near field enables one to use a small hollow optical fiber with a wavelength-scale hollow diameter for atom guidance. In fact, four kinds of micron-sized hollow fibers, whose hollow diameters are 7, 2, 1.4, and 0.3 μm, have been employed for guiding Rb atoms with the D_2 line whose wavelength is 780 nm. Each of these hollow optical fibers, which is similar to one with an outer diameter of 125 μm for communication uses, has a triple structure where a thin (2–4 μm) cylindrical core with propagating modes and a cladding are annularly arranged around a hollow region [17, 18, 25]. In this section, we analyze the light field in such a hollow fiber to examine the optical near field required for AQW. We first discuss the exact light-field modes, and then shift to a simplified expression of the light field based on the weakly guiding approximation (WGA) with linearly polarized (LP) modes. As a consequence, it is shown that the HE_{11} mode, or approximately the LP_{01} mode, is suitable for guiding atoms.

11.2.1 Exact Light-Field Modes in Hollow Optical Fibers

Let us consider the light field in a cylindrical coordinate system (ρ, ϕ, z), in which the z axis is taken along the center axis of a hollow optical fiber with a hollow diameter $2a$ and a core thickness d. When a light beam with a wave number k is coupled to the hollow fiber, the longitudinal component E_z of the electric field follows the next Maxwell equation [26, 27],

$$\frac{\partial^2 E_z}{\partial \rho^2} + \frac{1}{\rho}\frac{\partial E_z}{\partial \rho} + \frac{1}{\rho^2}\frac{\partial^2 E_z}{\partial \rho^2} + (n^2k^2 - \beta^2)E_z = 0 \tag{11.3}$$

where β is a propagation constant, and n is the refractive index. We get a similar equation for the longitudinal component H_z of the magnetic field. The solutions satisfying Eq. 11.3 are given by

$$E_z(\rho,\phi) = \begin{cases} A_1 \mathrm{I}_m(v\rho)\cos(m\phi+\varphi) & (\rho < a) \\ (A_2\mathrm{J}_m(u\rho) + A_3\mathrm{N}_m(u\rho))\cos(m\phi+\varphi) & (a \le \rho \le a+d) \\ A_4\mathrm{K}_m(w\rho)\cos(m\phi+\varphi) & (\rho > a+d) \end{cases} \tag{11.4}$$

where φ is an arbitrary phase constant. The four functions, $\mathrm{J}_m(u\rho)$, $\mathrm{N}_m(u\rho)$, $\mathrm{I}_m(v\rho)$, and $\mathrm{K}_m(w\rho)$ are the m-th order Bessel and modified Bessel functions of the first and second kinds, respectively. Assuming that the refractive indices of the hollow region, the core, and the cladding are 1 (vacuum), n_1, and n_2, respectively, we get three transverse characteristic constants $u = \sqrt{n_1^2k^2 - \beta^2}$, $v = \sqrt{\beta^2 - k^2}$, and $w = \sqrt{\beta^2 - n_2^2k^2}$. We also obtain similar solutions for the longitudinal magnetic component $H_z(\rho,\phi)$ by replacing $\cos(m\phi+\varphi)$ by $\sin(m\phi+\varphi)$. The other components $E_\rho(\rho,\phi)$, $E_\phi(\rho,\phi)$, $H_\rho(\rho,\phi)$, and $H_\phi(\rho,\phi)$ can be derived from the longitudinal ones $E_z(\rho,\phi)$ and $H_z(\rho,\phi)$. On the other hand, the four coefficients A_1, A_2, A_3, and A_4 are determined by the boundary conditions at $\rho = a$ and $\rho = a+d$, and the conservation law of the power transmitted through the hollow fiber.

The continuity conditions with respect to the tangential components $E_z(\rho,\phi)$, $E_\phi(\rho,\phi)$, $H_z(\rho,\phi)$, and $H_\phi(\rho,\phi)$ at $\rho = a$ and $\rho = b = a+d$ lead us to a secular equation. Then, the secular equation yields the following dispersion equation [17] describing the hybrid modes such as the HE and EH modes [26]:

$$\begin{aligned}
&[\mathrm{TE}][\mathrm{TM}] \\
+\;& \left\{\frac{1}{ab}\left(\frac{\beta m}{k}\right)^2\left(\frac{1}{u^2}+\frac{1}{v^2}\right)\left(\frac{1}{u^2}+\frac{1}{w^2}\right)\right. \\
& \left.[\mathrm{J}_m(ua)\mathrm{N}_m(ub) - \mathrm{J}_m(ub)\mathrm{N}_m(ua)]\right\}^2 \\
+\;& \frac{1}{ab}\left(\frac{\beta m}{k}\right)^2\left\{\frac{2n_1^2}{u^2}\left(\frac{1}{u^2}+\frac{1}{v^2}\right)\left(\frac{1}{u^2}+\frac{1}{w^2}\right)\right. \\
& [\mathrm{J}_m(ua)\mathrm{N}'_m(ua) - \mathrm{J}'_m(ua)\mathrm{N}_m(ua)][\mathrm{J}_m(ub)\mathrm{N}'_m(ub) - \mathrm{J}'_m(ub)\mathrm{N}_m(ub)] \\
-\;& \frac{a}{b}\left(\frac{1}{u^2}+\frac{1}{w^2}\right)^2 \\
& \left[\left(\frac{1}{u}\mathrm{N}'_m(ua) + \frac{1}{v}\frac{\mathrm{I}'_m(va)}{\mathrm{I}_m(va)}\mathrm{N}_m(ua)\right)\mathrm{J}_m(ub)\right.
\end{aligned}$$

$$
\begin{aligned}
& - \left(\frac{1}{u} \mathrm{J}'_m(ua) + \frac{1}{v} \frac{\mathrm{I}'_m(va)}{\mathrm{I}_m(va)} \mathrm{J}_m(ua) \right) \mathrm{N}_m(ub) \Bigg] \\
& \Bigg[\left(\frac{n_1^2}{u} \mathrm{N}'_m(ua) + \frac{1}{v} \frac{\mathrm{I}'_m(va)}{\mathrm{I}_m(va)} \mathrm{N}_m(ua) \right) \mathrm{J}_m(ub) \\
& - \left(\frac{n_1^2}{u} \mathrm{J}'_m(ua) + \frac{1}{v} \frac{\mathrm{I}'_m(va)}{\mathrm{I}_m(va)} \mathrm{J}_m(ua) \right) \mathrm{N}_m(ub) \Bigg] \\
- \quad & \frac{b}{a} \left(\frac{1}{u^2} + \frac{1}{v^2} \right)^2 \\
& \Bigg[\left(\frac{1}{u} \mathrm{N}'_m(ub) + \frac{1}{w} \frac{\mathrm{K}'_m(wb)}{\mathrm{K}_m(wb)} \mathrm{N}_m(ub) \right) \mathrm{J}_m(ua) \\
& - \left(\frac{1}{u} \mathrm{J}'_m(ub) + \frac{1}{w} \frac{\mathrm{K}'_m(wb)}{\mathrm{K}_m(wb)} \mathrm{J}_m(ub) \right) \mathrm{N}_m(ua) \Bigg] \\
& \Bigg[\left(\frac{n_1^2}{u} \mathrm{N}'_m(ub) + \frac{n_2^2}{w} \frac{\mathrm{K}'_m(wb)}{\mathrm{K}_m(wb)} \mathrm{N}_m(ub) \right) \mathrm{J}_m(ua) \\
& - \left(\frac{n_1^2}{u} \mathrm{J}'_m(ua) + \frac{n_2^2}{w} \frac{\mathrm{K}'_m(wb)}{\mathrm{K}_m(wb)} \mathrm{J}_m(ub) \right) \mathrm{N}_m(ua) \Bigg] \Bigg\} \\
= \quad & 0
\end{aligned} \tag{11.5}
$$

where prime denotes differentiation with respect to ρ, and the two factors [TE] and [TM] are given by

$$
\begin{aligned}
[\mathrm{TE}] = & \left[\frac{1}{u} \mathrm{J}'_m(ua) + \frac{1}{v} \frac{\mathrm{I}'_m(va)}{\mathrm{I}_m(va)} \mathrm{J}_m(ua) \right] \\
& \left[\frac{1}{u} \mathrm{N}'_m(ub) + \frac{1}{w} \frac{\mathrm{K}'_m(wb)}{\mathrm{K}_m(wb)} \mathrm{N}_m(ub) \right] \\
& \left[\frac{1}{u} \mathrm{N}'_m(ua) + \frac{1}{v} \frac{\mathrm{I}'_m(va)}{\mathrm{I}_m(va)} \mathrm{N}_m(ua) \right] \\
& \left[\frac{1}{u} \mathrm{J}'_m(ub) + \frac{1}{w} \frac{\mathrm{K}'_m(wb)}{\mathrm{K}_m(wb)} \mathrm{J}_m(ub) \right]
\end{aligned} \tag{11.6}
$$

and

$$
\begin{aligned}
[\mathrm{TM}] = & \left[\frac{n_1^2}{u} \mathrm{J}'_m(ua) + \frac{1}{v} \frac{\mathrm{I}'_m(va)}{\mathrm{I}_m(va)} \mathrm{J}_m(ua) \right] \\
& \left[\frac{n_1^2}{u} \mathrm{N}'_m(ub) + \frac{n_2^2}{w} \frac{\mathrm{K}'_m(wb)}{\mathrm{K}_m(wb)} \mathrm{N}_m(ub) \right] \\
& - \left[\frac{n_1^2}{u} \mathrm{N}'_m(ua) + \frac{1}{v} \frac{\mathrm{I}'_m(va)}{\mathrm{I}_m(va)} \mathrm{N}_m(ua) \right] \\
& \left[\frac{n_1^2}{u} \mathrm{J}'_m(ub) + \frac{n_2^2}{w} \frac{\mathrm{K}'_m(wb)}{\mathrm{K}_m(wb)} \mathrm{J}_m(ub) \right]
\end{aligned} \tag{11.7}
$$

In addition, two equations, [TE] = 0 and [TM] = 0, yield the TE and TM modes, respectively.

In the case of guiding Rb atoms, we use the D_2 line with a wavelength of 780 nm. As an example, let us take a silica-glass hollow fiber with $2a = 7\ \mu$m, $d = 3.8\ \mu$m, and $n_2 = 1.45$, in which the core is germanium-doped so that the relative refractive index difference $\Delta n = (n_1^2 - n_2^2)/2n_1^2$ is equal to 0.18%. From Eqs. 11.5–11.7, we see that six propagating modes, TE_{01}, TM_{01}, HE_{11}, HE_{21}, HE_{31}, and EH_{11}, can be excited for the wavelength 780 nm (see also Fig. 11.4). The lowest mode is the HE_{11} mode. The TE_{01}, TM_{01}, and HE_{21} modes have almost the same dispersion curve, so they make up the second group of propagating modes. In the same way, the EH_{11} and HE_{31} modes make up the third group.

We can also obtain cutoff frequencies of the hybrid modes from the next equation [17]:

$$\begin{aligned}
&\left[\left(\frac{1}{u}\mathrm{N}'_m(ua) + \frac{1}{v}\frac{\mathrm{I}'_m(va)}{\mathrm{I}_m(va)}\mathrm{N}_m(ua)\right)\mathrm{J}_m(ub)\right.\\
&\left.-\left(\frac{1}{u}\mathrm{J}'_m(ua) + \frac{1}{v}\frac{\mathrm{I}'_m(va)}{\mathrm{I}_m(va)}\mathrm{J}_m(ua)\right)\mathrm{N}_m(ub)\right]\\
&\left[\left(\frac{n_1^2}{u}\mathrm{N}'_m(ub) + \frac{1}{v}\frac{\mathrm{I}'_m(va)}{\mathrm{I}_m(va)}\mathrm{N}_m(ua)\right)\mathrm{J}_m(ub)\right.\\
&\left.-\left(\frac{n_1^2}{u}\mathrm{J}'_m(ua) + \frac{1}{v}\frac{\mathrm{I}'_m(va)}{\mathrm{I}_m(va)}\mathrm{J}_m(ua)\right)\mathrm{N}_m(ub)\right]\\
=\ &\frac{1}{a^2}\left(\frac{\beta m}{k}\right)^2\left(\frac{1}{u^2}+\frac{1}{v^2}\right)^2\\
&[\mathrm{J}_m(ua)\mathrm{N}_m(ub) - \mathrm{J}_m(ub)\mathrm{N}_m(ua)]^2
\end{aligned} \tag{11.8}$$

On the other hand, the two equations set to zero on the first and second factors of the left-hand side of Eq. 11.8 yield the cutoff frequencies of the TE and TM modes, respectively.

11.2.2 Approximate Light-Field Modes in Hollow Optical Fibers

As mentioned above, the exact light-field modes are divided into several groups. Therefore, each group of the exact modes can be approximately expressed with a proper mode. This results from the fact that the relative refractive index difference Δn between the core and the cladding is very small: the small relative refractive index difference guarantees the applicability of the WGA [26] to the present case. This approximation greatly simplifies the analysis of the light-field modes and the estimation of the optical potential mentioned in Sect. 11.3.

Under the WGA, the amplitude vector $\boldsymbol{E}_0$ of the electric field can be divided into two components, the longitudinal one $\boldsymbol{E}_z$ and the transverse one $\boldsymbol{E}_t$, in a cylindrical coordinate system (ρ, ϕ, z): $\boldsymbol{E}_0(\rho, \phi) = \boldsymbol{E}_z(\rho, \phi) + \boldsymbol{E}_t(\rho, \phi)$ at a position z. Moreover, the transverse component $\boldsymbol{E}_t$ can be written as either $(E_x, 0)$ or $(0, E_y)$ in accordance with the direction of the linear polarization. What we should note is that the WGA leads to the LP modes [26]. We then get a scalar equation for the transverse component E_i ($i = \mathrm{x}$ or y),

$$\nabla^2 E_i(\rho,\phi) + (n^2k^2 - \beta^2)E_i(\rho,\phi) = 0 \tag{11.9}$$

This equation yields a solution written as

$$E_i(\rho,\phi) = E(\rho)\cos(m\phi + \varphi) \tag{11.10}$$

where the radial part $E(\rho)$ follows the next Maxwell equation

$$\frac{\mathrm{d}^2E(\rho)}{\mathrm{d}\rho^2} + \frac{1}{\rho}\frac{\mathrm{d}E(\rho)}{\mathrm{d}\rho} + (n^2k^2 - \beta^2 - \frac{m^2}{\rho^2})E(\rho) = 0 \tag{11.11}$$

From Eq. 11.11 we get the radial part $E(\rho)$ as follows:

$$E(\rho) = \begin{cases} B_1\mathrm{I}_m(v\rho) & (\rho < a) \\ (B_2\mathrm{J}_m(u\rho) + B_3\mathrm{N}_m(u\rho)) & (a \le \rho \le a+d) \\ B_4\mathrm{K}_m(w\rho) & (\rho > a+d) \end{cases} \tag{11.12}$$

where the four coefficients B_1, B_2, B_3, and B_4 are determined by boundary conditions at $\rho = a$ and $\rho = a+d$, and the conservation law of the transmission power. We also get similar solutions for the magnetic field.

From a secular equation coming from continuity conditions at the boundaries $\rho = a$ and $\rho = b = a + d$, we derive a dispersion equation describing the LP modes as [18]

$$\begin{aligned} &\left[\frac{\mathrm{J}_m(ua)}{\mathrm{I}_m(va)} - \frac{u}{v}\frac{\mathrm{J}'_m(ua)}{\mathrm{I}'_m(va)}\right]\left[\frac{\mathrm{N}_m(ub)}{\mathrm{K}_m(wb)} - \frac{u}{w}\frac{\mathrm{N}'_m(ub)}{\mathrm{K}'_m(wb)}\right] \\ = &\left[\frac{\mathrm{N}_m(ua)}{\mathrm{I}_m(va)} - \frac{u}{v}\frac{\mathrm{N}'_m(ua)}{\mathrm{I}'_m(va)}\right]\left[\frac{\mathrm{J}_m(ub)}{\mathrm{K}_m(wb)} - \frac{u}{w}\frac{\mathrm{J}'_m(ub)}{\mathrm{K}'_m(wb)}\right] \end{aligned} \tag{11.13}$$

The dispersion equation 11.13 yields the LP_{m1} modes ($m = 0, 1, 2, \cdots$) for the hollow fibers considered here. This result reflects the situation that both the hollow diameter and the core thickness are small enough to be only a few times as large as a wavelength. In fact, from the numerical analysis of Eq. 11.13, we see that three LP modes can be excited in the 7-μm hollow fiber for a wavelength of 780 nm. Figure 11.4a shows the dispersion curves with respect to several lower modes. The solid circles indicate the three LP modes: the LP_{01} mode, the LP_{11} mode, and the LP_{21} mode. Comparing the results derived from the numerical analysis of the exact modes with Eqs. 11.5–11.7, we find that: (1) the LP_{01} mode is approximately equal to the HE_{11} mode, (2) the LP_{11} mode is made up of the TE_{01}, TM_{01}, and HE_{21} modes, and (3) the LP_{21} mode is made up of the EH_{11} and HE_{31} modes. It should be noted that in Fig. 11.4a, the points where each dispersion curve intersects the horizontal axis indicate the cutoff frequencies. According to Eq. 11.13, the 7- and 2-μm hollow fibers are multimode, while the 1.4- and 0.3-μm hollow fibers are single-mode.

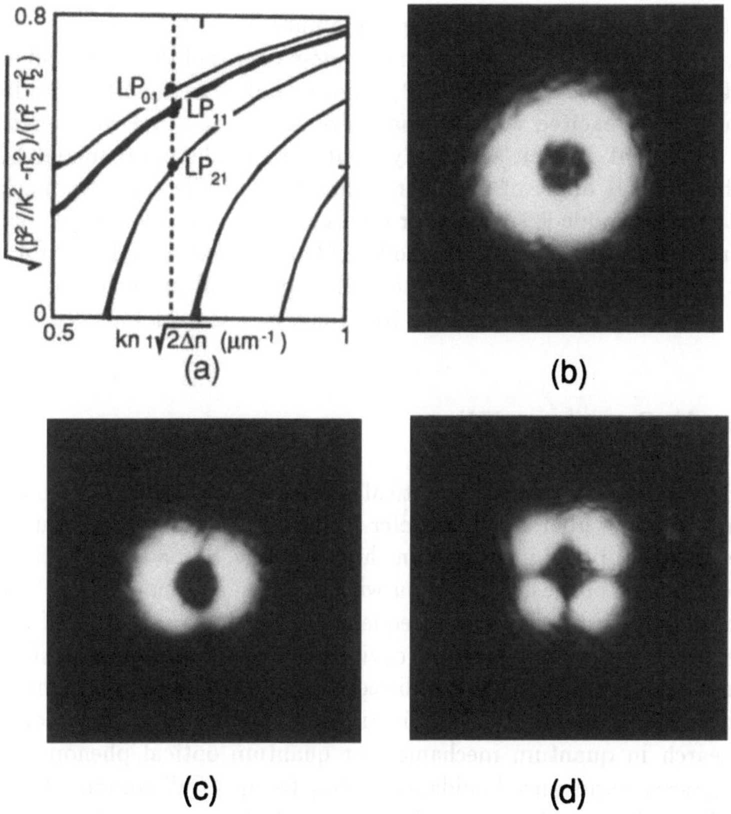

Fig. 11.4. **a** Dispersion curves of the propagating modes for a wavelength of 780 nm in a 7-μm hollow optical fiber with a core thickness of 3.8 μm and a relative refractive index difference of 0.18%. **b** LP_{01} mode, **c** LP_{11} mode, and **d** LP_{21} mode. These cross-sectional patterns are taken with a CCD camera

11.2.3 Field Intensity of the LP Modes

We can evaluate the field intensity $P(\rho, \phi)$ by calculating the z component of the time-averaged Poynting vector $\bar{S}_z$ defined as [26]

$$\overline{S_z} = \frac{1}{2}(\boldsymbol{E} \times \boldsymbol{H}^*) \cdot \boldsymbol{e}_z = \frac{1}{2}(E_\rho H_\phi^* - E_\phi H_\rho^*) \tag{11.14}$$

where $\boldsymbol{e}_z$ is a unit vector in the z axis. Substituting Eq. 11.10 and the corresponding equation for the magnetic field into Eq. 11.14, we get the transverse intensity profile $P(\rho, \phi)$ of the LP_{m1} mode as follows:

$$P(\rho, \phi) = \begin{cases} \alpha B_1^2 \mathrm{I}_m^2(v\rho) \cos^2(m\phi + \varphi) & (\rho < a) \\ \alpha \left(B_2 \mathrm{J}_m(u\rho) + B_3 \mathrm{N}_m(u\rho)\right)^2 \cos^2(m\phi + \varphi) & (a \le \rho \le a + d) \\ \alpha B_4^2 \mathrm{K}_m(w\rho)^2 \cos^2(m\phi + \varphi) & (\rho > a + d) \end{cases} \tag{11.15}$$

where $\alpha = \beta/2\omega_L\mu_0$ with the magnetic permiability μ_0 of the vacuum.

Figure 11.4b–d show the cross-sectional mode-patterns of the 7-μm hollow fiber: b is the LP_{01} mode, c is the LP_{11} mode, and d is the LP_{21} mode. These LP modes are excited with a Ti:sapphire laser tuned to a wavelength of 780 nm. Note that we can selectively excite one of these LP modes by adjusting the incident angle of the laser beam. These CCD camera images show that the LP_{01} mode is suitable for guiding atoms: the LP_{01} mode has no nodes around the inner wall of a hollow fiber, as shown in Fig. 11.4b. In the next section, we will estimate the optical potential produced by the blue-detuned optical near field to guide Rb atoms with Eq. 11.15.

11.3 Atomic Quantum Wires

This section deals with the near-field optical guidance of a thermal Rb atomic beam through a hollow fiber. First, we refer to the optical potential produced by the blue-detuned optical near field on the Rb atoms. Second, we describe the laser spectroscopy of the guided atoms with two-step photoionization. The photoionization spectrum shows the frequency dispersion properties of the dipole force. We will then touch on the cavity QED effects in a micron-sized hollow fiber which works as a kind of dielectric cylindrical cavity to confine atoms. This gives an example of applications of the AQW in unexplored fields of research in quantum mechanical or quantum optical phenomena. Finally, we present an advanced guidance scheme for an AQW composed of a slantwise-polished hollow fiber in which a guide light beam is coupled to the core of the hollow fiber with a 45°-cut edge directly from the side via total internal reflection. This enables one to use the AQW together with additional atom-optical devices such as an atomic funnel, discussed in Sect. 11.5.

11.3.1 Near-Field Optical Potential

In accordance with experimental purposes, various kinds of hollow optical fibers are available. In this and subsequent sections, four kinds of hollow fiber with different hollow diameters and core thicknesses are employed. As mentioned already, all of these hollow fibers support LP_{m1} modes for the guide-wavelength of 780 nm. From Eq. 11.15, the transverse intensity profile $P_{\mathrm{onf}}(\rho, \phi)$ of the optical near field with the LP_{m1} mode induced in the hollow region is given by

$$P_{\mathrm{onf}}(\rho, \phi) = \frac{\beta}{2\omega_L\mu_0} B_1^2 \mathrm{I}_m(v\rho)^2 \cos^2(m\phi) \tag{11.16}$$

where an arbitrary phase constant φ is taken as zero.

If the detuning of a laser frequency ω_L from an atomic resonant frequency ω_0 is large enough so that the spontaneous emission per bounce is negligible, the two-level atomic system can be applied to alkali-metal atoms such as Rb. In this case, the optical potential $U_{\mathrm{opt}}(\rho, \phi)$ of the dipole force on an atom with a longitudinal velocity v_z is written as [1, 8]

$$U_{\mathrm{opt}}(\rho, \phi) = \frac{1}{2}\hbar\Delta \ln\left\{1 + \frac{P_{\mathrm{onf}}(\rho, \phi)}{P_{\mathrm{sat}}}\frac{\Gamma^2}{4\Delta^2 + \Gamma^2}\right\} \tag{11.17}$$

where $\Delta = \omega_{\mathrm{L}} - \omega_0 - \beta v_z$ is a frequency detuning including an atomic Doppler shift kv_z along the center axis of the hollow fiber. The saturation power P_{sat} and the natural linewidth Γ are equal to 1.6 mW/cm^2 and $2\pi \times 6.1$ MHz, respectively, for the Rb D_2 line.

Substituting Eq. 11.16 into Eq. 11.17, we can estimate the optical potential $U_{\mathrm{opt}}^{m1}(\rho, \phi)$ of the LP_{m1} modes. Hereafter, the LP_{01} mode, which is suitable for atom guidance, is considered. Note that Eq. 11.16 yields a somewhat higher optical potential than that obtained from an exact mode such as the HE_{11} mode. This comes from the fact that the WGA is not good across the core–hollow interface because the refractive index difference between the core and the hollow (vacuum) is not necessarily small: the LP modes bring about a leak in the optical near field over a longer distance from the inner wall surface than in the case of the exact modes, so that the potential barrier is higher. Nevertheless, Eq. 11.16 is very useful for a rough estimate of the experimental results. In fact, as shown in Fig. 11.4, the experimental observation of the field intensity profile supports representation with the LP_{m1} modes for the present hollow fibers.

11.3.2 Laser Spectroscopy of Guided Atoms with Two-Step Photoionization

A collimated thermal atomic beam is a good source for a demonstration of the AQW. The incoming atoms have a mean longitudinal velocity of more than 100 m/s along a hollow fiber. On the other hand, the mean transverse velocity approaching the inner wall is small enough that many atoms can be reflected by a blue-detuned optical near field produced under feasible experimental conditions. However, the use of such a thermal atomic beam gives rise to difficulties with the detection of the guided atoms owing to the large atomic transmission velocity and the small atom flux. This difficulty can be resolved by ionization detection. There are two ionization methods: (1) photoionization, and (2) surface ionization. In this section, we focus on laser-spectroscopic experiments by way of two-step photoionization. In Sect. 11.4, we introduce a scheme of surface ionization with a hot wire to examine the spatial distribution of a guided atom flux.

Figure 11.5 schematically shows the experimental set-up employed for guiding a Rb atomic beam with a micron-sized hollow optical fiber [20]. In a vacuum chamber, a Rb atomic beam from a hot oven collimated through a few pinholes is introduced into a hollow optical fiber coaxially placed behind a

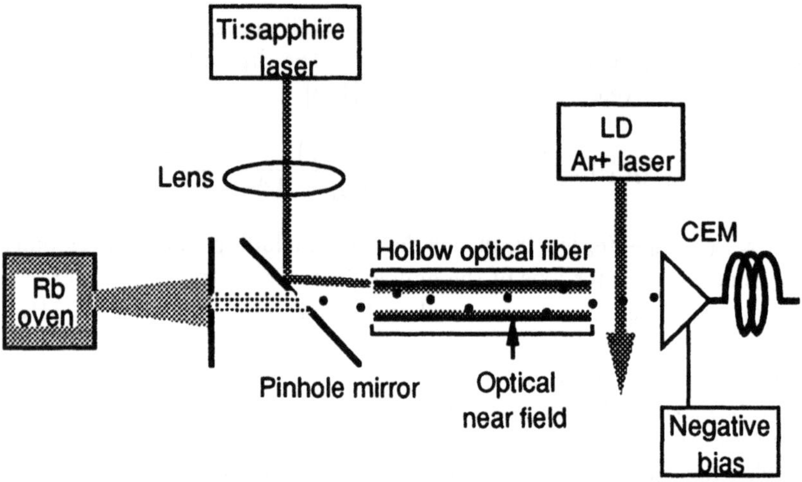

Fig. 11.5. Experimental set-up for two-step photoionizaion. A Rb atomic beam from a hot oven is introduced into a straight hollow fiber with a typical length of 3 cm, while a guide light beam from a Ti:sapphire laser with a wavelength of 780 nm is coupled to the core. The guided Rb atoms are detected with a channel electron multiplier (CEM) applying a high negative bias via two-step photoionization with a diode laser and an Ar-ion laser

pinhole mirror. At a typical oven temperature of 200°C, the mean longitudinal transmission velocity of the Rb atomic beam is about 300 m/s, while the mean transverse velocity is estimated to be about 30 cm/s [28]. In addition, the typical incident atom flux is of the order of 10^6 atom/s. Atoms not entering the hollow region are blocked by a plate with a pinhole placed just in front of the entrance facet of the hollow fiber so that they cannot be observed downstream of the fiber.

A guide-light beam from a Ti:sapphire laser tuned to the Rb D_2 line with a wavelength of 780 nm is also coupled to the thin core of the hollow fiber via the pinhole mirror. In this case, as mentioned in Sect. 11.2, the 7- and 2-μm hollow fibers support three LP modes, LP_{01}, LP_{11}, and LP_{21}, while the 1.4- and 0.3-μm hollow fibers support only the LP_{01} mode. In fact, we can selectively excite the LP_{01} mode to make it suitable for guiding atoms by adjusting the incident angle of the guide light beam and monitoring the mode pattern on the cross section of the exit facet with a CCD camera. The typical transmission light power with the LP_{01} mode is about 100 mW. From Eqs. 11.16 and 11.17, we can see that the excitation of the LP_{01} mode with a coupling power of 100 mW produces an optical potential of the order of 10 mK, corresponding to a transverse atomic velocity of about 1 m/s.

The Rb atoms coming out of the hollow fiber are detected with a CEM via two-step photoionization with a diode laser tuned to the Rb D_2 line and a high-power Ar-ion laser whose wavelength is 476.5 nm. The vacuum pressure is kept below 10^{-6} Pa in the detection region. Two light beams from the diode

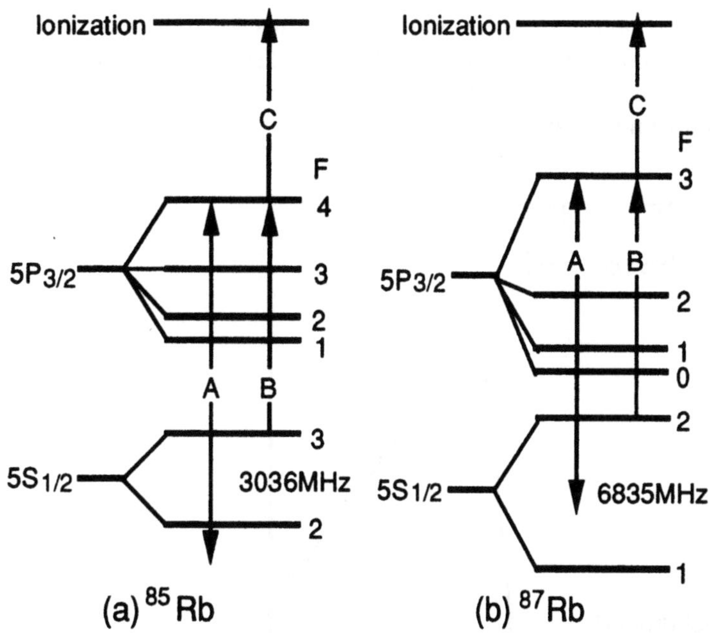

Fig. 11.6. Relevant energy levels of two stable Rb isotopes: **a** ^{85}Rb, and **b** ^{87}Rb. In the two cases of guiding and detecting, **a** the ^{85}Rb atoms in the $5S_{1/2}$, $F = 3$ upper ground state, and **b** the ^{87}Rb atoms in the $5S_{1/2}$, $F = 2$ upper ground state, arrows A, B, and C indicate the frequencies of the guide laser, the diode laser, and the Ar-ion laser, respectively. The ionization level is at 4.177 eV above the $5S_{1/2}$ ground state

laser and the Ar-ion laser are introduced from a direction perpendicular to the axis of the atomic beam and overlapped downstream of the hollow fiber. Note that this scheme leads to Doppler-free spectroscopy of the guided atoms. The CEM-applied negative bias of −3 kV, in which the quantum efficiency is 0.9, gathers the ionized atoms and sends out pulse signals. The total pulse count per second gives the guided atom flux.

The photoionization is carried out in two steps. (1) The diode laser excites the guided Rb atoms in one of the $5S_{1/2}$ hyperfine ground states (see Sect. 11.1.2 for the meaning of $5S_{1/2}$) to one of the $5P_{3/2}$ hyperfine excited states. (2) The Ar-ion laser excites the Rb atoms in the $5P_{3/2}$ state to the ionization level at an energy of 4.177 eV above the $5S_{1/2}$ ground state. Keep in mind that the atomic beam employed here includes two stable Rb isotopes with mass numbers 85 (^{85}Rb) and 87 (^{87}Rb); the natural abundance ratio of ^{85}Rb to ^{87}Rb is about 7 to 3 [29]. Figure 11.6 shows the relevant energy levels of (a) ^{85}Rb and (b) ^{87}Rb. In addition, as a typical case of guiding and detecting ^{85}Rb or ^{87}Rb atoms, the frequencies of the guide laser, the diode laser, and the Ar-ion laser are indicated by arrows A, B, and C, respectively. The advantage of this two-step photoionization consists in the species- and state-selective detection of the guided atoms: we can select the quantum energy state of the guided atoms by tuning the diode laser. Thus, this scheme leads to Doppler-free and state-selective spectroscopy of the guided atoms with high signal-to-noise ratio.

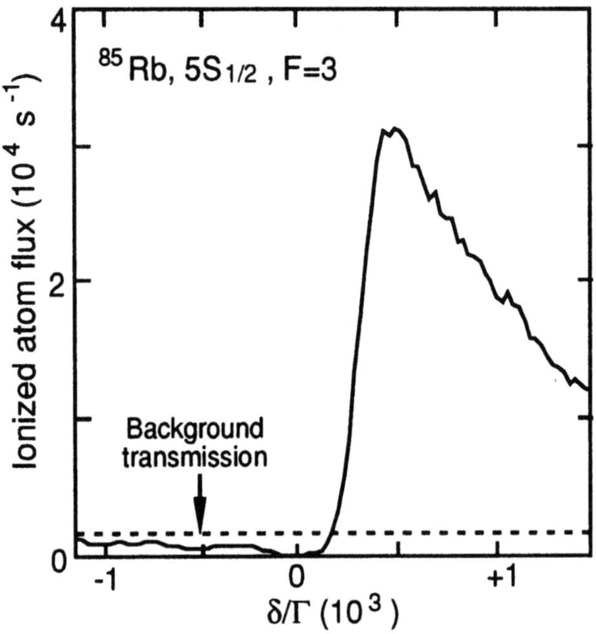

Fig. 11.7. Doppler-free two-step photoionization spectrum of the ^{85}Rb atoms in the $5S_{1/2}$, $F = 3$ upper ground state guided through a 7-μm hollow fiber over a distance of 3 cm. The *solid curve* shows the ionized atom flux plotted as a function of normalized blue-detuning δ/Γ with respect to the $5S_{1/2}$, $F = 3$ upper ground state. The *broken line* shows a background transmission level without the guide light

Figure 11.7 shows a typical photoionization spectrum of the ^{85}Rb atoms guided through a 7-μm hollow fiber over a distance of 3 cm as a function of the frequency detuning δ of the guide laser normalized to the natural linewidth Γ. Here, note that the frequency detuning $\delta = \omega_L - \omega_0$, which does not include the atomic Doppler shift, is measured with respect to the $5S_{1/2}$, $F = 3$ upper ground state. In this case, as shown in Fig. 11.6a, since the diode laser is tuned to the $5S_{1/2}$, $F = 3 \leftrightarrow 5P_{3/2}$, $F = 4$ transition, the photoionization signal comes from the ^{85}Rb atoms in the $5S_{1/2}$, $F = 3$ upper ground state.

As expected, the guided atom flux greatly increases in the blue-detuning region. In addition, the foot of the photoionization spectrum extends to the blue detuning over +20 GHz. Note that the broken line in Fig. 11.7 shows the background transmission level without the guide-light beam. When the atomic beam is shut off with a mechanical shutter, or when the diode laser is detuned from any hyperfine transition of the Rb D_2 line, the background level becomes almost zero very rapidly. This fact leads us to the conclusion that the background signal must come from the atoms ballistically flying through the hollow region. Comparing the maximum value of the guided atom flux with the background transmission, we get an enhancement factor of 20. From a similar experiment with a 2-μm hollow fiber, in which the axis of the hollow fiber is slightly tilted against that of the Rb atomic beam, a higher enhancement factor of 80 has been obtained [30]. On the other hand, the guided atom flux decreases in the red-detuning region. This may be because the atoms adhere to the surface of the inner wall due to the attractive dipole force from the red-detuned optical near field. As shown in this example, the state-selective photoionization spectrum of atoms guided through a hollow fiber directly reflects the frequency dispersion properties of the dipole interaction between an atom and an optical near field.

The condition for efficient two-step photoionization of a ground-state atom is given by [31]

$$P_{\rm ion} \sim P_{\rm sat} \frac{\sigma_0}{\sigma_{\rm ion}} \tag{11.18}$$

where $P_{\rm ion}$ is the light intensity required for ionizing the atom in an excited state, and $\sigma_{\rm ion}$ is the ionization cross section from the excited state to the ionization level. Taking the ratio of the Ar-ion laser intensity to the saturation ionization intensity $P_{\rm ion}$, we can roughly estimate the photoionization efficiency of the guided Rb atoms in the $5S_{1/2}$ ground state. In the case of a typical Ar-ion laser intensity of 0.5 GW/m^2, assuming the resonant excitation cross section $\sigma_0 = (3/2\pi)\lambda^2 = 3 \times 10^{-9}$ cm^2 of the $5S_{1/2} \to 5P_{3/2}$ transition [32] and $\sigma_{\rm ion} = 2.5 \times 10^{-17}$ cm^2 for a wavelength of 476.5 nm, we get a photoionization efficiency of about 30%, where we guess the value of $\sigma_{\rm ion}$ from the results given in Ref. [33]. The net detection efficiency is given by the photoionization efficiency times the quantum efficiency of the CEM. We can also perform two-step photoionization with Ar-ion laser wavelengths of both 488 nm and 459 nm, although the ionization efficiency is lower.

11.3.3 Observation of Cavity QED Effects in a Dielectric Cylinder

The dipole interaction between an atom and an optical near field occurs in extremely close vicinity to a material surface. On the other hand, it is well known that the van der Waals force acts on an atom near a dielectric surface [34]. This indicates that we cannot ignore the influence of such a force on atom manipulation with an optical near field. In fact, as mentioned in Sect. 11.1, the van der Waals force on a ground-state atom near a dielectric plane has been examined with an AQS [14]. In the case where an atom is in a small cavity, the result is generalized from the cavity QED effect [35]. The cavity QED effect, which results from the interaction between an atom and the vacuum modified by the cavity [36], yields two kinds of attractive force:

the first is the well-known van der Waals force, and the second is the so-called Casimir–Polder force [37]. The latter gives rise to position-dependent Lamb shifts of atomic quantum energy levels.

Although study of the cavity QED effect is very important for a deeper understanding of the nature of a vacuum, few experimental results have been reported except for some simple cases of a plane and a planar cavity, although there has been much theoretical work [35]. This is mainly because the cavity QED effect is very small, so that observation is difficult. In fact, observations of the van der Waals force and the Casimir–Polder force have been made in a cavity composed of two parallel conductor planes [38]. Recently, an elaborate experimental study of the Casimir–Polder force was performed [39]. On the other hand, cavity QED effects near a spherical surface [40] and inside a microsphere [41, 42] have been theoretically examined. In particular, the cavity QED effect near a dielectric surface [40, 43–45] is a problem of great consequence for near-field optical atom manipulation. In this subsection, as an application of the AQW to the study of unsolved quantum mechanical or quantum optical phenomena, we touch on observations of the cavity QED effect inside a hollow glass fiber regarded as a sort of dielectric cylindrical cavity [46].

The cavity QED effect gives rise to an energy shift of an atomic quantum state. In general, the formulae of the atomic energy shifts in the case of curved dielectric surfaces are very complicated, so that their analysis requires considerable computational time. Here, for simplicity, we will apply an approximate method with a formula for a planar conductor cavity to analyze the experimental results. An exact formula for an atomic energy shift inside a dielectric cylinder has recently been obtained [47].

The atomic energy shift depends on the atomic position, so that it produces an attractive cavity potential. The cavity potential $U_{\mathrm{cav}}(\rho)$ on an atom inside a planar conductor cavity with a space interval $2a$ is written in a simple analytical form as [35, 38]

$$U_{\mathrm{cav}}(\rho) = -\sum_{\mathrm{e}} \frac{\pi |d_{\mathrm{eg}}|^2}{48\varepsilon_0 a^3} \int_0^\infty \frac{r^2 \cosh(\pi r \rho/a)}{\sinh(\pi r)} \tan^{-1}\left(\frac{r\lambda_{\mathrm{eg}}}{4a}\right) \mathrm{d}r \qquad (11.19)$$

where d_{eg} and λ_{eg} are a matrix element of a dipole moment and a wavelength of an atomic dipole transition from the ground state, respectively.

Let us consider the case where an atom is guided through a hollow optical fiber with a hollow diameter of $2a$ by the blue-detuned optical near field with the LP_{01} mode. Then, the total potential $U_{\mathrm{tot}}(\rho)$ is given by the summation of the optical potential $U_{\mathrm{opt}}^{01}(\rho)$ and the modified cavity potential $f \cdot U_{\mathrm{cav}}(\rho)$: $U_{\mathrm{tot}}(\rho) = U_{\mathrm{opt}}^{01}(\rho) + f \cdot U_{\mathrm{cav}}(\rho)$, where we introduce a scaling coefficient f to approximate the dielectric cylinder case. In addition, we assume that the scaling coefficient f can be the product of a dielectric factor ξ and a geometric factor η. Moreover, since the inner wall is considered as approximately a

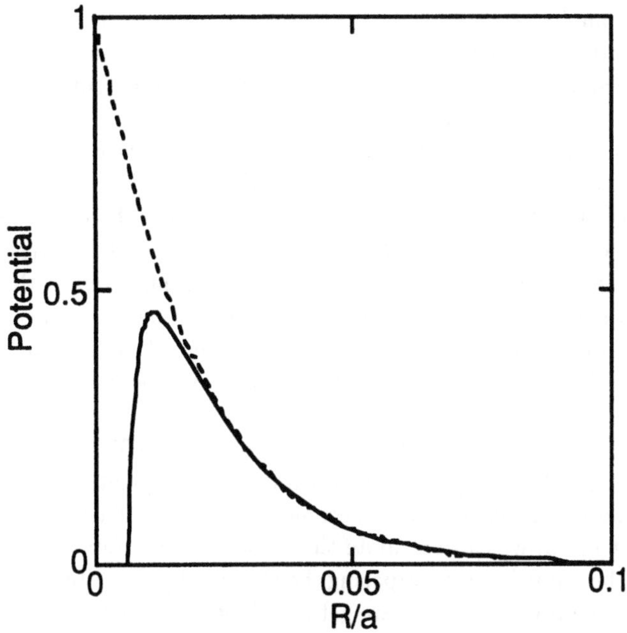

Fig. 11.8. Cross-sectional profile of the potential barrier produced near the inner wall of a hollow fiber as a function of the normalized distance R/a from the surface. The *solid curve* shows the total potential $U_{tot}(R) = U_{opt}^{01}(R) + U_{cav}(R)$, while the *broken curve* shows the optical potential $U_{opt}^{01}(R)$ of the LP_{01} mode. These potentials are normalized to a maximum value of $U_{opt}^{01}(0)$

plane for an atom near the surface, we can write the dielectric factor ξ as $\xi = (n_1^2 - 1)/(n_1^2 + 1)$ [16, 34], where n_1 is a refractive index of the core.

Figure 11.8 shows a potential barrier to reflect ^{85}Rb atoms plotted as a function of the normalized distance R/a from the inner-wall surface. The solid curve shows a total potential $U_{tot}(R)$, which is composed of an optical potential $U_{opt}^{01}(R)$ under the excitation of the LP_{01} mode with a power of 100 mW and a blue detuning of +1 GHz, and a cavity potential $U_{cav}(R)$ with $f = 1$. The broken curve shows the pure optical potential $U_{opt}^{01}(R)$. Both potentials are normalized to the maximum value $U_{opt}^{01}(0)$ of the optical potential on the inner-wall surface. As shown in Fig. 11.8, the potential barrier is greatly reduced due to the cavity potential near the surface.

The optical potential is in proportion to the light power coupled to the hollow fiber at a fixed blue-detuning. On the other hand, the cavity potential becomes larger as the hollow diameter becomes smaller. This indicates that the repulsive optical potential produced by a weak blue-detuned optical near field can be canceled by the attractive cavity potential if we employ a small hollow fiber with a hollow diameter comparable to a wavelength. As a consequence, atoms cannot be successfully transmitted through the hollow fiber. This situation can be confirmed by the observation of the threshold power in the atom guidance. For this purpose, two-step photoionization experiments have been carried out, in which the power of the guide light beam was changed at a large blue-detuning.

Figure 11.9a shows the result in the case of a 0.3-μm hollow fiber: the two-step photoionization signal on the ^{87}Rb atoms in the $5S_{1/2}$, $F = 2$ upper ground state is plotted as a function of the transmission power of the guide light. Here, the blue detuning δ of +4 GHz with respect to the ^{85}Rb, $5S_{1/2}$, $F = 3$ upper ground state is chosen so that the optical pumping between the hyperfine ground states due to the spontaneous transition can be suppressed. The magnified figure in the low-power region is shown in Fig. 11.9b, in which we can find a threshold of the guiding of ^{87}Rb atoms at a power of about 2.6 mW.

The threshold yields $U_{\text{tot}} = 0$. In this case, the scaling coefficient f is given by $f = |U^{01}_{\text{opt}}/U_{\text{cav}}|$. For simplicity, we estimate f at a position $R = R_0$ where the total potential with $f = 1$ becomes maximum. Then, from Eqs. 11.16, 11.17, and 11.19, we can calculate the value of f as a function of the coupled power of the guide light. Comparing the threshold power obtained experimentally with the numerical result, we get a scaling coefficient $f \simeq 1.3$, which leads to a geometric factor $\eta \simeq 3.7$. This is a rough estimation including uncertainty, but is a reasonable value: qualitative theoretical analysis says

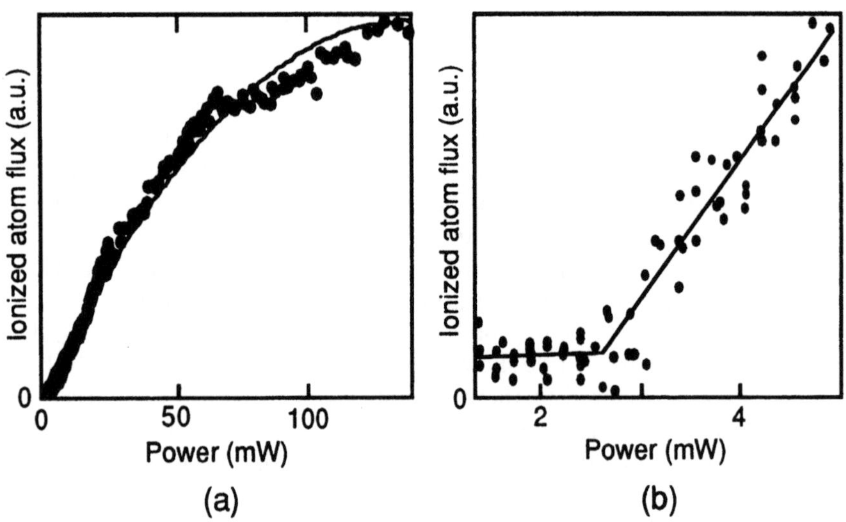

Fig. 11.9. **a** Two-step photoionization signal of ^{87}Rb atoms in the $5S_{1/2}$, $F = 2$ upper ground state guided through a 0.3-μm hollow fiber as a function of the transmission power of the guide light, and **b** the magnified figure in the low-power region. The frequency of the guide light is fixed at a blue-detuning of +4 GHz with respect to the ^{85}Rb, $5S_{1/2}$, $F = 3$ upper ground state

that the geometric factor should be between 1 and 10. Analysis with the exact formula given in Ref. [47] is now in progress.

Two kinds of force contribute to the cavity potential given by Eq. 11.19. The van der Waals force is dominant in a region below a wavelength, but the Casimir–Polder force is dominant in a region above a wavelength. Since the wavelength used here is 0.78 μm, the case with a 0.3-μm hollow fiber corresponds to the van der Waals case. To examine the Casimir–Polder case, we can use a 1.4-μm hollow fiber. In fact, a similar result has previously been obtained [46].

11.3.4 Atomic Quantum Wires with a Light Coupled Sideways

The AQW experiments mentioned earlier employ a thermal atomic beam, in which a guide light beam is introduced into a hollow fiber from the front via a pinhole mirror, as shown in Fig. 11.5. Note that one of our goal is to realize atom manipulation with extremely high spatial accuracy. This requires the guiding of cold atoms. In this connection, there is a plan to develop an atomic funnel with a blue-detuned optical near field that efficiently introduces cold atoms from an MOT into a hollow fiber (see Sect. 11.5 for details). However, the above scheme for guiding atoms is not suitable for the attachment of additional atom-optical devices such as an atomic funnel to the AQW. In this subsection, to solve this problem, we present a new scheme for AQW [48] in which a guide light beam is coupled to the core directly from the side via total internal reflection at an edge. This scheme enables one to combine the AQW with some atom-optical devices without extra optics.

The side coupling of a guide light beam is carried out with a hollow fiber with a slanting polished facet. In the case of a silica-glass fiber with a refractive index of 1.45, the critical angle at which total internal reflection occurs is 43.5° for a wavelength of 780 nm. Therefore, we can use a hollow fiber with a 45°-cut edge that is easily fabricated. The fabrication process for a 45°-cut hollow fiber consists of four steps: (1) fixing a hollow fiber to a device to keep the fiber axis at an angle of 45° to the plane, (2) rough grinding, (3) fine grinding and polishing with several kinds of abrasives, and (4) washing and baking to remove any remnants. The mean roughness of the polished edge surface can be below 50 nm.

The experiments of guiding Rb atoms with 7- and 2-μm hollow fibers have been made in a lateral light-coupling scheme. Figure 11.10 schematically shows the experimental set-up with two-step photoionization. In these experiments, results similar to those from the front light-coupling scheme have been obtained [48]. In addition, surface-ionization experiments in a lateral light-coupling scheme have also been performed with a platinum hot wire. We will describe such a surface-ionization experiment in relation to the observation of the spatial distribution of the guided atom flux in Sect. 11.4.

There are several advantages in this method in addition to the guiding of cold atoms. First, the scheme improves the removal of unwanted light such

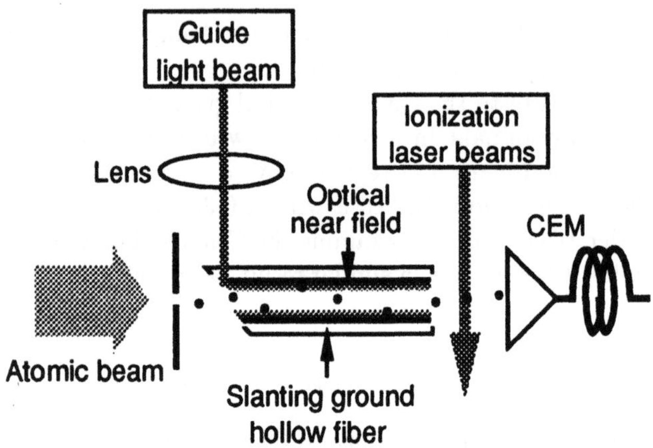

Fig. 11.10. Set-up for the two-step photoionization experiment with an AQW with light from the side. A guide light beam is coupled to the core via total internal reflection at the 45°-cut edge

as propagating modes in the hollow region. This leads to the suppression of spontaneous emission events induced by the absorption of photons from the extra light fields. Second, this scheme prevents the leak of incident light at the entrance of the AQW, so that heating and optical pumping of atoms near the entrance can be avoided. Moreover, this scheme also enables one to couple the guide light beam from the rear direction. Therefore, the lateral light-coupling scheme is very useful for a detailed study of AQW: for example, research into the influence of spontaneous emission due to absorption of near-field photons, and the atomic Doppler effects inside the hollow fiber, etc. This research will give us a deeper understanding of the dipole interaction between atoms and the optical near field, including the cavity QED effects in a dielectric cylinder.

11.4 Optically Controlled Atomic Deposition

For many years, fabrication of atomic-scale matter such as quantum dots has attracted considerable attention. In consequence of the rapid progress of atom-optical techniques, hopes of nano-fabrication based on atom manipulation with laser light have increased in recent years. For example, it has been possible to make small structures with fine parallel-line patterns of metal atoms on a substrate [5, 49–51]. In these experiments, optical standing waves which play the role of an atomic lens or an atomic grating are often employed for focusing of an atomic beam and generating a line-pattern [52]. In addition to atom lithography [53], atom holography has also been demonstrated [54]. However, atom-optical methods with far-field light are affected by the diffraction of light waves. The diffraction effect puts a limit on the size of the fabricated structure. Moreover, it is not easy to make an arbitrary pattern with such methods: although it may not necessarily be impossible, it usually requires elaborate apparatus.

To overcome difficulties in the traditional techniques with far-field light, we consider the application of an optical near field to nano-fabrication. To this end, in this section, we discuss the possibility of fabricating an arbitrary pattern using an AQW. This will lead us to a novel technique of optically controlled atomic deposition.

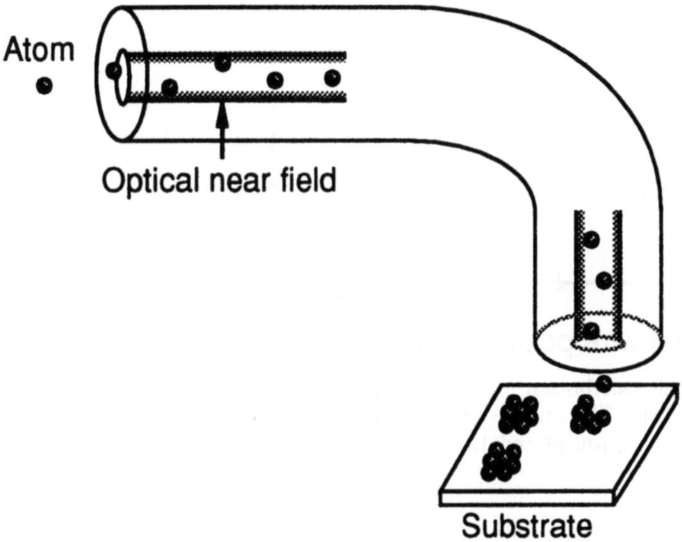

Fig. 11.11. Sketch of atom deposition with an AQW on a substrate

Figure 11.11 is a sketch of the creation of a dot-shaped structure with a hollow fiber close to a blue-detuned optical near field. Since an optical fiber is flexible, we can carry atoms aiming at any point on a substrate by using a bent hollow fiber. The curvature of the AQW is determined by the ratio of the potential barrier to the atomic kinetic energy. Assuming laser-cooled atoms, we can estimate the radius of curvature to be less than 1 cm.

11.4.1 Spatial Distribution of Guided Atoms

Let us move on to the measurement of the spatial distribution of atoms guided through a hollow fiber escorted by a blue-detuned optical near field [30]. This illustrates the feasibility of forming a dot-shaped pattern with an AQW. The experimental set-up is similar to that in Fig. 11.5 except for the detection scheme. In this case, surface ionization with a fine hot-wire is employed instead of laser photoionization: a tungsten (W) wire, a rhenium (Re) wire, a platinum (Pt) wire, and so on, can be used for the surface ionization of alkali–metal atoms according to the work function of each atomic species.

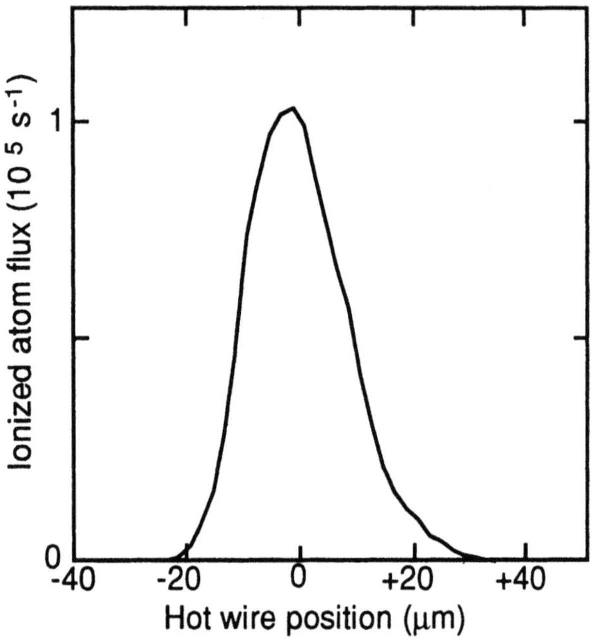

Fig. 11.12. Spatial distribution of Rb atoms guided through a 7-μm hollow fiber over a distance of 3 cm. The surface-ionization signal is obtained with a Pt hot wire having a diameter of 10 μm placed 12 mm downstream from the exit facet

Here, a Pt hot-wire with a radius of 10 μm is used for the ionization and detection of the guided Rb atoms.

Figure 11.12 shows a surface-ionization signal on Rb atoms guided through a 7-μm hollow fiber over a distance of 3 cm. This spatial distribution is obtained from a cross-sectional scan of the guided atom flux with the Pt hot-wire at a distance of 12 mm downstream of the exit facet of the hollow fiber. The guide-laser frequency is blue-detuned at an optimal value of +3 GHz with respect to the ^{85}Rb, $5S_{1/2}$, $F = 3$ upper ground state. Remember that the atomic beam contains two stable isotopes, ^{85}Rb and ^{87}Rb. Therefore, the surface-ionization signal involves contributions from all of the four $5S_{1/2}$ hyperfine ground states of two Rb isotopes: the ^{85}Rb atoms in the $5S_{1/2}$, $F = 3$ upper ground state is dominant in the thermal atomic beam.

The spatial distribution shown in Fig. 11.12 has the full width of 20 μm at the half-maximum. In addition, considering that the quantum efficiency of the CEM is 0.9 and the diameter of the hot-wire is 10 μm, we can estimate the guided Rb flux Φ to be 1.7×10^5 atom/s measured above the background level. Here, we suppose that the surface-ionization efficiency is almost 1: it has been reported that in practice the surface-ionization efficiency reaches up to 0.8 [55].

Taking into account the detecting position of 12 mm downstream of the hollow fiber and the mean longitudinal atomic velocity of 360 m/s, we figure the divergence 2θ of the guided Rb flux to be 1.1 mrad, which leads to a transverse atomic velocity of 19 cm/s. On the other hand, the mean transverse velocity of the incident Rb beam can be estimated to be 20 cm/s in the current collimation scheme. This result implies that the transverse atomic velocity is constant through the short-distance guiding. This is because the dipole-force guiding of atoms is energy-conserving when the frequency detuning is large enough that spontaneous emission per bounce is negligible. However, if a long hollow fiber is employed, it is possible to reduce the transverse atomic velocity based on the laser cooling induced by the optical near field. In this connection, Sect. 11.5 deals with a near-field optical atomic funnel.

Let us estimate the deposition rate in this case. At a position x downstream of the exit facet, the flux density $d(\Phi, x)$ is given by $d(\Phi, x) = \Phi/\pi(x\theta + a)^2$, where a is the hollow radius of a hollow fiber. Then, assuming a density of 1.53 g/cm^3 for Rb metal, we get the supply rate $\eta(\Phi, x) = 3.2 \times 10^{-7}\text{nm} \cdot \Phi/(x/\text{mm} + 6.5)^2$ (units nm/s). As a result, in the case of a substrate placed at $x = 1$ mm, the inverse rate η^{-1} is estimated to be 17 min/nm. Since the number of Rb atoms not fixed to a substrate can be greatly reduced by cooling the substrate or processing it with selective etching, the supply rate η can be regarded as the deposition rate. The above value of the deposition rate is suitable for the creation of an atomic-scale structure on a substrate.

11.4.2 Precise Control of Deposition Rate

The guided atom flux Φ, or the deposition rate η, depends on oven temperature, hollow diameter, guide-laser intensity, blue detuning, etc. In particular, by changing the blue detuning, we can accurately control the deposition rate. This is one crucial advantage of atomic deposition with an AQW. To illustrate the feasibility of frequency control of the deposition rate [30], we will go back to the state-selective two-step photoionization of guided Rb atoms.

Figure 11.13 shows a photoionization spectrum of ^{87}Rb atoms in the $5S_{1/2}$, $F = 2$ upper ground state guided through a 1.4-μm hollow fiber. The solid curve shows the ionized atom flux plotted as a function of the normalized blue-detuning δ/Γ with respect to the ^{87}Rb, $5S_{1/2}$, $F = 2$ upper ground state. The broken line shows the background transmission without the guide light. In this case, the oven temperature is low enough that the incident atom flux is limited to a rather low level. It should be emphasized that the state-selective photoionization scheme enables one to observe a small number of guided atoms without background noise.

Here we are interested in the ratio of the atom-flux change $\Delta\Phi$ to the light-frequency change $\Delta\omega_L$, i.e., $\Delta\Phi/\Delta\omega_L$. From the central slope of the solid curve in Fig. 11.13, we can estimate $\Delta\Phi/\Delta\omega_L$ to be about 50 s^{-1}/GHz. Since the frequency fluctuation of laser light can easily be suppressed below 1 MHz, this result implies that the deposition rate can be controlled with an accuracy of 1 atom/s.

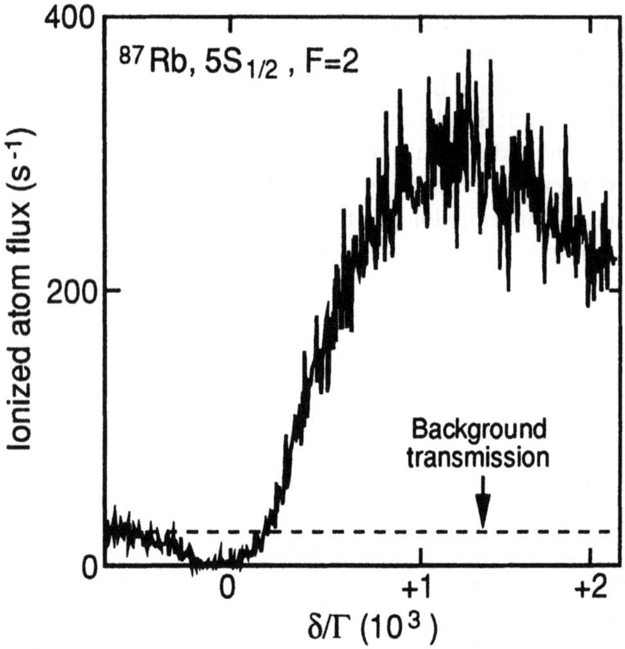

Fig. 11.13. State-selective two-step photoionization spectrum of the ^{87}Rb atoms in the $5S_{1/2}$, $F = 2$ upper ground state guided through a 1.4-μm hollow fiber at a low oven temperature. The *solid curve* shows the ionized atom flux as a function of the normalized blue-detuning δ/Γ with respect to the $5S_{1/2}$, $F = 2$ upper ground state. The *broken line* shows a background level without the guide light

In the present case, the thermal fluctuation of the guided atoms becomes a crucial issue. In addition, as shown by the broken line, there is a residual flux if a straight hollow fiber is employed. For precise control of the deposition rate, thermal fluctuation and background transmission should be suppressed. The thermal fluctuation will be greatly improved by the use of laser-cooled atoms. At the same time, this enables one to use a bent hollow fiber that blocks the ballistic atoms. Manipulation of cold atoms with an optical near field is presented in Sects. 11.5 and 11.6.

11.4.3 In-line Spatial Isotope Separation

Atomic deposition with an AQW also results in greater purity, because of the species- and state-selective guidance based on the frequency-dispersion properties of the dipole force. The strong point of this atomic deposition scheme is very important for atomic-scale crystal growth.

It should be noted that the species- and state-selectivity of the dipole interaction between an atom and an optical near field appears in its application to isotope separation. The first demonstration of such isotope separation was made with an AQS composed of a planar optical near field [10]. The AQW

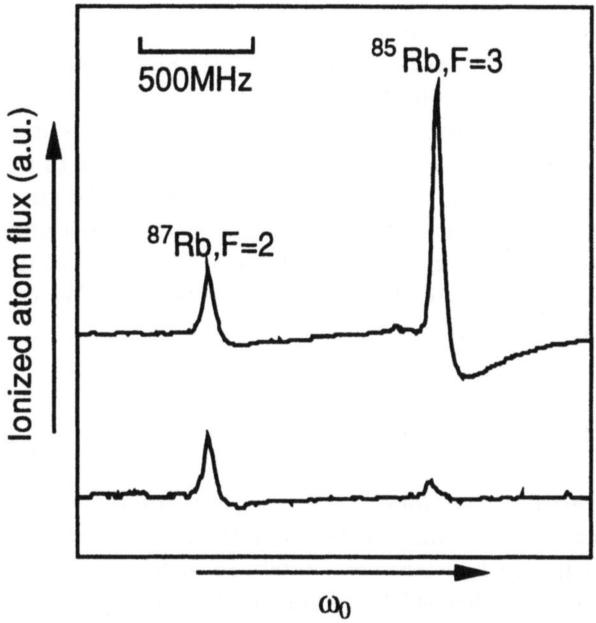

Fig. 11.14. In-line spatial separation between ^{85}Rb in the $5S_{1/2}$, $F = 3$ state and ^{87}Rb in the $5S_{1/2}$, $F = 2$ state with a 3-cm-long hollow optical fiber with an inside diameter of 7 μm. The *upper curve* shows the case where the guide laser is blue-detuned for both isotopes, while the *lower curve* shows the case where the guide laser is blue-detuned for ^{87}Rb but red-detuned for ^{85}Rb

greatly improves the isotope separation: it leads to in-line spatial isotope separation [20]. Since the AQW scheme of isotope separation can efficiently produce high-purity atomic samples, it will also be very useful for industrial applications.

As shown in Fig. 11.7, a state-selective two-step photoionization spectrum has a frequency dispersion shape with respect to an atomic resonant frequency. This leads us to isotope separation with AQW. Let us consider the case where two kinds of atoms having slightly different resonant frequencies are contained in an atomic guidance source. If the frequency of the guide light is higher than the resonant frequency of one atom but lower than that of another, the former can be successfully guided through the hollow fiber, while the latter is attracted by the red-detuned optical near field so that it will be trapped on the inner-wall surface.

An experiment on in-line spatial separation between the ^{85}Rb atoms in the $5S_{1/2}$, $F = 3$ upper ground state and the ^{87}Rb atoms in the $5S_{1/2}$, $F = 2$ upper ground state has been performed with a 3-cm-long hollow optical fiber [20]. The experimental set-up is the same as that shown in Fig. 11.5. Figure 11.14 shows the result with a 7-μm hollow fiber: the two-step photoionization signals are plotted as a function of the diode-laser frequency, i.e., the atomic resonant frequency ω_0. In each photoionization signal, the frequency of the diode laser is swept at a fixed blue-detuning of the guide light. Note that the $5S_{1/2}$, $F = 2$ state of ^{87}Rb is 1 GHz higher than the $5S_{1/2}$, $F = 3$ state of ^{85}Rb (see also Fig. 11.6).

The upper curve of Fig. 11.14 shows the case where the guide laser is blue-detuned for both isotopes. In this case, both isotopes are guided through the hollow fiber. The lower curve of Fig. 11.14 shows the case where the guide laser is blue-detuned for ^{87}Rb atoms but nearly red-detuned for ^{85}Rb atoms. In this case, the ^{87}Rb atoms are still guided through the hollow fiber, but the transmission of the ^{85}Rb atoms is greatly suppressed. Thus, the AQW can be regarded as an atomic-state filter.

11.5 Near-Field Optical Atomic Funnels

So far we have concentrated on describing the control of thermal atoms. It goes without saying that there is a limit to the use of thermal atoms whose mean velocity is several hundred meters per second. To manipulate atoms with high spatial accuracy, even beyond the diffraction limit of light waves, we need cold atoms whose mean velocity is low enough that they can be trapped in a narrow region for a long time. Therefore, we will now turn to the subject of guiding and manipulating cold atoms with an optical near field.

A dense ensemble of cold atoms, which is called an optical molasses, can be produced in the center of an MOT [2, 3, 56]. The MOT carries out deceleration and cooling of atoms at the same time. Atoms in the MOT are decelerated with Doppler cooling, which involves spatial modulation of atomic Zeeman sublevels so that they form a cold cloud with a very narrow velocity distribution around zero velocity: the temperature of the cold ensemble is below 100 μK. Moreover, using evaporative cooling in a pure magnetic trap that follows the MOT [57], we can cool the dense atomic ensemble still further to below 100 nK. This technique realizes the so-called Bose–Einstein condensation of an alkali–metal gas. For example, Bose–Einstein condensation has been observed on Rb [58], Na [59], and lithium (Li) [60]. Recently, an output-coupler of a Bose–Einstein condensed gas has been developed [61]. This can be regarded as an atom laser.

Although we can now generate an atomic ensemble cold enough for near-field optical manipulation with a standard MOT, it is not easy to take a large number of cold atoms out of the MOT: when the MOT is turned off, the atoms divergently escape from the MOT in random directions according to each one's momentum. Therefore, even if the hollow fiber is placed below the MOT, many atoms are lost without entering the hollow fiber. It should also be noted that the effective diameter of an optical molasses, which is usually about 1 mm, is much larger than the size of near-field optical devices such as an AQW composed of a micron-sized hollow fiber. In this section, we present an AQS with a conical optical near field to couple the cold atoms

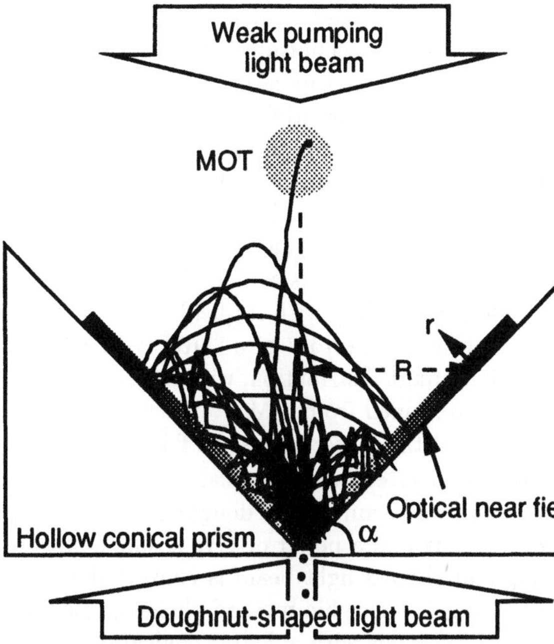

Fig. 11.15. Atomic funnel composed of an optical near field induced over the inner-wall surface of an inverted hollow conical prism with a small exit hole at the bottom. A blue-detuned doughnut-shaped light beam is shone upward to make an AQS, while an additional light beam is shone downward to induce the Sisyphus cooling of atoms released from an MOT. A cross-sectional trajectory of a ^{85}Rb atom randomly sampled from the MOT is also shown based on a Monte Carlo simulation

efficiently from an MOT to an AQW. The conical AQS works as an atomic funnel [62].

11.5.1 Atomic Funnel with Atomic Quantum Sheet

As mentioned in Sect. 11.1, the sheet of a blue-detuned optical near field works as an atomic mirror. Let us consider the case where a blue-detuned light beam is shone upward from the bottom of an inverted hollow conical prism with a small exit hole, as shown in Fig. 11.15. The blue-detuned light beam induces an AQS to reflect atoms on the inner-wall surface of the hollow conical prism via total internal reflection. Assume that an MOT is produced over the open face of the inverted hollow conical prism. If the MOT is turned off, the cold atoms fall into the funnel and then bounce on the optical near field.

However, this scheme does not work well for funneling cold atoms as it is. Because of a strong gravitational pull, as a result of the fall, the cold atoms released from the MOT are extremely accelerated in only a few millimeters. The accelerated atoms escape from the open face soon after a few bounces even if a powerful AQS is made by the intense light beam. This indicates that the atomic funnel requires a cooling mechanism to compensate for the atomic acceleration due to gravity. In fact, we can carry out recooling of the falling atoms inside the atomic funnel by irradiating a weak pumping light beam downward. The cooling takes place in the atomic reflection process involving optical pumping events. Since the atoms lose their kinetic energies as a consequence of the multiple up-and-down motion on the potential hill, the cooling process is called Sisyphus cooling, a name derived from Greek mythology. In the next subsection, we explain this cooling. Sisyphus cooling is also very popular in traditional laser cooling with far-field light [2].

Figure 11.15 shows a cross-sectional trajectory of a ^{85}Rb atom released from an MOT with a Maxwell–Boltzmann distribution, with a mean temperature of 10 μK, based on a result obtained from a Monte Carlo simulation developed below. Here we assume that the center of the MOT is placed 1 mm above the open face of a 4-mm-high inverted hollow conical prism with a 0.1-mm hole at the bottom. In addition, we assume that a doughnut-shaped light beam with a blue-detuning of +1 GHz and a power of 1 W is shone upward to make an AQS. Such a doughnut-shaped light beam is converted from a Gaussian light beam not only using holographic methods [63–65], but also using a couple of simple geometric methods with a double-conical prism [62] or a hollow fiber [66]. The doughnut-shaped light beam prevents heating or pumping due to extra light fields from the recooled atoms coming from the funnel as well as saving the light power.

As shown in the trajectory, due to the near-field optical cooling inside the funnel, the atoms released from the MOT are going down and bouncing off the inner wall many times before going out of the small exit hole. Therefore, in contrast to other types of atomic funnel [67–69], this atomic funnel serves the function of converting a cold but spatially extended atomic sample from an MOT to a collimated cold atomic beam. Although another scheme for forming a cold atomic beam from an MOT has been demonstrated [70], this type of atomic funnel is convenient as a source of a cold atomic beam for an AQW. Such a cold atomic beam is also useful for atomic reflection with a localized optical near field (cf. Sect. 11.6).

11.5.2 Sisyphus Cooling Induced by Optical Near Field

Sisyphus cooling can be understood based on a dressed-atom picture [71–73]. The dressed states are eigenstates of the Hamiltonian of the *atom + light field* system in the electric dipole representation [74]. However, the application of such a dressed-atom scheme, which was developed for far field, to near field is not necessarily self-evident, since the dressed-atom picture involves the quantization of light field, in which the photon number is well defined. The issue of the quantization of optical near field is still open, particularly that of a nanometric optical near field. Nevertheless, we apply the dressed-atom picture to the atoms interacting with an optical near field of the planar type: with minor alternations, the far-field dressed-atom picture will approximately describe the atomic quantum states modulated by the optical near field produced via total internal reflection of propagating light. In fact, a cooling experiment with a planar AQS has been demonstrated [72].

The near-resonant dipole interaction between an atom and a light field gives rise to a change of atomic quantum states, which is well known as light shifts [74]. As mentioned in Sect. 11.3, the dipole force of an optical near field on a two-level atom has the optical potential given in Eq. 11.17. Provided that the saturation parameter s, defined by the ratio of the Rabi frequency $\Omega(r) = \sqrt{P(r)/2P_{\mathrm{sat}}}\Gamma$ to the blue detuning $\delta = \omega_{\mathrm{L}} - \omega_0$, is small, i.e.,

$$s(r) = \frac{\Omega(r)^2}{2\delta^2} \ll 1 \tag{11.20}$$

the optical potential $U_{\mathrm{opt}}(r)$ is approximately equal to the position-dependent light shift given by $\hbar\Omega(r)^2/4\delta$ [74].

To demonstrate an atomic funnel supported by a blue-detuned optical near field, we also use alkali–metal atoms, including Rb. As shown in Fig. 11.6, the alkali–metal atoms have two hyperfine ground states. Note that we have so far supposed two-level atoms under a large frequency detuning. However, Sisyphus cooling involves spontaneous emission events that follow the absorption of photons of the optical near field and additional pumping light, so that the transition between the two hyperfine ground states via the excited states occurs in the funneling process. Therefore, we must now consider three-level atoms with Λ configuration [73] that consist of two hyperfine ground states and a group of excited states.

We assume that a three-level atom has two ground states, represented by kets $|g1, N\rangle$ and $|g2, N\rangle$, and a group of degenerated excited states represented by a ket $|e, N-1\rangle$ outside an optical near field, where the $|g2, N\rangle$ state is spaced at a frequency of δ_{hfs} above the $|g1, N\rangle$ state, and N is a photon number. Then, the dipole interaction between the three-level atom and the optical near field alters the three free atomic states $|g1, N\rangle$, $|g2, N\rangle$, and $|e, N-1\rangle$ into the three dressed-states $|1, N-1\rangle$, $|2, N-1\rangle$, and $|3, N-1\rangle$, respectively. On this occasion, as a straightforward expansion of the two-level configuration, the optical potential $U^1_{\mathrm{opt}}(r)$ for the lower atomic state and the optical potential $U^2_{\mathrm{opt}}(r)$ for the upper atomic state are given by [71]

$$U^1_{\mathrm{opt}}(r) = \frac{2}{3}\frac{\hbar\Omega(r)^2}{4\delta} \tag{11.21}$$

$$U^2_{\mathrm{opt}}(r) = \frac{2}{3}\frac{\hbar\Omega(r)^2}{4(\delta + \delta_{\mathrm{hfs}})} \tag{11.22}$$

where the factor of 2/3 associated with the Clebsch–Gordan coefficient stems from the sum of the transition strengths over the excited state hyperfinelevels.

Let us consider the case where a three-level atom is reflected by a blue-detuned optical near field produced near a dielectric surface. Figure 11.16

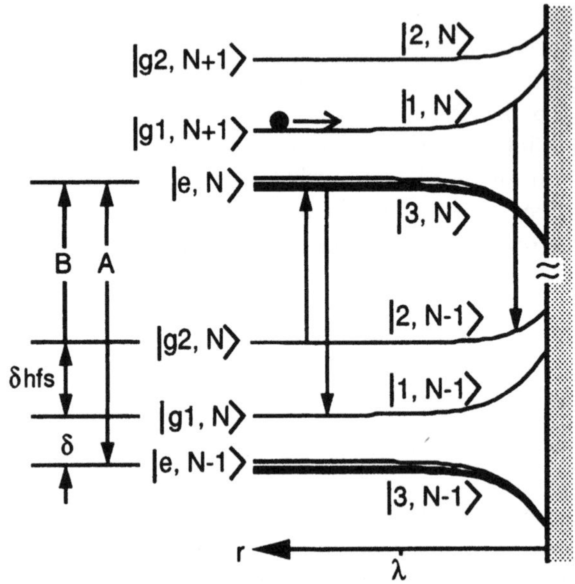

Fig. 11.16. Dressed-atom energy levels of a three-level atom near a surface with a blue-detuned optical near field. An atom in the lower state makes a spontaneous transition to the upper state on the hill of the optical potential produced by the optical near field (A) with a detuning of δ, and it is then repumped to the lower state by a pumping light (B). The potential barrier is induced in a region below a wavelength λ

shows the variations of the atomic quantum states as a function of the distance r from the surface, in which the blue-detuning δ is taken with respect to the lower ground state $|g1, N\rangle$. Arrows A and B indicate the frequencies of the light supporting the optical near field and the pumping light, respectively. A set of the three free states ($|g1, N\rangle$, $|g2, N\rangle$, $|e, N-1\rangle$) or a set of the three dressed states ($|1, N-1\rangle$, $|2, N-1\rangle$, $|3, N-1\rangle$) forms a manifold labeled with a photon number N. Moreover, a set of manifolds forms a ladder structure. The three-level atom can be transferred from a manifold to a lower manifold via spontaneous emission, or to a higher manifold via absorption of a photon. It should be noted that both dressed states $|1, N-1\rangle$ and $|2, N-1\rangle$ partially include the excited state $|e, N-1\rangle$, so that the atom in these dressed states can cause spontaneous emission.

Important to Sisyphus cooling is the fact that the optical potential for the upper state is smaller than that for the lower state because the blue-detuning for the upper state is larger by the hyperfine splitting $\delta_{\rm hfs}$ than that for the lower state. Therefore, if an atom in the lower state is spontaneously transferred to the upper state in the process of reflection it loses some kinetic energy, and the amount that loss is the difference between both optical potentials. Moreover, soon after reflection, if the atom is repumped to the lower state with a weak pumping light, the energy-loss event is repeated, so that the atom continues to lose its kinetic energy until the heating effects, including the scattering of photons, balance the cooling effects.

11.5.3 Monte Carlo Simulations

There are some factors that determine the collection efficiency of an atomic funnel, these are blue detuning, laser power, field intensity distribution, initial atomic temperature, release height, etc. Monte Carlo simulations show the dependence of these parameters on the atomic funnel [62]. As an example, to collect cold Rb atoms efficiently, we assume a specific scheme with an inverted hollow conical prism and a doughnut-shaped light beam, as mentioned above (see also Fig. 11.15). In addition, we take into account the van der Waals cavity potential in the Monte Carlo simulations.

We now consider the case where the intensity distribution of the optical near field over the inner wall surface of a 4-mm-high inverted hollow conical prism, with a slope angle $\alpha = 45^\circ$ and a refractive index $n_\mathrm{p} = 1.45$, is produced by a doughnut-shaped light beam with a ring diameter $2R_0 = 4$ mm. The specific conical hollow prism is chosen so that the collection efficiency becomes higher under feasible experimental conditions. On the other hand, the doughnut-shaped light beam can be converted from a Gaussian light beam with a waist $2w_0 = 4$ mm through a proper double-conical prism [62]. In this case, as a function of the distance R_d from the center axis, the intensity $P(R_\mathrm{d})$ of the doughnut-shaped light beam is given approximately by

$$P(R_\mathrm{d}) = \frac{Q_0}{\pi w_0^2}\left[\exp\left\{-\frac{2(R_\mathrm{d}-R_0)^2}{w_0^2}\right\} + \exp\left\{-\frac{2(R_\mathrm{d}+R_0)^2}{w_0^2}\right\}\right] \quad (11.23)$$

where Q_0 is the incident light power.

The optical potential $U^1_\mathrm{opt}(r)$ for the lower dressed state of a Rb atom at a distance r from the inner-wall surface of the prism is given by

$$U^1_\mathrm{opt}(r,R) = \frac{1}{12}\frac{\hbar\Gamma^2}{\delta}\frac{P(R)}{P_\mathrm{sat}}\exp\left(-\frac{2r}{\Lambda}\right) \quad (11.24)$$

where the decay length $\Lambda = (\lambda/2\pi)/\sqrt{n_\mathrm{p}^2\sin^2\alpha - 1}$ is nearly equal to 0.7λ, and the natural linewidth Γ and the saturation intensity P_sat are equal to $2\pi \times 6.1$ MHz and 1.6 mW/cm^2, respectively, for the Rb D_2-line with a wavelength λ of 780 nm. In addition, we omit small Doppler shifts of cold atoms. We can write the optical potential for the upper dressed state by replacing the detuning δ with $\delta + \delta_\mathrm{hfs}$ in a similar way. In Eq. 11.24, the coordinate R denotes the distance from the center axis of a position on the inner wall surface (see Fig. 11.15).

As mentioned in Sect. 11.3, the attractive potential due to cavity QED effects greatly reduces the potential barrier to reflecting atoms near the surface.

In this atomic funnel, we consider the van der Waals force. The cavity potential $U_{\rm vdw}(r)$ with the van der Waals force is given approximately by [34, 44]

$$U_{\rm vdw}(r) = -\frac{1}{16r^3}\frac{n_{\rm p}^2-1}{n_{\rm p}^2+1}\sum_{\rm e}\frac{\hbar\Gamma_{\rm eg}}{k_{\rm eg}^3} \tag{11.25}$$

where $\Gamma_{\rm eg}$ and $k_{\rm eg}$ are the natural linewidth and wave number, respectively, of an allowed dipole transition from the atomic ground state. As a result, the total potential $U_{\rm tot}(r,R)$ is the sum of the repulsive optical potential $U_{\rm opt}(r,R)$ and the attractive van der Waals cavity potential $U_{\rm vdw}(r)$: $U_{\rm tot}(r,R) = U_{\rm opt}(r,R) + U_{\rm vdw}(r)$.

Two kinds of transition from the lower dressed state occur in the process of reflection: $|1,N\rangle \to |1,N-1\rangle$ and $|1,N\rangle \to |2,N-1\rangle$. The transition $|1,N\rangle \to |3,N-1\rangle$ to the third dressed state is negligible compared with the other transitions because it has a very small transition probability. Assuming a branching ratio q, we get the transition rates $\Gamma_{11}(r,R) \sim q\Gamma s(r,R)$ and $\Gamma_{12}(r,R) \sim (1-q)\Gamma s(r,R)$ for each transition, and then the total transition rate $\Gamma_1(r,R) = \Gamma_{11}(r,R) + \Gamma_{12}(r,R) \sim \Gamma s(r,R)$ from the lower dressed state. In consequence of this, the total probability $P(v_\perp)$ of the spontaneous transition from the lower dressed state is given by [71, 75]

$$P(v_\perp) = 1 - \exp\left(-\frac{mv_\perp \Lambda\Gamma}{\hbar\delta}\right) \tag{11.26}$$

where m and $v_\perp$ are the atomic mass and atomic velocity perpendicular to the inner surface, respectively.

The point of the Monte Carlo simulation procedure is as follows. When an atom in the lower ground state enters the near-field region after a free fall from an MOT, we compare the transition probability $P(v_\perp)$ with a random number r_1 between 0 and 1. If $r_1 < P(v_\perp)$, the transition from the lower dressed state takes place. In this case, we add to the atom a momentum $\hbar n_{\rm p} k \sin\alpha$ of an optical near-field photon as well as a momentum $\hbar k$ of a spontaneous photon. Then we compare the branching ratio q ($= 0.75$ for the Rb D_2 line [71, 73]) with a random number r_2. If $r_2 > q$, we choose the first transition $|1,N\rangle \to |1,N-1\rangle$, but otherwise we choose the second transition $|1,N\rangle \to |2,N-1\rangle$. When the second transition $|1,N\rangle \to |2,N-1\rangle$ occurs in the process of reflection, the energy loss corresponding to the amount of difference $\Delta U_{\rm tot}(r,R)$ between both potentials is subtracted from the atomic kinetic energy. The expectation value $\langle \Delta K\rangle$ of the energy loss per bounce is given by [71]

$$\Delta K = -\frac{2}{3}\cdot\frac{\delta_{\rm hfs}}{\delta+\delta_{\rm hfs}}\cdot K_\perp \tag{11.27}$$

where $K_\perp$ is the atomic kinetic energy perpendicular to the inner-wall surface. Moreover, when an atom in the upper dressed state comes out of the near-field region and has an upward velocity $v_z > 0$, the atom absorbs a photon of

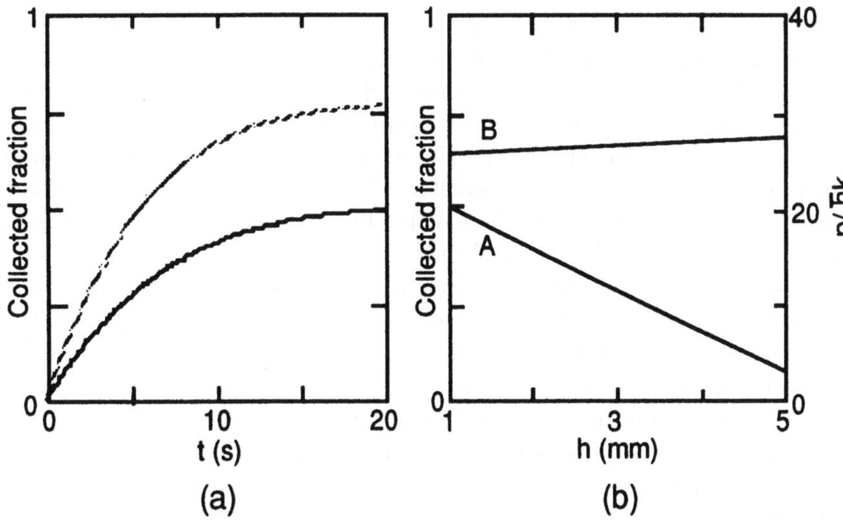

Fig. 11.17. Monte Carlo simulations of an atomic funnel composed of a 4-mm-high inverted hollow conical prism with a 100-μm exit hole, where a doughnut-shaped light beam with a power of 1 W and a blue detuning of +1 GHz irradiates the prism. **a** Temporal evolution of the collected fraction of the 10^4 ^{85}Rb atoms released from an MOT with a mean temperature of 10 μK placed at 1 mm above the open face: the *solid* and *broken curves* show the cases with and without the van der Waals cavity potential, respectively. **b** Collected fraction (A) and normalized momentum $p/\hbar k$ (B) averaged over the collected atoms at $t = 20$ s as a function of the height h of the MOT

the weak pumping light beam so that it returns to the lower ground state. It should be noted that there is an additional cooling effect due to the absorption of photons of the pumping light beam [71]. In this process, the atoms hitting the inner-wall surface are removed and counted as the lost fraction.

According to the above procedure, let us simulate the number of ^{85}Rb atoms coming out of the exit hole with a diameter of 100 μm. Figure 11.17a shows the collected fraction as a function of the time t passed since the cold ^{85}Rb atoms were released from an MOT with an effective diameter of 1 mm, in which the incident power Q_0 and the blue-detuning δ of the doughnut-shaped light beam are taken as 1 W and +1 GHz, respectively. Here we assume that the MOT consists of an ensemble of 10^4 atoms which follows a Maxwell–Boltzmann distribution with a mean temperature of 10 μK, and the center of the MOT is placed at 1 mm above the open face of the inverted hollow conical prism. The solid and broken curves in Fig. 11.17a show the cases with and without the van der Waals cavity potential, respectively.

We can see in Fig. 11.17a that the collected fraction approaches a constant value and both curves are expressed as $C_1\{1-\exp(-C_2t)\}$ with two constants C_1 and C_2. Therefore, if multiple loading of the funnel is carried out, the collection rate can be defined by the asymptotic value C_1 of the collected fraction. For example, assuming a loading number of Φ s^{-1} from the MOT, we obtain a collection rate 0.5Φ atom/s from the solid curve, with $C_1 = 0.5$ in Fig. 11.17a. Thus, the multiple loading of an atomic funnel with an MOT results in the creation of a cold atomic beam. In the present case, the fact that a cold atomic ensemble with an effective diameter of 1 mm is compressed into a small region with an effective diameter of 100 μm leads us to the conclusion that the conversion efficiency into an cold atomic beam is about 10^4 times as large as that without the atomic funnel.

Because of large gravitational acceleration, the collection efficiency of the atomic funnel depends greatly on the height of the MOT. Figure 11.17b shows the collected fraction (A) and the momentum p (B) averaged over the collected ^{85}Rb atoms as a function of the height h of the MOT center, in which they were obtained at $t = 20$ s, and the averaged momentum p was normalized to the recoil momentum $\hbar k$. In Fig. 11.17b, while the collected fraction decreases from 0.5 at $h = 1$ mm to 0.09 at $h = 5$ mm, the averaged momentum is almost constant. This is because only atoms recooled below a low temperature reach down to the bottom and then go out of the funnel. The average momentum is estimated to be about 26 times as large as the recoil momentum $\hbar k$, which corresponds to 80 μK in terms of temperature. If an MOT is produced inside the atomic funnel, the collection efficiency will be greatly increased. In fact, a scheme of trapping atoms inside a hollow pyramidal prism or a hollow conical prism has been demonstrated with a single circularly polarized light beam [76].

11.6 Atomic Quantum Dots

AQWs with a blue-detuned optical near field have greatly increased the spatial accuracy of atom manipulation. However, the optical near field employed in the AQW is produced by way of the total internal reflection of propagating light modes, so that the spatial accuracy is still characterized by the wavelength used. Therefore, we will turn to the subject of atom manipulation with an optical near field on a sharpened optical fiber, i.e., the AQD. To start, we introduce a Yukawa-type field intensity profile of the optical near field. Then, assuming a phenomenological expression for the optical potential, we describe the atomic reflection with a blue-detuned optical near field induced over the tip of a sharpened fiber. In addition, we present a novel form of atom trap with a sharpened fiber. Finally, the possibility of a three-dimensional atom trap with a set of sharpened fibers is discussed.

11.6.1 Phenomenological Approach to the Interaction Between Atoms and the Localized Optical Near Field

The theory of dipole force developed for an optical far field has been applied to the AQS and the AQW. However, in the nanometer region where the wavelength cannot be well defined, the far-field theory is not necessarily suitable for a description of the mechanical interaction between an atom and a localized optical near field. In particular, the quantization of the nanometric optical near field is now an open question. Under these circumstances, we have treated the interaction semiclassically so far, except for the atomic funnel with Sisyphus cooling. In this section, we adopt a phenomenological approach based on a Yukawa-type field intensity profile to describe the dipole force of the optical near field produced near the nanometric tip of a sharpened fiber (cf. Sect. 12.2). This approach greatly simplified the description of the dipole interaction without complicated numerical calculations of the optical near field with the original Maxwell equations.

Experimental observation shows that the decay length of the optical near field localized near nanometric matter depends on the size of the matter: it is almost equal to the inverse of the curvature of the surface (cf. Sects. 2.1 and 12.2). In this section, for simplicity, we consider a sharpened fiber with a hemispherical tip whose radius of curvature is a. Let us take the center of the sphere as the origin of a radial coordinate r. Now, to express the intensity profile of the optical near field in a simple analytical form, we introduce a Yukawa-type scalar potential $\varphi(r)$ given by [24, 77]

$$\varphi(r) = \int\int_A \frac{\exp(-r/\bar{\lambda}_c)}{r} \mathrm{d}S \tag{11.28}$$

where $\bar{\lambda}_c$ is the characteristic decay length of the optical near field, which corresponds to the Compton wavelength, and A denotes the integration region.

This scalar potential generates the intensity profile $P_{\mathrm{onf}}(r)$, given by

$$P_{\mathrm{onf}}(r) = P_0 \frac{\mathcal{H}(r)}{\mathcal{H}(a)} \tag{11.29}$$

where P_0 is the intensity on the surface of the hemispherical tip, and $\mathcal{H}(r)$ is the field-distribution function given by [77]

$$\mathcal{H}(r) = |\nabla\varphi(r)|^2 + \frac{1}{\bar{\lambda}_c^2}|\varphi(r)|^2 \tag{11.30}$$

The initial intensity profile P_0 can be estimated approximately based on the Bethe theory. According to this theory, the power Q_{onf} of the optical near field is given by [78]

$$Q_{\mathrm{onf}} = \frac{8}{\pi} k^2 a^4 P_{\mathrm{in}} \tag{11.31}$$

Using the incident field intensity $P_{\mathrm{in}} = Q_0/\pi b_c^2$ coupled to a sharpened fiber with an incident power Q_0 and a core radius b_c, we can evaluate the field intensity $P_0 = Q_{\mathrm{onf}}/2\pi a^2$ on a hemisphere. For example, in the case of $a =$ 50 nm, $2b_c = 5$ μm, $Q_0 = 1$ mW, assuming a wavelength of 780 nm, we get $Q_{\mathrm{onf}} \sim 50$ nW and $P_0 \sim 3 \times 10^5$ mW/cm^2.

Assuming $\bar{\lambda}_c \simeq a$, we can write down approximations of the scalar potential $\varphi(r)$ and the gradient $\nabla\varphi(r)$ on the polar axis as follows:

$$\varphi(r) = \frac{2\pi a^2}{r}\left\{\exp\left(-\frac{r-a}{a}\right) - \exp\left(-\frac{\sqrt{r^2+a^2}}{a}\right)\right\} \quad (11.32)$$

$$\begin{aligned}\nabla\varphi(r) = & -2\pi\left\{\left(\frac{a}{r}+\frac{a^2}{r^2}\right)\exp\left(-\frac{r-a}{a}\right)\right. \\ & \left. -\left(\frac{a}{\sqrt{r^2+a^2}}+\frac{a^2}{r^2}\right)\exp\left(-\frac{\sqrt{r^2+a^2}}{a}\right)\right\} \end{aligned} \quad (11.33)$$

For simplicity, we here take into account the contributions over all the hemisphere, that is, we take the hemisphere as the integral region A. Substituting Eqs. 11.29–11.33 into Eq. 11.17, we can estimate the optical potential $U_{\mathrm{opt}}(r)$ produced by the localized optical near field. In what follows we ignore the influence of far-field components on atoms by taking a large frequency detuning.

11.6.2 Atom Deflection

Using a sharpened fiber escorted by a blue-detuned optical near field, we can deflect atoms near the tip, as shown in Fig. 11.3a. In this case, the total potential $U_{\mathrm{tot}}(r)$ to reflect an atom can be written as the sum of the optical potential $U_{\mathrm{opt}}(r)$ and the cavity potential $U_{\mathrm{cav}}(r)$. The cavity QED effect near a nanometric sphere is expected to be very large. Although the exact formula of the cavity QED effect in the nano-space has not been obtained, we will use the result calculated in the case of a microsphere [40]. This is a good approximation if we consider only the van der Waals force in the immediate vicinity of the surface. Then, the cavity potential $U_{\mathrm{cav}}(r)$ can be replaced by the van der Waals potential $U_{\mathrm{vdw}}(r)$ [34, 40, 45], which is similar to Eq. 11.25:

$$U_{\mathrm{vdw}}(r) = -\frac{1}{16(r-a)^3}\frac{n_1^2-1}{n_1^2+1}\sum_{\mathrm{e}}\frac{\hbar\Gamma_{\mathrm{eg}}}{k_{\mathrm{eg}}^3} \quad (11.34)$$

where n_1 is the refractive index of the core. Here we also assume that the surface seen by an atom is almost planar in the van der Waals region.

To illustrate the feasibility of atomic deflection, we consider a specific case using a sharpened fiber with a hemispherical tip whose radius of curvature and core diameter are 50 nm and 5 μm, respectively. These values are chosen so that a potential barrier which is suitable for the reflection of Rb atoms is produced under feasible experimental conditions, and at the same time the radius of curvature is as small as possible: from the numerical estimation of Eq. 11.34, including the Rb D lines, we can see that the van der Waals force becomes very large in the region below 50 nm from the surface.

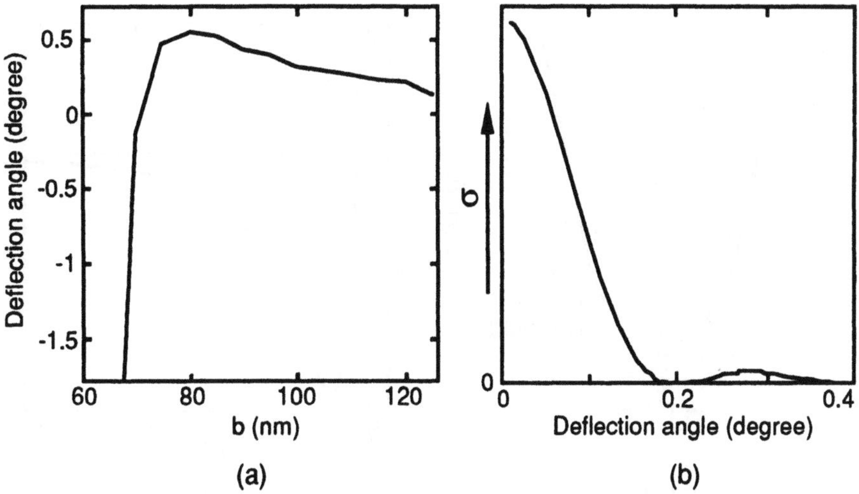

Fig. 11.18. Numerical calculation of the deflection of slow ^{85}Rb atoms with a sharpened fiber escorted by a blue-detuned optical near field: a deflection angle θ as a function of the impact parameter b; b differential cross section σ calculated with the first Born approximation as a function of the deflection angle θ

Let us estimate the deflection angle θ. As is well known, the deflection angle θ is given within classical mechanics by [79]

$$\theta = \pi - 2b \int_{r_t}^{\infty} \frac{\mathrm{d}r}{r^2} \left(1 - \frac{b^2}{r^2} - \frac{U_{\mathrm{tot}}(r)}{K_{\mathrm{A}}} \right)^{-1/2} \tag{11.35}$$

where b is the impact parameter, and K_{A} is the atomic kinetic energy. The turning point r_{t} follows the equation, $1-b^2/r_{\mathrm{t}}^2-U(r_{\mathrm{t}})/K_{\mathrm{A}}=0$. Figure 11.18a shows the deflection angle for a ^{85}Rb atom with $K_{\mathrm{A}}/\hbar\Gamma = 10^3$ as a function of the impact parameter b, in which we assume $Q_0 = 1$ mW and $\Delta/2\pi = +1.5$ GHz with respect to the ^{85}Rb, $5\mathrm{S}_{1/2}$, $F = 2$ lower ground state. Such a slow Rb atom with an equivalent temperature of 200 mK can easily be obtained from a laser-cooled atomic beam. As shown in Fig. 11.18a, a rather large deflection angle $\theta \sim 0.5°$ is expected at $b \sim 80$ nm. From this, we get a deviation of about 0.9 mm from the incident atomic path at a distance 10 cm downstream of the sharpened fiber. On the other hand, we see a negative deflection angle in the region below $b \sim 70$ nm. This is because the Rb atom is attracted to the sharpened fiber by the van der Waals force.

Actually, an atomic beam would be employed for this kind of experiment. In this case, the cross section is a well-defined observable because the atom flux has a known velocity distribution. Assuming a spherical symmetric potential, we can write the differential cross section $\sigma(\theta)$ quantum-mechanically with the first Born approximation as follows [79]:

$$\sigma(\theta) \sim \left| \frac{2m}{\hbar^2} \int_0^{\infty} U_{\mathrm{tot}}(r) \frac{\sin(qr)}{q} r\mathrm{d}r \right|^2 \tag{11.36}$$

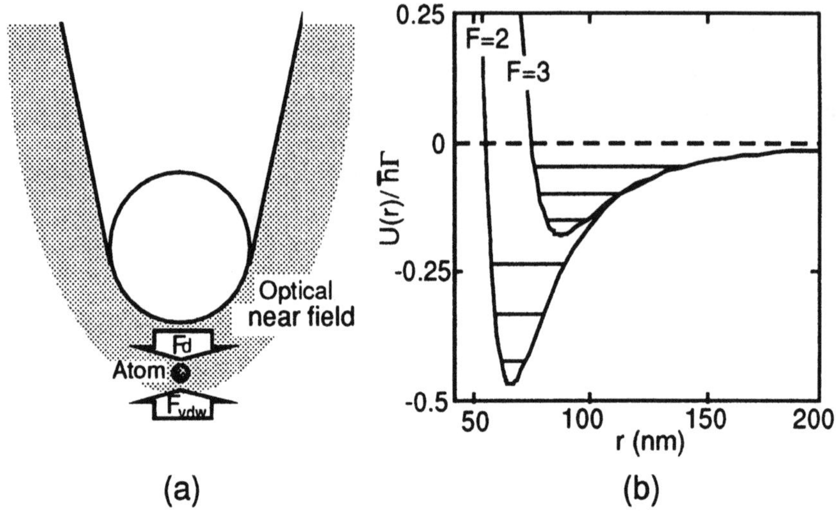

Fig. 11.19. Atom trap with a sharpened fiber placed in the vertical direction. **a** Sketch of an optical van der Waals hybrid trap in which an atom is one-dimensionally trapped by the repulsive dipole force F_{d} and the attractive van der Waals force F_{vdw}, and **b** normalized potentials $U_{\mathrm{tot}}(r)/\hbar\Gamma$ composed of the repulsive optical potential and the attractive cavity potential for both ^{85}Rb, $5S_{1/2}$ ground states near the tip as a function of the distance r from the surface together with quantized vibrational energy levels

where the momentum transfer q is given by $2k_{\mathrm{A}}\sin(\theta/2)$ with an atomic wave number k_{A}. Figure 11.18b shows the differential cross section $\sigma(\theta)$ for an ^{85}Rb atomic beam with a mean kinetic energy of $K_{\mathrm{A}}/\hbar\Gamma = 10^3$ as a function of the deflection angle θ under the same conditions as the above classical calculation. As is expected, the cross section is large in a region $\theta < 0.2°$. The total cross section is estimated to be twice as large as πa^2.

11.6.3 Atom Trap with a Sharpened Optical Fiber

An atom trap composed of a sharpened optical fiber has been proposed [24]. This atom trap requires a subtle balance among three forces, the attractive dipole force, the enhanced radiation pressure, and the centrifugal force. In this scheme, an atom can be trapped on a circular orbit near the tip. However, the balance condition is more complicated if the cavity QED effects are taken into account. In addition, as the AQD for the precise control of atomic motion, a static trap is preferable. In fact, a static atom trap with a sharpened fiber can be produced by a combination of the dipole force and the cavity forces. Here we consider only the van der Waals force, which is dominant in the immediate vicinity of the surface.

Figure 11.19a is a sketch of a one-dimensional atom trap near the hemispherical tip of a sharpened fiber placed in the vertical direction. If a proper radius of curvature (a) of the tip is chosen under a proper blue-detuning and a proper light-field intensity, as shown in Fig. 11.19a, an atom is pushed back by the repulsive dipole force F_d when it approaches the tip, while it is pulled back by the attractive van der Waals force F_{vdw} when it escapes from the tip.

Similarly to the case of atomic deflection, the total potential $U_{tot}(r)$ is made up of the repulsive optical potential $U_{opt}(r)$ and the attractive van der Waals cavity potential $U_{cav}(r)$. Then, the zone where the total potential $U_{tot}(r)$ is minimum is produced around the tip. Figure 11.19b shows an example of the potential to trap ^{85}Rb atoms, in which two normalized total potentials $U_{tot}(r)/\hbar\Gamma$ for both ^{85}Rb, $5S_{1/2}$, $F = 2$ and 3 ground states are plotted as a function of the distance r from the hemispherical surface. The radius of curvature a is taken to be 20 nm, while the incident power Q_0 and the blue detuning $\delta/2\pi$ are 40 mW and +1 GHz, respectively, with respect to the ^{85}Rb, $5S_{1/2}$, $F = 2$ lower ground state.

In this case, the ^{85}Rb atom and the dielectric tip form a sort of diatomic molecule, so that the atom is trapped in one of the quantized vibrational energy levels with an order n produced within the potential curve. In Fig. 11.19b, the vibrational energy levels with $n = 0$, 1, and 2 are drawn for both potentials in the harmonic oscillator approximation. The ^{85}Rb atom can make the transition between these vibrational levels via absorption and spontaneous emission of photons. However, according to the Frank–Condon principle [80], the ^{85}Rb atom trapped, for example, in the $n = 0$ level of the $F = 2$ lower ground state has little probability of transition to the $F = 3$ upper ground state because the wavefunction of the ^{85}Rb atom in the $F = 2$, $n = 0$ energy level does not overlap much with the other wavefunctions in the $F = 3$ energy levels. As a result, the atom will stay at the same energy level for rather a long time.

In the nano-scale region, gravity is less effective at confining atoms. In fact, the depth of the gravitational potential is only 20 nK in terms of temperature for a height change of 100 nm. Therefore, in the above scheme of an atom trap, no effective potential to confine an atom is produced along the tip surface, so that the atom will escape from the direction sooner or later. On the other hand, a two-dimensional trap in a horizontal plane can be produced in a similar way to a whispering gallery atom [81]. In this case, an atom will be dynamically trapped while rotating around the tip.

11.6.4 Three-Dimensional Atom Trap

Basically, an atom trap composed of a single sharpened fiber cannot three-dimensionally confine atoms in a nanometer region, in which the effect of a gravitational trap is extremely small: the stable minimum point of a potential to trap atoms cannot be three-dimensionally produced near the nanometric tip of a sharpened fiber with a simple structure. In addition, even if a nanometric sharpened fiber with a more complicated structure is employed, the tunnel effect due to the matter wave character of cold atoms will be crucial for their confinement. In the case of cold-atom manipulation, the spatial accuracy will be determined by the diffraction limit of the atomic de Broglie waves. From these circumstances, we need a moderate confined space to trap cold atoms tightly with matter wave characteristics. This need also results from the fact that the cavity QED effects, including the van der Waals force on atoms, are prominent in the near-field region. A feasible method for a three-dimensional atom trap is the use of multiple sharpened fibers.

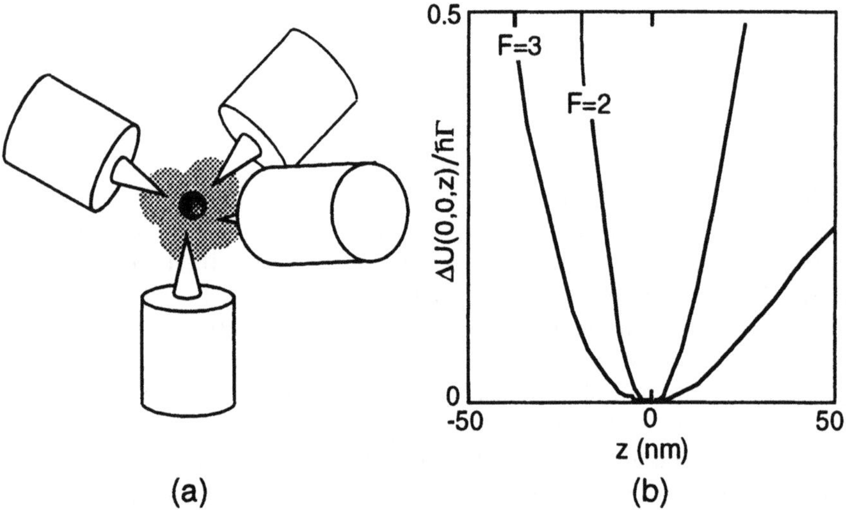

Fig. 11.20. Three-dimensional atom trap: a sketch of an atom trap with four sharpened fibers arranged tetrahedrally; b normalized cross-sectional profile $\Delta U(0,0,z)/\hbar\Gamma$ in the vertical direction of the potential composed of the repulsive optical potentials and the attractive van der Waals cavity potentials on ^{85}Rb as a function of the displacement z

Figure 11.20a shows a three-dimensional atom trap made up of four sharpened fibers arranged tetrahedrally. In this trap, an atom is captured at the center where the potential is minimum. The total potential $U^i_{\text{tot}}(\boldsymbol{r})$ produced by each sharpened fiber with a blue-detuned optical near field is the sum of the repulsive optical potential $U^i_{\text{opt}}(\boldsymbol{r})$ and the attractive van der Waals cavity potential $U^i_{\text{vdw}}(\boldsymbol{r})$: $U^i_{\text{tot}}(\boldsymbol{r}) = U^i_{\text{opt}}(\boldsymbol{r}) + U^i_{\text{vdw}}(\boldsymbol{r})$ with $i = 1, 2, 3$, and 4. Then, the three-dimensional trap potential is given by $U(\boldsymbol{r}) = \sum_{i=1}^{4} U^i_{\text{tot}}(\boldsymbol{r})$.

As an example, let us consider the case where the center of the hemispherical tip of each sharpened fiber is placed at the center of gravity of each triangular face of a tetrahedron with a side L. The four centers of gravity are at $\text{P}_1(0,0,-L/2\sqrt{6})$, $\text{P}_2(L/6\sqrt{3},-L/6,L/6\sqrt{6})$, $\text{P}_3(L/6\sqrt{3},L/6,L/6\sqrt{6})$, and $\text{P}_4(-L/3\sqrt{3},0,L/6\sqrt{6})$, if the origin of the coordinate is taken to the center of gravity of the tetrahedron. Figure 11.20b shows the cross-sectional profile

$\Delta U(0,0,z)$ in the vertical direction (the z-axis) of the potential change on a ^{85}Rb atom normalized to $\hbar\Gamma$ plotted as a function of the z displacement. Here, the length of a side is taken to be $L/a = 11$, with $a = 50$ nm, while the incident light power Q_0 and the blue-detuning δ are taken to be 2 mW and $2\pi \times 1$ GHz, respectively, with respect to the ^{85}Rb, $5S_{1/2}$, $F = 2$ lower ground state.

As shown in Fig. 11.20b, the potential with the minimum at the origin can be produced for both ^{85}Rb hyperfine ground states, although the potential for the upper ground state is somewhat shallow. A cold atom can be trapped in a quantized vibrational level. We also get a similar horizontal profile of the potential change. Since the atom stays in the light field for a long time, spontaneous transition can occur between both hyperfine states. If an atom trapped in the $F = 2$ lower state is transferred to the $F = 3$ upper state with a higher potential, it has the possibility of being heated due to the motion in the deeper potential curve. Then, if the atom is turned back to the lower state, it may escape from the shallower potential. Thus, the stability of the atom trap depends on the spontaneous transition events between the atomic states dressed by the optical near field. In other words, the stability will depend on the blue detuning, the field intensity, and so on.

11.7 Future Outlook

The study of atom manipulation with optical near field has been developing rapidly. For example, some theoretical work has been done by a Russian group [82]. Basically, this kind of atom manipulation deals with the dipole interaction between atoms and optical near field. In this case, the picture of atoms dressed by the optical near field is expected to be a good description of the mechanical interaction. This naturally leads us to the quantization of the optical near field. However, as has already been pointed out, field quantization meets some theoretical difficulties involving the handling of the vacuum–matter interface in the nanometric region. How should we define the photon number or the creation and annihilation operators of the optical near field? How should we express the field operator, including the light frequency and the quantization volume? How should we write the Lagrangian or the Hamiltonian of the *atom + optical near-field* system? Thus, the quantum mechanical treatment of the optical near field, at least regarding the interaction with atoms, is one of important issues that must be resolved.

In this connection, the atom deflection experiments with the scheme described in Sect. 11.6 will give us useful information on the atom–optical near-field interaction. The profile of optical near field has usually been investigated through destructive observation with a fiber probe. However, using atoms as a probe, we can investigate the optical near field nondestructively. Therefore, this scheme enables one to examine the interaction as a scattering problem or an inverse scattering problem. At the same time, we can also check the pertinence of the phenomenological Yukawa model through the study. This is very important for an understanding of the short-range electromagnetic interaction.

The cavity QED effect in the near-field region is also an important factor, but one that is full of open questions. In fact, an experimental result shows that the optical near-field space is strongly modified by vacuum–matter interaction due to the zero-point energy [83]. The hollow fiber and the sharpened fiber escorted by the optical near field are powerful tools for the experimental study of cavity QED effects. The scheme of atom deflection described in Sect. 11.6 can be applied to the investigation of attractive forces through comparison with the dipole force.

The realization of precise control of atomic motion gives rise to fruitful studies in the related area. The use of a bent hollow fiber enables one to carry atoms aiming at any point on a substrate. For this purpose, an AQW with a tapered hollow fiber has been proposed [84]. An AQW with a bent hollow fiber can also be used as a novel form of gravitational trap [85]. A hollow fiber escorted by a blue-detuned optical near field can be regarded not only as a waveguide of atomic de Broglie waves [16], but also as a cavity of an atom laser [86]. Moreover, a sharpened fiber with an optical near field can be used as quantum tweezers for an atom as well as for a nanometric particle [87].

The geometric study of optical near field is also very interesting. As shown in Sect. 11.6, the geometric arrangement of the optical near field is crucial for atom traps. The differential geometric or topological characters may lead us to a new field of research in optics.

The techniques of AQWs and AQDs also have industrial applications. The manipulation of silicon, which is one of the most important materials in the semiconductor industry, is the center of interest. For example, we will be able to carry out crystal growth of silicon on the atomic scale with near-field optical devices if a laser tuned to the guide-wavelength of 252 nm is developed. This technique is, of course, also useful in surface science research. In addition, near-field optical devices for atom manipulation are great tools for isotope separation, as seen in Sect. 11.4. These enable one to carry out isotope separation as neutral atoms while being in control of the number and the purity. It should be emphasized that it may be possible to manipulate molecules with the techniques developed here. Molecule manipulation also has wider applications, including isotope separation of carbon and oxygen that are important in the medical industry.

Thus, these advanced techniques of near-field optical atom manipulation will open new areas of scientific research, including quantum mechanical and quantum optical subjects, and industrial applications including nanofabrication.

References

1. C. Cohen-Tannoudji, in *Fundamental systems in quantum optics*, J. Dalibard, J. M. Raimond, J. Zinn-Justin (eds.), pp. 1–164 (Elsevier, Amsterdam, 1992)
2. P. J. Meystre, S. Stenholm (eds.), J. Opt. Soc. Am. B **2**, 1706 (1985); S. Chu, C. Wieman (eds.), J. Opt. Soc. Am. B **11**, 2019 (1989)
3. E. L. Raab, M. Prentiss, A. Cable, S. Chu, D. E. Pritchard, Phys. Rev. Lett. **59**, 2631 (1987)
4. M. H. Anderson, J. R. Ensher, M. R. Matthews, C. E. Wieman, E. A. Cornell, Science **269**, 198 (1995)
5. J. Mlynek, V. Balykin, P. Meystre (eds.), Appl. Phys. B **54**, 319 (1992)
6. V. I. Balykin, V. S. Letokhov, *Atom optics with laser light*, (Harwood Academic, Chur, 1995)
7. H. Haken, H. C. Wolf, *The physics of atoms and quanta*, 5th edn. (Springer-Verlag, Berlin, 1996)
8. J. P. Dowling, J. Gea-Banacloche, in *Advances in atomic, molecular, and optical physics*, Vol. 37, B. Bederson, H. Walther (eds.), pp. 1–94, (Academic Press, San Diego, 1996)
9. S. Feron, J. Reinhardt, M. Ducloy, O. Gorceix, S. Nic Chormaic, Ch. Miniatura, J. Robert, J. Baudon, V. Lorent, H. Haberland, Phys. Rev. A **49**, 4733 (1994)
10. V. I. Balykin, V. S. Letokhov, Yu. B. Ovchinnikov, A. I. Sidorov, Phys. Rev. Lett. **60**, 2137 (1988)
11. C. G. Aminoff, A. M. Steane, P. Bouyer, P. Desbiolles, J. Dalibard, C. Cohen-Tannoudji, Phys. Rev. Lett. **71**, 3083 (1993)
12. G. J. Liston, S. M. Tan, D. F. Wallis, Appl. Phys. B **60**, 211 (1995)
13. A. Landragin, G. Labeyrie, C. Henkel, R. Kaiser, N. Vansteenkiste, C. I. Westbrook, A. Aspect, Opt. Lett. **21**, 1591 (1996)
14. A. Landragin, J.-Y. Courtois, G. Labeyrie, N. Vansteenkiste, C. I. Westbrook, A. Aspect, Phys. Rev. Lett. **77**, 464 (1996)
15. T. Esslinger, M. Weidemüller, A. Hemmerich, T. W. Hänsch, Opt. Lett. **18**, 450 (1993)
16. C. M. Savage, S. Marksteiner, P. Zoller, in *Fundamentals of quantum optics III*, F. Ehlotzky (ed.) (Springer-Verlag, Berlin, 1993); S. Marksteiner, C. M. Savage, P. Zoller, S. L. Rolston, Phys. Rev. A **50**, 2680 (1994)
17. H. Ito, K. Sakaki, T. Nakata, W. Jhe, M. Ohtsu, Opt. Commun. **115**, 57 (1995)
18. H. Ito, K. Sakaki, T. Nakata, W. Jhe, M. Ohtsu, Ultramicroscopy **61**, 91 (1995)
19. M. J. Renn, E. A. Donley, E. A. Cornell, C. E. Wieman, D. Z. Anderson, Phys. Rev. A **53**, R648 (1996)
20. H. Ito, T. Nakata, K. Sakaki, M. Ohtsu, K. I. Lee, W. Jhe, Phys. Rev. Lett. **76**, 4500 (1996)

21. W. Jhe, M. Ohtsu, H. Hori, S. R. Friberg, Jpn. J. Appl. Phys. **33**, L1680 (1994)
22. M. A. Ol'Shanii, Yu. B. Ovchinnikov, V. S. Letokhov, Opt. Commun. **98**, 77 (1993)
23. M. J. Renn, D. Montgomery, O. Vdovin, D. Z. Anderson, C. E. Wieman, E. A. Cornell, Phys. Rev. Lett. **75**, 3253 (1995); M. J. Renn, A. A. Zozulya, E. A. Donley, E. A. Cornell, D. Z. Anderson, Phys. Rev. A **55**, 3684 (1997)
24. H. Hori, in *Near-field optics*, D. W. Pohl, D. Courjon (eds.), pp. 105–114 (Kluwer, Dordrecht, 1993); M. Ohtsu, S. Jiang, T. Pangaribuan, M. Kozuma, in *Near-field optics*, D. W. Pohl, D. Courjon (eds.), pp. 131–139 (Kluwer, Dordrecht, 1993)
25. S. Sudo, I. Yokohama, H. Yasaka, Y. Sakai, T. Ikegami, IEEE Photon. Technol. Lett. **2**, 128 (1990)
26. D. Marcuse, *Light transmission optics*, 2nd edn. (Robert E. Krieger, Malabar, 1989); *Theory of dielectric optical waveguides*, 2nd edn. (Academic Press, San Diego, 1991)
27. G. P. Agrawal, *Nonlinear fiber optics*, 2nd edn. (Academic Press, San Diego, 1995)
28. N. F. Ramsey, *Molecular beams* (Oxford University Press, London, 1956)
29. W. Happer, Rev. Mod. Phys. **44**, 169 (1972)
30. H. Ito, K. Sakaki, W. Jhe, M. Ohtsu, Appl. Phys. Lett. **70**, 2496 (1997)
31. V. S. Letokhov, *Laser photoionization spectroscopy* (Academic Press, San Diego, 1987)
32. G. W. Drake (ed.), *Atomic, molecular, and optical physics handbook* (AIP Press, New York, 1996)
33. T. P. Dinneen, C. D. Wallace, Kit-Yan N. Tan, P. L. Gould, Opt. Lett. **17**, 1706 (1992)
34. M. Chevrollier, M. Fichet, M. Oria, G. Rahmat, D. Bloch, M. Ducloy, J. Phys. II France **2**, 631 (1992)
35. P. R. Berman (ed.), *Cavity quantum electrodynamics* (Academic Press, San Diego, 1994)
36. P. W. Milonni, *The quantum vacuum* (Academic Press, Boston, 1994)
37. H. B. G. Casimir, D. Polder, Phys. Rev. **73**, 360 (1948)
38. V. Sandoghdar, C. I. Sukenik, E. A. Hinds, S. Haroche, Phys. Rev. Lett. **68**, 3432 (1992); C. I. Sukenik, M. G. Boshier, D. Cho, V. Sandoghdar, E. H. Hinds, Phys. Rev. Lett. **70**, 560 (1993)
39. S. K. Lamoreaux, Phys. Rev. Lett. **78**, 5 (1997)
40. W. Jhe, J. W. Kim, Phys. Rev. A **51**, 1150 (1995); Phys. Lett. A **197**, 192 (1995)
41. K. Jiang, W. Jhe, Phys. Rev. A **53**, 1126 (1996)
42. V. V. Klimov, M. Ducloy, V. S. Letokhov, J. Mod. Opt. **43**, 549 (1996)
43. J. M. Wylie, J. E. Sipe, Phys. Rev. A **30**, 1185 (1984)
44. J.-Y. Courtois, J.-M. Courty, J. C. Mertz, Phys. Rev. A **53**, 1862 (1996)

45. H. Nha, W. Jhe, Phys. Rev. A **54**, 3505 (1996)
46. H. Ito, K. Sakaki, W. Jhe, M. Ohtsu, in *Atom optics*, M. G. Prentiss, W. D. Phillips (eds.), Proceedings SPIE 2995, 138 (1997)
47. H. Nha, W. Jhe, Phys. Rev. A **56**, 2213 (1997)
48. H. Ito, K. Sakaki, W. Jhe, M. Ohtsu, Opt. Commun. **141**, 43 (1997)
49. G. Timp, R. E. Behringer, D. M. Tennant, J. E. Cunningham, M. Prentiss, K. K. Berggren, Phys. Rev. Lett. **69**, 1636 (1992)
50. J. J. McClelland, R. E. Scholten, E. C. Palm, R. J. Celotta, Science **262**, 877 (1993); R. Gupta, J. J. McClelland, Z. J. Jabbour, R. J. Celotta, Appl. Phys. B **67**, 1378 (1995)
51. R. W. McGowan, D. M. Giltner, S. A. Lee, Opt. Lett. **20**, 2535 (1995)
52. K. S. Johnson, A. Chu, T. W. Lynn, K. K. Berggren, M. S. Shahriar, M. Prentiss, Opt. Lett. **20**, 1310 (1995); K. S. Johnson, A. P. Chu, K. K. Berggren, M. Prentiss, Opt. Commun. **126**, 326 (1996); A. P. Chu, K. S. Johnson, M. Prentiss, Opt. Commun. **134**, 105 (1997)
53. K. K. Berggren, A. Bard, J. L. Wilbur, J. D. Gillaspy, A. G. Helg, J. J. McClelland, S. L. Rolston, W. D. Phillips, M. Prentiss, G. M. Whitesides, Science **269**, 1255 (1995); K. S. Johnson, K. K. Berggren, A. Black, C. T. Black, A. P. Chu, N. H. Dekker, D. C. Raiph, J. H. Thywissen, R. Younkin, M. Tinkham, M. Prentiss, G. M. Whitesides, Appl. Phys. Lett. **69**, 2773 (1996)
54. J. Fujita, M. Morinaga, T. Kishimoto, M. Yasuda, S. Matsui, F. Shimizu, Nature **380**, 691 (1996); M. Morinaga, M. Yasuda, T. Kishimoto, F. Shimizu, J. Fujita, S. Matsui, Phys. Rev. Lett. **77**, 802 (1996)
55. G. Scoles (ed.), *Atomic and molecular beam methods* (Oxford University Press, London, 1988)
56. C. Monroe, W. Swann, H. Robinson, C. Wieman, Phys. Rev. Lett. **65**, 1571 (1990)
57. K. B. Davis, M. O. Mewes, M. A. Joffe, M. R. Andrews, W. Ketterle, Phys. Rev. Lett. **74**, 5202 (1995)
58. M. R. Anderson, J. R. Ensher, M. R. Matthews, C. E. Wieman, E. A. Cornell, Science **269**, 198 (1995)
59. K. B. Davis, M.-O. Mewes, M. R. Andrews, N. J. van Druten, D. S. Durfee, D. M. Kurn, W. Ketterle, Phys. Rev. Lett. **75**, 3969 (1995)
60. C. C. Bradley, C. A. Sackett, J. J. Tollett, R. G. Hilet, Phys. Rev. Lett. **75**, 1687 (1995)
61. M. R. Andrews, C. G. Townsend, H.-J. Miesner, D. S. Durfee, D. M. Kurn, W. Ketterle, Science **275**, 637 (1997); M.-O. Mewes, M. R. Andrews, D. M. Kurn, D. S. Durfee, C. G. Townsend, W. Ketterle, Phys. Rev. Lett. **78**, 582 (1997)
62. H. Ito, K. Sakaki, W. Jhe, M. Ohtsu, Phys. Rev. A **56**, 712 (1997)
63. H. S. Lee, B. W. Stewart, K. Choi, H. Fenichel, Phys. Rev. A **49**, 4922 (1994)
64. K. T. Gahagan, G. A. Swartzlander, Jr., Opt. Lett. **21**, 827 (1996)

65. T. Kuga, Y. Torii, N. Shiokawa, T. Hirano, Y. Shimizu, H. Sasada, Phys. Rev. Lett. **78**, 4713 (1997)
66. J. Yin, H.-R. Noh, K.-I. Lee, K.-H. Kim, Y.-Z. Wang, W. Jhe, Opt. Commun. **138**, 287 (1997)
67. E. Riis, D. S. Weiss, K. A. Moler, S. Chu, Phys. Rev. Lett. **64**, 1658 (1990)
68. J. Yu, J. Djemaa, P. Nosbaum, P. Pillet, Opt. Commun. **112**, 136 (1994)
69. T. B. Swanson, N. J. Silva, S. K. Mayer, J. J. Maki, D. H. McIntyre, J. Opt. Soc. Am. B **13**, 1833 (1996)
70. Z. T. Lu, K. L. Corwin, M. J. Renn, M. H. Anderson, E. A. Cornell, C. E. Wieman, Phys. Rev. Lett. **77**, 3331 (1996)
71. J. Söding, R. Grimm, Yu. B. Ovchinnikov, Opt. Commun. **119**, 652 (1995)
72. P. Desbiolles, M. Arndt, P. Szriftgiser, J. Dalibard, Phys. Rev. A **54**, 4292 (1996)
73. H. Nha, W. Jhe, Phys. Rev. A **56**, 729 (1997)
74. C. Cohen-Tannoudji, J. Dupont-Roc, G. Grynberg, *Atom-photon interactions* (Wiley-Interscience, New York, 1992)
75. M. A. Kasevich, D. S. Weiss, S. Chu, Opt. Lett. **15**, 607 (1990)
76. K. I. Lee, J. A. Kim, H. R. Noh, W. Jhe, Opt. Lett. **21**, 1177 (1996); J. A. Kim, K. I. Lee, H. R. Noh, W. Jhe, M. Ohtsu, Opt. Lett. **22**, 117 (1997)
77. M. Ohtsu, H. Hori, *Near-field nano-optics* (Plenum, New York, 1998)
78. U. Dürig, D. W. Pohl, F. Rohner, J. Appl. Phys. **59**, 3318 (1986)
79. R. G. Newton, *Scattering theory of waves and particles*, 2nd edn. (Springer-Verlag, New York, 1982)
80. L. D. Landau, E. M. Lifshitz, *Quantum mechanics – Non-relativistic theory*, 3rd edn. (Pergamon, Oxford, 1975)
81. H. Mabuchi, H. J. Kimble, Opt. Lett. **19**, 749 (1994); D. W. Vernooy, H. J. Kimble, Phys. Rev. A **55**, 1239 (1997)
82. V. V. Klimov, V. S. Letokhov, Opt. Commun. **121**, 130 (1995); J. Mod. Opt. **42**, 1485 (1995); JETP **81**, 49 (1995); Laser Phys. **6**, 475 (1996)
83. T. Saiki, M. Ohtsu, K. Jang, W. Jhe, Opt. Lett. **21**, 674 (1996)
84. M. V. Subbotin, V. I. Balykin, D. V. Laryushin, V. S. Letokhov, Opt. Commun. **139**, 107 (1997)
85. D. J. Harris, C. M. Savage, Phys. Rev. A **51** (1995) 3967
86. G. M. Moy, J. J. Hope, C. M. Savage, Phys. Rev. A **55**, 3631 (1997)
87. L. Novotny, R. X. Bian, X. S. Xie, Phys. Rev. Lett. **79**, 645 (1997)

Chapter 12

Related Theories

12.1 Comparison of Theoretical Approaches

Several approaches have been proposed for optical near-field problems. These have been used for analyzing experimental results as well as in trying to understand the phenomena observed in the experiments and their underlying physics.

Tables 12.1 categorize and summarize such approaches in terms of their theoretical framework, method, and viewpoint. As emphasized in Chap. 1, here one should keep in mind that sample–probe interaction via the optical near field is essential for near-field optical microscopy and spectroscopy, and that a three-dimensional treatment is essential. As pointed out in Chaps. 2 and 6, resolution, contrast, and sensitivity are also very closely related to the probe's apex size and the cone angle, the approach distance, illumination conditions (such as the polarization of incident light), and the scanning method (such as constant-height mode or constant-distance mode). Thus, the most desirable theoretical approach is one that can take account of all these considerations consistently, as well as provide physical insights.

Macroscopic approaches require electromagnetic boundary conditions in order to solve the equations, but it is generally difficult in macroscopic approaches to set boundary conditions for arbitrary three-dimensional objects, instead of using numerical point-to-point matching. Owing to the high cost of such approaches in computational time, they are usually applied only to two-dimensional problems. Moreover, they cannot systematically provide any spectroscopic information on nanometric/mesoscopic systems; nor can they handle quantum systems such as single atoms/molecules, quantum dots, and internal degrees of freedom such as spin/angular momentum. Semi-microscopic and microscopic approaches, on the other hand, can do these things, although they also require numerical computation that takes a time which is roughly proportional to the third power of the space to be considered.

We will therefore concentrate on the semi-microscopic and microscopic approaches in the following sections. Readers interested in a specific approach listed in Tables 12.1, are recommended to refer to the papers cited, review articles [1–3], and books [4–6], all of which are publicly available.

Table 12.1. Comparison of theoretical approaches

Category	Method	Object analyzed				
		Probe or sample	Probe–sample system			
		Field distribution	Field distribution	Scan intensity	Spectroscopy	Force
Macroscopic (classical electrodynamics)	**Diffraction theory**	Rayleigh[7] Sommerfeld[8] Born[9] Jackson[10] Bethe[11] Bouwkamp[12] Leviatan[13] Roberts[14]				
	MMP(2D) **MMP(3D)**	Novotny[15] Novotny[18]		Pohl[16] Novotny [17]		Novotny[19]
	Mie scattering	Mie[20] Hulst[21] Newton[22] Prasad[24]		Barchiesi[23]		
	Grating theory (perturbation)	Petit[25] Labeke[26] Sentenac[27] Goudonnet[28] Barchiesi[29]		Labeke [30] Wang[31]		
	BEM(2D)	Tanaka[32]				Kawata [33]
	Green dyadic (2D)		Dereux[34]			
	Non-global	Depasse[35]		Jang [36] Zvyagin [37,38]		
	FDTD (2D) **FDTD (3D)**		Christensen[39] Kann[40]	Kann[41] Furukawa[42]	Baian [43]	
	Intuitive	Denk[44]		Banno[45]		

[cont'd]

Table 12.1. cont'd

Category	Method	Object analyzed				
		Probe or sample	Probe–sample system			
		Field distribution	Field distribution	Scan intensity	Spectroscopy	Force
Fourier optic	**Angular spectrum**	Weyl[46] Wolf[47] Sheppard[48] Madrazo[49]		Ohtsu[6] Berntsen [50] Inoue[51]		
	Spectrum analysis			Uma[52]		
Semi-microscopic	**Coupled dipole**	Lukosz[53] Taubenblatt[54]		Novotny [56]	Novotny [55]	
	Image dipole			Jhe[57]	Ruppin [58]	
	Linear response	Kubo[92]		Kobayashi [59-61]		
Microscopic	**Self-consistent field**	Agarwal[62]		Girard [63,65] Labani[64] Keller[66] Li[69]	Girard [67]	Girard [68]
	Iterative	Martin [70-72]	Martin [73]	Hanewinkel[91]		
	Semi-quantum theory	Cho[74]	Cho[75]		Girard[76] Cho[77] Kobayashi[78]	
	Intuitive	Vigoureux [79]		Hori[80] Ohtsu[6]		Hori [80] Ohtsu [2]

12.2 Semi-microscopic and Microscopic Approaches

12.2.1 Basic Equations

In semi-microscopic and microscopic approaches, an optical field is represented by an electric field satisfying the Maxwell equations (classical), while a material system (probe and sample) is described in either classical or quantum mechanical terms. On the assumption that there are no external charges and currents, and that the time–dependence is the homogeneous one, $\exp(-i\omega t)$, the Maxwell equations read [8–10, 22]

$$\nabla \cdot [\boldsymbol{E}(\boldsymbol{r},\omega) + 4\pi \boldsymbol{P}(\boldsymbol{r},\omega)] = 0 \tag{12.1}$$

$$\nabla \cdot \boldsymbol{H}(\boldsymbol{r},\omega) = 0 \tag{12.2}$$

$$\nabla \times \boldsymbol{E}(\boldsymbol{r},\omega) = i\left(\frac{\omega}{c}\right)\boldsymbol{H}(\boldsymbol{r},\omega) \tag{12.3}$$

$$\nabla \times \boldsymbol{H}(\boldsymbol{r},\omega) = -i\left(\frac{\omega}{c}\right)[\boldsymbol{E}(\boldsymbol{r},\omega) + 4\pi \boldsymbol{P}(\boldsymbol{r},\omega)] \tag{12.4}$$

The vector wave equation for the electric field $\boldsymbol{E}(\boldsymbol{r},\omega)$ is readily obtained by taking a rotation of Eq. 12.3, and with the help of Eq. 12.4:

$$\nabla \times \nabla \times \boldsymbol{E}(\boldsymbol{r},\omega) = \left(\frac{\omega}{c}\right)^2 [\boldsymbol{E}(\boldsymbol{r},\omega) + 4\pi \boldsymbol{P}(\boldsymbol{r},\omega)] = \left(\frac{\omega}{c}\right)^2 \boldsymbol{D}(\boldsymbol{r},\omega) \tag{12.5}$$

that is,

$$\nabla \times \nabla \times \boldsymbol{E}(\boldsymbol{r},\omega) - \left(\frac{\omega}{c}\right)^2 \boldsymbol{E}(\boldsymbol{r},\omega) = 4\pi \left(\frac{\omega}{c}\right)^2 \boldsymbol{P}(\boldsymbol{r},\omega) \tag{12.6}$$

or

$$\nabla[\nabla \cdot \boldsymbol{E}(\boldsymbol{r},\omega)] - \left[\Delta + \left(\frac{\omega}{c}\right)^2\right] \boldsymbol{E}(\boldsymbol{r},\omega) = 4\pi \left(\frac{\omega}{c}\right)^2 \boldsymbol{P}(\boldsymbol{r},\omega) \tag{12.7}$$

Alternatively, we have

$$\boldsymbol{D}(\boldsymbol{r},\omega) = \epsilon(\boldsymbol{r},\omega)\boldsymbol{E}(\boldsymbol{r},\omega), \quad V(\boldsymbol{r},\omega) \equiv \left(\frac{\omega}{c}\right)^2 [\epsilon(\boldsymbol{r},\omega) - \epsilon_{\text{ref}}] \tag{12.8}$$

$$\nabla \times \nabla \times \boldsymbol{E}(\boldsymbol{r},\omega) - \left(\frac{\omega}{c}\right)^2 \epsilon_{\text{ref}} \boldsymbol{E}(\boldsymbol{r},\omega) = V(\boldsymbol{r},\omega)\boldsymbol{E}(\boldsymbol{r},\omega) \tag{12.9}$$

Mathematically, Eqs. 12.6 and 12.9 have the same structure, namely, an inhomogeneous equation with the source term (right-hand side); the only difference is in their viewpoint. This might be easier to understand if the equations are transformed into the integral form as

$$\begin{aligned}
\boldsymbol{E}(\boldsymbol{r},\omega) &= \boldsymbol{E}_0(\boldsymbol{r},\omega) + \int \mathbb{T}(\boldsymbol{r},\boldsymbol{r}',\omega) \cdot \boldsymbol{P}(\boldsymbol{r}',\omega)\,\mathrm{d}^3 r' && (12.10)\\
&= \boldsymbol{E}_0(\boldsymbol{r},\omega) + \int \mathbb{G}(\boldsymbol{r},\boldsymbol{r}',\omega) \cdot V(\boldsymbol{r}',\omega)\boldsymbol{E}(\boldsymbol{r}',\omega)\,\mathrm{d}^3 r' && (12.11)
\end{aligned}$$

in terms of the field propagator $\mathbb{T}(\boldsymbol{r},\boldsymbol{r}',\omega)$ in free space. Here $\boldsymbol{E}_0(\boldsymbol{r},\omega)$ is a solution of the homogeneous equation, or incident field, and the field propagator satisfies the following equations:

$$\nabla\times\nabla\times\mathbb{T}(\boldsymbol{r},\boldsymbol{r}',\omega)-\left(\frac{\omega}{c}\right)^2\mathbb{T}(\boldsymbol{r},\boldsymbol{r}',\omega)=4\pi\left(\frac{\omega}{c}\right)^2\delta(\boldsymbol{r}-\boldsymbol{r}') \quad (12.12)$$

$$\mathbb{T}(\boldsymbol{r},\boldsymbol{r}',\omega)=4\pi\left(\frac{\omega}{c}\right)^2\mathbb{G}(\boldsymbol{r},\boldsymbol{r}',\omega) \quad (12.13)$$

Equation 12.10 emphasizes the optical response of the infinitesimal fluctuating volume of polarized matter $\boldsymbol{P}(\boldsymbol{r},\omega)$, or approximately the dipole moments, while Eq. 12.11 treats it as light scattering from a mesoscopic/nanometric probe apex–sample system perturbed by $V(\boldsymbol{r},\omega)$. If we note the optical linear response of the material with susceptibility $\chi(\boldsymbol{r},\omega)$,

$$\boldsymbol{P}(\boldsymbol{r},\omega)=\chi(\boldsymbol{r},\omega)\,\boldsymbol{E}(\boldsymbol{r},\omega) \quad (12.14)$$

we can rewrite Eq. 12.10 as

$$\boldsymbol{E}(\boldsymbol{r},\omega)=\boldsymbol{E}_0(\boldsymbol{r},\omega)+\int\mathbb{T}(\boldsymbol{r},\boldsymbol{r}',\omega)\cdot\chi(\boldsymbol{r}',\omega)\,\boldsymbol{E}(\boldsymbol{r}',\omega)\,\mathrm{d}^3r' \quad (12.15)$$

It follows from Eqs. 12.11 and 12.15 that there is no need to impose the additional electromagnetic boundary conditions required for the macroscopic approaches, and that the interactions between probe and sample are self-consistently taken into consideration. The latter can be seen more explicitly if the susceptibility is set as

$$\chi(\boldsymbol{r},\omega)=\sum_{i=\text{probe}}^{\text{sample}}\alpha_i(\omega)\,\delta(\boldsymbol{r}-\boldsymbol{r}_i) \quad (12.16)$$

by using the point-like polarizability $\alpha_i(\omega)$; the effective field in the sample–probe system is then given by

$$\boldsymbol{E}(\boldsymbol{r},\omega)=\boldsymbol{E}_0(\boldsymbol{r},\omega)+\sum_{i=\text{probe}}^{\text{sample}}\mathbb{T}(\boldsymbol{r},\boldsymbol{r}_i,\omega)\cdot\alpha_i(\omega)\,\boldsymbol{E}(\boldsymbol{r}_i,\omega) \quad (12.17)$$

This is especially preferable from the atomic/molecular viewpoint. As will be shown in Sect. 12.2.4, one can solve optical near-field problems in either a classical or a hybrid way by evaluating the susceptibility with the help of quantum mechanics. Moreover, there are several possible ways of obtaining the effective field numerically by discretizing in

1. coordinate (r-) space,
2. Fourier (q-) space, or
3. partial wave (l-) space.

We will work in r–space in the following sections, because we intend to focus on sub-wavelength phenomena.

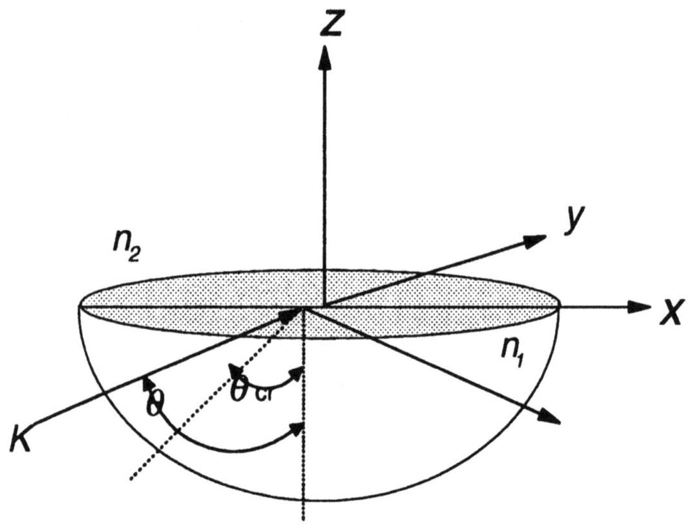

Fig. 12.1. Modeling coordinate system

12.2.2 Example of an Evanescent Field

In the previous section we described the basic equations that we rely on, and outlined how to approach optical near-field problems from the semi-microscopic and microscopic points of view. Before going into detail, let us give an example of an evanescent field generated at a flat interface by total internal reflection (see Fig. 12.1). This corresponds to $\boldsymbol{E}_0(\boldsymbol{r},\omega)$ in Eqs. 12.10 and 12.11.

According to the Fresnel formula [9, 10], a p-polarized electric field (parallel to the plane of incidence) can be expressed as

$$\boldsymbol{E}_0(\boldsymbol{r},\omega) = E_p \boldsymbol{e}_p \exp\left[i\left(\boldsymbol{K}\cdot\boldsymbol{r} - \omega t\right)\right] \tag{12.18}$$

$$\boldsymbol{e}_p = \left[-i\sqrt{\left(n_1 \sin\theta/n_2\right)^2 - 1}, 0, \sin\theta/n_2\right] N_p \tag{12.19}$$

$$E_p = \frac{2n_1\cos\theta}{n_2\cos\theta + in_1\sqrt{\left(n_1/n_2\right)^2\sin^2\theta - 1}} \tag{12.20}$$

while an s-polarized field (perpendicular to the plane of incidence) can be written in the form

$$\boldsymbol{E}_0(\boldsymbol{r},\omega) = E_s \boldsymbol{e}_s \exp\left[i\left(\boldsymbol{K}\cdot\boldsymbol{r} - \omega t\right)\right] \quad \boldsymbol{e}_s = [0,1,0] \tag{12.21}$$

$$E_s = \frac{2n_1\cos\theta}{n_1\cos\theta + in_2\sqrt{\left(n_1/n_2\right)^2\sin^2\theta - 1}} \tag{12.22}$$

Here the incident angle θ exceeds the critical angle $\theta_{cr} = \sin^{-1}(n_2/n_1)$, and the complex wave number is given by

$$\boldsymbol{K} = \frac{2\pi n_1}{\lambda}\left[\sin\theta, 0, i\sqrt{\sin^2\theta - (n_2/n_1)^2}\right] \tag{12.23}$$

In Eq. 12.19, the normalization constant is designated N_p. It follows from the above equations that the amplitude of the evanescent field drops off exponentially in a direction normal to the surface (z-direction) as it penetrates the less dense medium (n_2), whose length is given by $\left|K_z^{-1}\right|$.

12.2.3 Direct and Indirect Field Propagators

In Sect. 12.2.1, we used the field propagator $\mathbb{T}(\boldsymbol{r}, \boldsymbol{r}', \omega)$ in free space to obtain the optical effective field. If we include the reflection effect from the substrate that supports the sample, we have to introduce another field propagator $\mathbb{T}_{\text{indirect}}(\boldsymbol{r}, \boldsymbol{r}', \omega)$. To avoid confusion, the former will be denoted as $\mathbb{T}_{\text{direct}}(\boldsymbol{r}, \boldsymbol{r}', \omega)$, and thus $\mathbb{T}(\boldsymbol{r}, \boldsymbol{r}', \omega)$ in Sect. 12.2.1 should be replaced by the sum of both propagators, as follows:

$$\mathbb{T}(\boldsymbol{r}, \boldsymbol{r}', \omega) = \mathbb{T}_{\text{direct}}(\boldsymbol{r}, \boldsymbol{r}', \omega) + \mathbb{T}_{\text{indirect}}(\boldsymbol{r}, \boldsymbol{r}', \omega) \tag{12.24}$$

Let us now show the explicit form of Eq. 12.24. Noting that the field propagator given by

$$\mathbb{G}_0(\boldsymbol{r}, \boldsymbol{r}') = \frac{\exp\left(iq\left|\boldsymbol{r} - \boldsymbol{r}'\right|\right)}{|\boldsymbol{r} - \boldsymbol{r}'|} \tag{12.25}$$

is a solution of the equation

$$\left(\nabla^2 + q^2\right)\mathbb{G}_0(\boldsymbol{r}, \boldsymbol{r}') = -4\pi\delta(\boldsymbol{r} - \boldsymbol{r}') \tag{12.26}$$

we have

$$\mathbb{T}_{\text{direct}}(\boldsymbol{r}, \boldsymbol{r}', \omega) = \left(q^2\mathbb{I} + \nabla\nabla\right)\cdot\mathbb{G}_0(\boldsymbol{r}, \boldsymbol{r}'), \quad q = \left(\frac{\omega}{c}\right) \tag{12.27}$$

Here the unit matrix is denoted as $\mathbb{I}$. By simple algebraic calculation, we can verify that Eq. 12.27 satisfies Eq. 12.12, which is what we want. By setting $\boldsymbol{R} = \boldsymbol{r} - \boldsymbol{r}'$ and $R = |\boldsymbol{r} - \boldsymbol{r}'|$, and using the relation

$$\nabla_R \cdot \frac{\exp(iqR)}{R}\boldsymbol{P}(\boldsymbol{r}', \omega) = \boldsymbol{P}(\boldsymbol{r}', \omega)\cdot\nabla_R\frac{\exp(iqR)}{R} \tag{12.28}$$

$$= \boldsymbol{P}(\boldsymbol{r}', \omega)\cdot\left\{(iq)\frac{\exp(iqR)}{R} - \frac{\exp(iqR)}{R^2}\right\}\left(\frac{\boldsymbol{R}}{R}\right) \tag{12.29}$$

we can obtain the following explicit form of the direct part:

$$\mathbb{T}_{\text{direct}}(\boldsymbol{r}, \boldsymbol{r}', \omega) = \left\{\mathbb{T}_3(\boldsymbol{r}, \boldsymbol{r}') + \mathbb{T}_2(\boldsymbol{r}, \boldsymbol{r}') + \mathbb{T}_1(\boldsymbol{r}, \boldsymbol{r}')\right\}\exp(iqR) \tag{12.30}$$

with

$$\mathbb{T}_3(\boldsymbol{r},\boldsymbol{r}') \equiv \frac{3\boldsymbol{R}\boldsymbol{R}\cdot - R^2\mathbb{I}}{R^5} \tag{12.31}$$

$$\mathbb{T}_2(\boldsymbol{r},\boldsymbol{r}') \equiv (-iq)\left(\frac{3\boldsymbol{R}\boldsymbol{R}\cdot - R^2\mathbb{I}}{R^4}\right) \tag{12.32}$$

$$\mathbb{T}_1(\boldsymbol{r},\boldsymbol{r}') \equiv q^2\left(\frac{R^2\mathbb{I} - \boldsymbol{R}\boldsymbol{R}\cdot}{R^3}\right) \tag{12.33}$$

This consists of three terms, each of which is responsible for quasi-static, radiative, and far-field propagation, respectively.

Now let us turn to the indirect part. Assume a point dipole moment $\boldsymbol{p}(\omega)$ at $\boldsymbol{r}_0$ above the infinite flat plane which divides the space into [I] $z > 0$ and [II] $z < 0$. In order to find the electric field at $\boldsymbol{r}$ (close to the source point $\boldsymbol{r}_0$) as $\boldsymbol{\epsilon}^{(+)}(\boldsymbol{r},\omega) = \mathbb{T}_{\text{indirect}}(\boldsymbol{r},\boldsymbol{r}_0,\omega)\cdot\boldsymbol{p}(\omega)$, we will expand the optical field in regions [I] and [II] in terms of the plane wave basis [62], using what is called angular spectrum representation.

$$\boldsymbol{E}^{(1)}(\boldsymbol{r},\omega) = \left[q^2\mathbb{I} + \nabla(\boldsymbol{p}(\omega)\cdot\nabla)\right]\mathbb{G}_0(\boldsymbol{r},\boldsymbol{r}') + \int \boldsymbol{\epsilon}^{(+)}(u,v;\omega)\exp(i\boldsymbol{K}_0\cdot\boldsymbol{r})\,\mathrm{d}u\mathrm{d}v \tag{12.34}$$

$$\boldsymbol{E}^{(2)}(\boldsymbol{r},\omega) = \int \boldsymbol{\epsilon}^{(2)}(u,v;\omega)\exp(i\boldsymbol{K}'\cdot\boldsymbol{r})\,\mathrm{d}u\mathrm{d}v \tag{12.35}$$

The first term of Eq. 12.34 represents the radiation field due to the dipole moment $\boldsymbol{p}(\omega)$. The second term and the right-hand side of Eq. 12.35 indicate the fields reflected from, and transmitted to, the interface. It is also convenient to expand $\mathbb{G}_0(\boldsymbol{r},\boldsymbol{r}')$ similarly,

$$\mathbb{G}_0(\boldsymbol{r},\boldsymbol{r}') = \frac{i}{2\pi}\int\frac{\mathrm{d}u\mathrm{d}v}{w_0}\exp\left[iu(x-x') + iv(y-y') + iw_0|z-z'|\right] \tag{12.36}$$

in order to satisfy the continuity conditions of both electric and magnetic fields at $z = 0$. After expressing $\boldsymbol{\epsilon}^{(+)}(u,v;\omega)$ in terms of $\boldsymbol{p}(\omega)$ and $\boldsymbol{K}_{\parallel} = (u,v)$, we can obtain $\mathbb{T}_{\text{indirect}}$ in (u,v) space. If we take a limit as $c \to \infty$, neglecting the retardation effect, we obtain the compact form of the indirect part:

$$\mathbb{T}_{\text{indirect}}(\boldsymbol{r},\boldsymbol{r}',\omega) \to \frac{1}{2\pi}\frac{\epsilon(\omega)-1}{\epsilon(\omega)+1} \cdot \int\frac{\mathrm{d}u\mathrm{d}v}{K_{\parallel}}\exp\left[i\boldsymbol{K}_{\parallel}\cdot\left(\boldsymbol{r}_{\parallel}-\boldsymbol{r}'_{\parallel}\right) - K_{\parallel}(z+z')\right] \times \begin{pmatrix} u^2 & uv & -iK_{\parallel}u \\ uv & v^2 & -iK_{\parallel}v \\ iK_{\parallel}u & iK_{\parallel}v & K_{\parallel}^2 \end{pmatrix} \tag{12.37}$$

$$= \frac{\epsilon(\omega)-1}{\epsilon(\omega)+1}\mathbb{T}_3(\boldsymbol{r},\boldsymbol{r}'_{\mathrm{M}})\cdot\begin{pmatrix}-1 & 0 & 0\\ 0 & -1 & 0\\ 0 & 0 & 1\end{pmatrix} \tag{12.38}$$

This gives us a very clear physical interpretation, showing the field propagation from the image point $\boldsymbol{r}'_{\mathrm{M}} = (x', y', -z')$ of the dipole location $\boldsymbol{r}' = (x', y', z')$ to the observation point $\boldsymbol{r} = (x, y, z)$.

12.2.4 Electric Susceptibility of Matter

In this section we discuss the electric susceptibility of matter with the help of quantum theory. Following the usual linear response theory [82, 89], we initially assume that a fixed external optical field $\boldsymbol{E}(\boldsymbol{r}, t)$ is imposed on an electron many-particle system within the first order in $\boldsymbol{E}(\boldsymbol{r}, t)$. The total Hamiltonian for each electron (charge e, mass M, momentum $\boldsymbol{p}$) is given in the minimal coupling [88]

$$H = H_0 + H_1 \tag{12.39}$$

$$H_0 \equiv \frac{\boldsymbol{p}^2}{2M} + U(\boldsymbol{r}), \quad H_1 \equiv -\frac{e}{2Mc}(\boldsymbol{p}\cdot\boldsymbol{A} + \boldsymbol{A}\cdot\boldsymbol{p}) \tag{12.40}$$

where the potential energy is denoted as $U(\boldsymbol{r})$. Using a Fourier transform of the external field $\boldsymbol{E}(\boldsymbol{r}, t)$ and vector potential $\boldsymbol{A}(\boldsymbol{r}, t)$ with Coulomb gauge,

$$\boldsymbol{E}(\boldsymbol{r}, t) = \boldsymbol{E}(\boldsymbol{r}) \int_{-\infty}^{\infty} \mathrm{d}\omega \tilde{E}(\omega) \exp(-i\omega t) \tag{12.41}$$

$$\begin{aligned}\boldsymbol{A}(\boldsymbol{r}, t) &= -c\int_{-\infty}^{t} \mathrm{d}t' \boldsymbol{E}(\boldsymbol{r}, t') \\ &= c\boldsymbol{E}(\boldsymbol{r}) \int_{-\infty}^{\infty} \mathrm{d}\omega \tilde{E}(\omega)\left(\frac{-i}{\omega + i\eta}\right) \exp(-i\omega t)\end{aligned} \tag{12.42}$$

we obtain the Fourier transform of H_1 as

$$\tilde{H}_1(\boldsymbol{r}, \omega) = \frac{ie\hbar}{2M}(\nabla\cdot\boldsymbol{E}(\boldsymbol{r}) + \boldsymbol{E}(\boldsymbol{r})\cdot\nabla)\tilde{E}(\omega)\left(\frac{-i}{\omega + i\eta}\right) \tag{12.43}$$

Since the expectation value of any one-body operator O_1 evolving according to the time-dependent Schrödinger equation can be expressed as $\langle O_1 \rangle = \mathrm{Tr}\{\rho(t) O_1\}$, the expectation value of the current (density $n(\boldsymbol{r})$ and velocity $\boldsymbol{v}$),

$$\boldsymbol{j}(\boldsymbol{r}, t) = \frac{e[n(\boldsymbol{r})\boldsymbol{v} + \boldsymbol{v}n(\boldsymbol{r})]}{2} \tag{12.44}$$

$$= \left(\frac{e}{2M}\right)\left\{n(\boldsymbol{r})\left[\boldsymbol{p} - \frac{e}{c}\boldsymbol{A}(\boldsymbol{r},t)\right] + \left[\boldsymbol{p} - \frac{e}{c}\boldsymbol{A}(\boldsymbol{r},t)\right] n(\boldsymbol{r})\right\}$$

$$= \left(\frac{e}{2M}\right)[n(\boldsymbol{r})\boldsymbol{p} + \boldsymbol{p}n(\boldsymbol{r})] - \left(\frac{e^2}{Mc}\right)\boldsymbol{A}(\boldsymbol{r},t)\,n(\boldsymbol{r}) \quad (12.45)$$

$$\equiv \boldsymbol{j}_0(\boldsymbol{r}) + \boldsymbol{j}_1(\boldsymbol{r},t) \quad (12.46)$$

is obtained by keeping terms up to the first order in $\boldsymbol{E}$ in the form

$$\langle \boldsymbol{j}(\boldsymbol{r},t)\rangle = \mathrm{Tr}[\rho_0 \boldsymbol{j}_1] + \mathrm{Tr}[\rho_1 \boldsymbol{j}_0] \quad (12.47)$$

Here the density operator $\rho(t)$, satisfying the Liouville equation $i\hbar \mathrm{d}\rho(t)/\mathrm{d}t = [H_0 + H_1, \rho(t)]$ is also kept up to the first order in the form

$$\rho(t) = \rho_0 + \rho_1(t) \quad (12.48)$$

where ρ_0 is constructed from the orthonormal basis $\{\psi_\alpha\}$ of eigenstates with the eigenvalue ϵ_α of the unperturbed Hamiltonian H_0 as

$$\rho_0 = \sum_\alpha f(\epsilon_\alpha)\,|\alpha\rangle\langle\alpha|, \quad f(\epsilon_\alpha) \simeq \theta(\epsilon_F - \epsilon_\alpha) \quad (12.49)$$

In order to obtain the explicit form of the current response, we transform the Liouville equation into Fourier ω-space,

$$\hbar\omega\tilde{\rho}_1 = [H_0, \tilde{\rho}_1] + \left[\tilde{H}_1, \rho_0\right] \quad (12.50)$$

and taking matrix elements between states $\langle\alpha|$ and $|\beta\rangle$, we get

$$\langle\alpha|\,\tilde{\rho}_1\,|\beta\rangle = \frac{f(\epsilon_\beta) - f(\epsilon_\alpha)}{\epsilon_{\beta\alpha} + \hbar\omega + i\eta}\langle\alpha|\,\tilde{H}_1\,|\beta\rangle, \quad \epsilon_{\beta\alpha} \equiv \epsilon_\beta - \epsilon_\alpha \quad (12.51)$$

Thus the first term (diamagnetic term) in Eq. 12.47 can be reduced to

$$\mathrm{Tr}[\rho_0 \boldsymbol{j}_1] = -\frac{e^2}{Mc}\boldsymbol{A}(\boldsymbol{r},\omega)\sum_\alpha f(\epsilon_\alpha)\,|\psi_\alpha(\boldsymbol{r})|^2 \equiv -\frac{e^2}{Mc}\boldsymbol{A}(\boldsymbol{r},\omega)\,n(\boldsymbol{r}) \quad (12.52)$$

Noting that $\boldsymbol{E} = (i\omega/c)\,\boldsymbol{A}$ and $\boldsymbol{j} = -i\omega\boldsymbol{P} = -i\omega\chi\boldsymbol{E}$, we can find the diamagnetic susceptibility as

$$\chi(\boldsymbol{r},\boldsymbol{r}',\omega) = -\frac{e^2}{M\omega^2}n(\boldsymbol{r})\,\delta(\boldsymbol{r}-\boldsymbol{r}')\,\mathbb{I} \quad (12.53)$$

For the second term (paramagnetic term) in Eq. 12.47, with the help of Eq. 12.51, we have

$$\mathrm{Tr}[\tilde{\rho}_1 \boldsymbol{j}_0] = \sum_{\alpha,\beta}\frac{f(\epsilon_\beta) - f(\epsilon_\alpha)}{\epsilon_{\beta\alpha} + \hbar\omega + i\eta}\left(\frac{i}{\omega}\right) \cdot \left[\int \mathrm{d}^3r'\,\langle\alpha|\,\boldsymbol{j}_0(\boldsymbol{r}')\cdot\boldsymbol{E}(\boldsymbol{r}')\,|\beta\rangle\right]\langle\beta|\,\boldsymbol{j}_0(\boldsymbol{r})\,|\alpha\rangle \quad (12.54)$$

With the help of $\boldsymbol{\mu}(\boldsymbol{r},\omega) = (i/\omega)\boldsymbol{j}_0(\boldsymbol{r})$, the paramagnetic susceptibility in nonlocal form can be obtained as

$$\chi(\boldsymbol{r},\boldsymbol{r}',\omega) = \sum_{\alpha,\beta} \frac{f(\epsilon_\beta) - f(\epsilon_\alpha)}{\epsilon_{\beta\alpha} + \hbar\omega + i\eta} \langle\alpha|\boldsymbol{\mu}(\boldsymbol{r}')|\beta\rangle\langle\beta|\boldsymbol{\mu}(\boldsymbol{r})|\alpha\rangle \tag{12.55}$$

Before moving on to the next section, let us remark on the classical counterpart of the susceptibility derived above. If we consider a region that is small in relation to the wavelength, where both the driving field and material response are spatially uniform, we may approximate nonlocal susceptibility as locally uniform susceptibility $\chi(\omega)$. It can then be related to the macroscopic (classical) dielectric function $\epsilon(\omega)$ as follows:

$$\chi(\omega) = \frac{3}{4\pi}\frac{\epsilon(\omega) - 1}{\epsilon(\omega) + 2} \tag{12.56}$$

including the depolarization effect (Lorentz–Lorenz effect). Alternatively, we can regard it as a polarizable sphere of radius r whose polarizability is

$$\alpha(\omega) = \left(\frac{4\pi}{3}\right) r^3 \chi(\omega) = \frac{\epsilon(\omega) - 1}{\epsilon(\omega) + 2} r^3 \tag{12.57}$$

12.3 Numerical Examples

In accordance with the above formulation, we will now numerically describe typical signal behavior in near-field optical microscopy and spectroscopy. Let us recall how a fiber probe is functionalized and fabricated in practice, as described in detail in Chaps. 2 and 3. As the idealized limit of a nanometric protruded probe with a metal coating, we approximate it as a nanometric sphere with radius a. One or a series of nanometric spheres with radius b are also assumed as an idealized sample embedded in a homogeneous medium with refractive index n_2 on a flat substrate with refractive index n_1, $(n_1 > n_2)$. Figure 12.2 illustrates such a modeling configuration.

In the collection mode (c-mode), an optical evanescent field $\boldsymbol{E}_0(\boldsymbol{r},\omega)$ (with a monochromatic frequency ω) is initially generated at the interface by total internal reflection. A p- or s-polarized plane wave incidence, with unit amplitude and incident angle θ, is assumed, as explained in Sect. 12.2.2. The effective field to be used for signal calculation is a solution of Eq. 12.17 with a field propagator (cf. Sect. 12.2.3) and susceptibility (cf. Sect. 12.2.4), and the parameters used in the simulation are explicitly mentioned in each relevant subsection.

12.3.1 Weak vs. Strong Coupling

Before going into numerical details, it may be helpful to make an order estimate of the absolute value of the second term in Eq. 12.17 compared with the

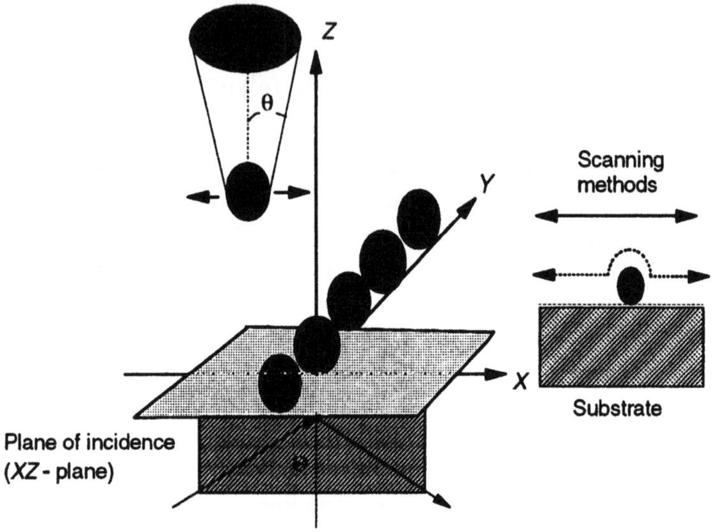

Fig. 12.2. Modeling configuration

first term. Since the second term represents the interaction between probe and sample, it may be said that the coupling of the probe and sample is weak if the second term is much smaller than the first term. Conversely, it may be said that the coupling is strong if the opposite is true. When a weakly coupled system is considered, one may be justified in using a perturbative method [37,59], that is, in replacing $\boldsymbol{E}(\boldsymbol{r},\omega)$ in the second term by $\boldsymbol{E}_0(\boldsymbol{r},\omega)$, which saves a lot of computational time. As the system becomes strongly coupled, we have to solve the equation precisely by, for example, the matrix inversion method, which takes much more time than the perturbative method. Such a case, which is related to resonance phenomena, will be discussed in more detail in Sect. 12.3.4.

Let us look at an example of weak coupling: Fig. 12.3 depicts near-field signals defined by $|\boldsymbol{E}(\boldsymbol{r}=\boldsymbol{R}_p,\omega)|^2$, the square modulus of the effective field at the probe center $\boldsymbol{R}_p$, as a function of the scanning position. The radii of the probe and sample sphere are both 20 nm, and the dielectric constants are $\epsilon_p = 2.25$ and $\epsilon_s = 4.0$ for the wavelength of incident light $\lambda = 780$ nm. P-polarized light (the electric field in the incident xz plane) is incident at an angle $\theta = 45°$ from the substrate side $n_1 = 1.5$ to the air side $n_2 = 1.0$. Dotted and solid lines represent the results for perturbative calculation and exact calculation, respectively, and both curves show the same characteristic peak in the signal for p-polarization at the center of the sample. Judging from the figure, the perturbative approach seems reasonable with respect to the relevant sizes of the probe and sample spheres, and the approach distance.

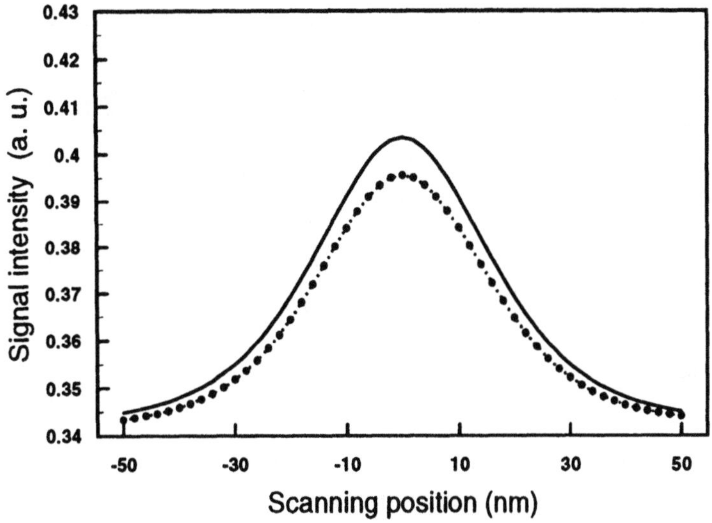

Fig. 12.3. Near-field signals as a function of the scanning position (p-polarization). Dotted and solid lines represent the results for perturbative calculation and exact calculation, respectively

Note that the signals coming from the z component of the effective field are about ten times larger than those from the x component.

Figure 12.4 shows the incident s-polarization case (in which the electric field is perpendicular to the incident xz plane). The edge behavior depends on the scanning direction, but basically the signal for s-polarization has a dip at the center of the sample. The experimental results described in previous chapters (Figs. 6.6 and 6.9 in Chap. 6, Figs. 7.6 and 7.9 in Chap. 7, and Fig. 8.11 in Chap. 8) show the characteristics.

The above characteristics for the interpretation of experimental images can be qualitatively understood from the signs of the polarizability of the samples and the field propagator. Noting that the signal modulation mainly comes from the quasi-static dipole term $\mathbb{T}_3(\boldsymbol{r}, \boldsymbol{r}')$ in the near-field range, we examine the sign of the quasi-static term. When both have the same sign, the second term in the perturbative version of Eq. 12.17 is added in phase with the first term, and a signal peak occurs. On the other hand, when they have opposite signs, a signal dip occurs. From Eq. 12.57, the polarizability $\alpha(\omega)$ is positive (negative) when the dielectric constant (assumed to be real, for simplicity) of the sample is larger (smaller) than unity. To the incident field for s-polarization, the initial evanescent field has only the y component, and $\mathbb{T}_3(\boldsymbol{r}, \boldsymbol{r}')$ becomes negative on the z axis, and thus a signal dip manifests itself in $\epsilon(\omega) > 1$. To the incident field for p-polarization, the initial evanescent

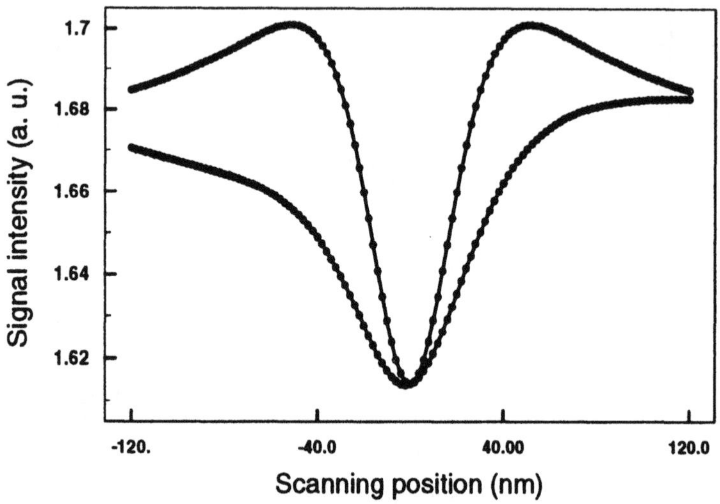

Fig. 12.4. Near-field signals as a function of the scanning position (s-polarization). *Solid lines* with *filled circles* and *open circles* represent the results for the x−scan and y−scan, respectively

field has a dominant z component and $\mathbb{T}_3(\boldsymbol{r},\boldsymbol{r}')$ becomes positive on the z axis. Therefore the signal of a sample with $\epsilon(\omega) > 1$ has a peak at the center.

Finally, we comment on the different edge behavior for s-polarization. Qualitatively, this can result from the existence of any dielectric constant difference in the direction of the electric field vector at a scanning point. When the probe is scanned parallel to the direction of the electric field vector, there is a gap in the dielectric constant near the edges of the sample. On the other hand, there is continuity in the dielectric constant when the probe is scanned vertically to the electric field vector. This difference produces the different signal behavior near the edges.

12.3.2 Near-Field- and Far-Field-Propagating Signals

We begin by defining the signal intensity $I_{\mathrm{far}}(\theta)$, which from now on we will call the far-field-propagating signal, as integration of $|\boldsymbol{E}_{\mathrm{far}}(\boldsymbol{R}+\boldsymbol{R}_{\mathrm{p}},\omega)|^2$ over the solid angle $\mathrm{d}\Omega = \sin\theta\mathrm{d}\theta\mathrm{d}\phi$,

$$I_{\mathrm{far}}(\theta) = \int R^2 \mathrm{d}\Omega \, |\boldsymbol{E}_{\mathrm{far}}(\boldsymbol{R}+\boldsymbol{R}_{\mathrm{p}},\omega)|^2 . \tag{12.58}$$

Here, $\boldsymbol{E}_{\mathrm{far}}(\boldsymbol{R}+\boldsymbol{R}_{\mathrm{p}},\omega)$ designates the field propagated to the point $\boldsymbol{R} = (R\sin\theta\cos\phi, R\sin\theta\sin\phi, R\cos\theta)$ in the far zone, originating in the effective near field at the center of the probe $\boldsymbol{R}_{\mathrm{p}}$.

$$\boldsymbol{E}_{\mathrm{far}}\left(\boldsymbol{R}+\boldsymbol{R}_{\mathrm{p}},\omega\right)=\mathbb{T}_{1}\left(\boldsymbol{R}+\boldsymbol{R}_{\mathrm{p}},\boldsymbol{R}_{\mathrm{p}}\right)\cdot\alpha_{\mathrm{p}}\left(\omega\right)\boldsymbol{E}\left(\boldsymbol{R}_{\mathrm{p}},\omega\right)\tag{12.59}$$

Using the explicit representation of $\mathbb{T}_1(\boldsymbol{r},\boldsymbol{r}')$ in Eq. 12.33, we can write

$$\boldsymbol{E}_{\mathrm{far}}\left(\boldsymbol{R}+\boldsymbol{R}_{\mathrm{p}},\omega\right)=-\frac{q^{2}\alpha_{p}\left(\omega\right)}{R}\left(s_{x},s_{y},s_{z}\right)\equiv-\frac{q^{2}\alpha_{p}\left(\omega\right)}{R}\boldsymbol{s}\tag{12.60}$$

$$\boldsymbol{s}=\begin{pmatrix}\sin^{2}\theta\cos^{2}\phi-1 & \sin^{2}\theta\sin\phi\cos\phi & \sin\theta\cos\theta\cos\phi\\ \sin^{2}\theta\sin\phi\cos\phi & \sin^{2}\theta\sin^{2}\phi-1 & \sin\theta\cos\theta\sin\phi\\ \sin\theta\cos\theta\cos\phi & \sin\theta\cos\theta\sin\phi & \cos^{2}\theta-1\end{pmatrix}\cdot\boldsymbol{E}\left(\boldsymbol{R}_{\mathrm{p}},\omega\right)\tag{12.61}$$

Thus, $I_{\mathrm{far}}(\theta)$ can be rewritten as

$$\begin{aligned}I_{\mathrm{far}}(\theta) &= q^{4}\left|\alpha_{\mathrm{p}}(\omega)\right|^{2}\int_{0}^{2\pi}\mathrm{d}\phi\int_{0}^{\theta}\sin\theta\mathrm{d}\theta\left(\left|s_{x}\right|^{2}+\left|s_{y}\right|^{2}+\left|s_{z}\right|^{2}\right) &(12.62)\\ &= q^{4}\left|\alpha_{\mathrm{p}}(\omega)\right|^{2}\left[\left(\left|E_{x}\left(\boldsymbol{R}_{\mathrm{p}},\omega\right)\right|^{2}+\left|E_{y}\left(\boldsymbol{R}_{\mathrm{p}},\omega\right)\right|^{2}\right)\right.\\ &\quad\cdot\left(16-15\cos\theta-\cos3\theta\right)\\ &\quad\left.+\left|E_{z}\left(\boldsymbol{R}_{\mathrm{p}},\omega\right)\right|^{2}\left(16-18\cos\theta+2\cos3\theta\right)\right] &(12.63)\end{aligned}$$

If the far-field propagation is perfectly coupled to a fiber mode, the angle θ corresponds to the cone angle of the probe described in Chaps. 2 and 3, and $I_{\mathrm{far}}(\theta)$ is the physical quantity to be compared with the experimentally measured signal intensity. It is worth noting that the near-field signal considered in the previous section, $I_{\mathrm{near}}=\left|\boldsymbol{E}\left(\boldsymbol{R}_{\mathrm{p}},\omega\right)\right|^{2}$, is a special case of $I_{\mathrm{far}}(\theta=90°)$ apart from the coefficient $16\,q^{4}\left|\alpha_{\mathrm{p}}(\omega)\right|^{2}$, and that the far-field-propagating signal for p-polarization has a stronger dependence on the cone angle θ than that for s-polarization.

We give a clear example of how the cone angle of the probe affects the far-field-propagating signal [60,61]. Let both probe and sample spheres have a radius of 15 nm, and let the probe be scanned constantly on the $z=60$ nm plane from the substrate surface. We assume an argon laser beam with a wavelength of 488 nm, directed onto the glass prism under the total internal reflection condition (c-mode). Figure 12.5 illustrates the fact that the signal intensity for p-polarization has a peak at the center of the sample (as before) when θ is larger than about 45°, while it has peaks at the edges of the sample when a smaller value of θ is used. The signal intensity for s-polarization, not shown in the figure, has similar behavior to that described in the previous section (see Fig. 12.4). When a smaller value of θ is used, the width between the peaks for p-polarization is broader than that for s-polarization. For a larger value of θ, the far-field-propagating signal is similar to the near-field signal. This simulated behavior coincides with the experimental findings discussed in Chaps. 7 and 8.

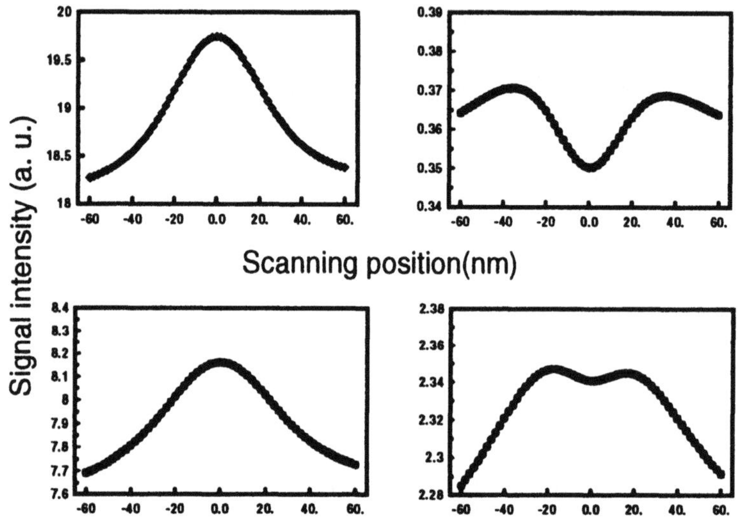

Fig. 12.5. Far-field-propagating signals for p-polarization as a function of the scanning position.The results for $\theta = 15°$, $30°$, $45°$, and $60°$ are shown clockwise from the top-right

The θ dependence can be interpreted as indicating that the z component of the dipole field excited by p-polarization light propagates less in the z-direction of the far-field zone.

The contrast of the scanning intensity (normalized intensity modulation) varies according to the cone angle θ of the probe and the incident polarization. From the figures, it follows that the contrast for p-polarization is about one and a half times as large as that for s-polarization (see Fig. 12.4). This is because the signal intensity for s-polarization, namely, the denominator of the normalization factor, is larger than that for p-polarization when a smaller value of θ is employed. The contrast and signal behavior might vary according to the scanning method used for the probe. We will examine this point in more detail in the next section.

12.3.3 Scanning Methods

We now demonstrate how the far-field-propagating signal intensities and contrast (or visibility) as functions of the cone angle θ of the probe depend on two fundamental linear polarizations of incident light and three scanning methods used for the probe: constant-height, constant-distance, and constant-intensity. Using the same parameters as in the previous subsections, we evaluate the visibility or contrast

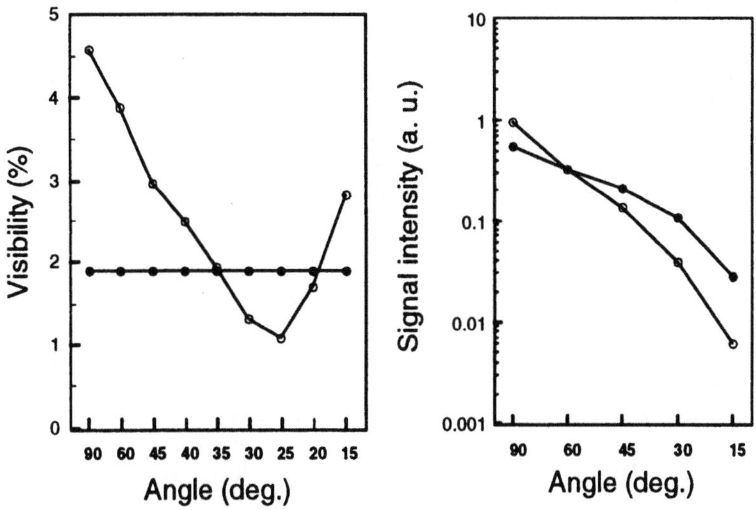

Fig. 12.6. Polarization-dependent visibility and normalized far-field-propagating signals as functions of the cone angle θ (constant-height and constant-intensity modes). *Solid lines* with *open circles* and *filled circles* represent the results for p-polarization and s-polarization, respectively

$$v(\theta) = \frac{I_{\text{far}}^{\max}(\theta) - I_{\text{far}}^{\min}(\theta)}{I_{\text{far}}^{\max}(\theta) + I_{\text{far}}^{\min}(\theta)} \tag{12.64}$$

as a function of the cone angle θ, where the terms max and min represent the maximum and minimum of $I_{\text{far}}(\theta)$ defined in Eq. 12.62 for each scan. We use the term "constant-height mode" for the situation in which the probe is scanned horizontally while the z component of the probe apex position remains constant, and the term "constant-distance mode" for the situation in which the probe is scanned along the sample geometry so that the distance either between the probe apex and the surface of the substrate or between the probe apex and the sample surface remains constant. In the "constant-intensity mode," the probe is scanned so that the trajectory gives a contour map of the intensity.

Figure 12.6 shows the far-field-propagating signal normalized by that for $\theta = 90°$ and p-polarization, as well as the polarization-dependent visibility, in both constant-height and constant-intensity modes. It follows from the figures that p-polarization is preferable for a probe with a cone angle of more than about 60°, in order to maximize the contrast (visibility) and maintain a high signal intensity. For s-polarization, a probe with a cone angle of around 30° would be preferable.

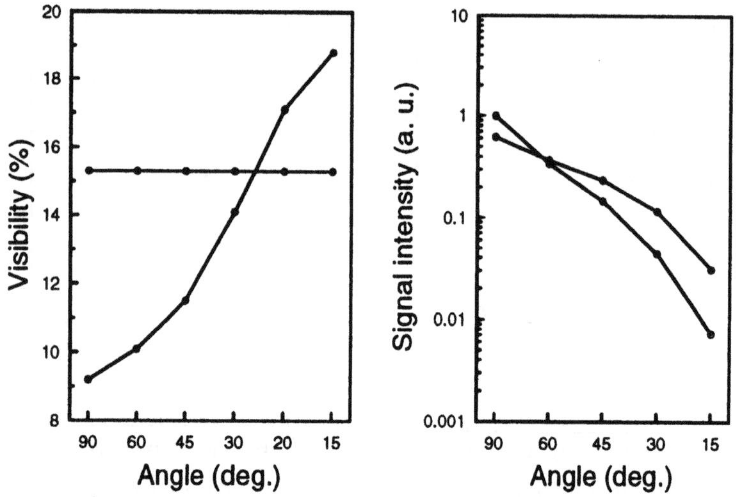

Fig. 12.7. Polarization-dependent visibility and normalized far-field-propagating signals as functions of the cone angle θ (constant-distance mode). *Solid lines* with *open circles* and *filled circles* represent the results for p-polarization and s-polarization, respectively

The visibility becomes higher in the constant-distance mode than in the constant-height mode for both p- and s-polarization, because the probe apex is scanned closer to the surface of the substrate, and the baseline of the signal intensity rises. As can be seen in Fig. 12.7, the visibility for s-polarization is better than for p-polarization, unlike in the constant-height and constant-intensity modes.

For each polarization and scanning method, as we have seen, the cone angle of the probe has an optimal value for maximizing the contrast and maintaining a high signal intensity. These numerical results support the assertion that the cone angle of the fiber probe is one of the critical parameters in experimental resolution and contrast, as emphasized in Chaps. 2, 6, 7, and 9.

12.3.4 Possibility of Spin-Polarization Detection

In the preceding subsections, the sample–probe systems were regarded as weak-coupling systems that allow the use of a perturbative method. Resonance phenomena, however, such as local plasmon resonance and proximity resonance, lead systems toward the limit of strong coupling. We therefore focus on this limit in order to study some aspects of the potential of near-

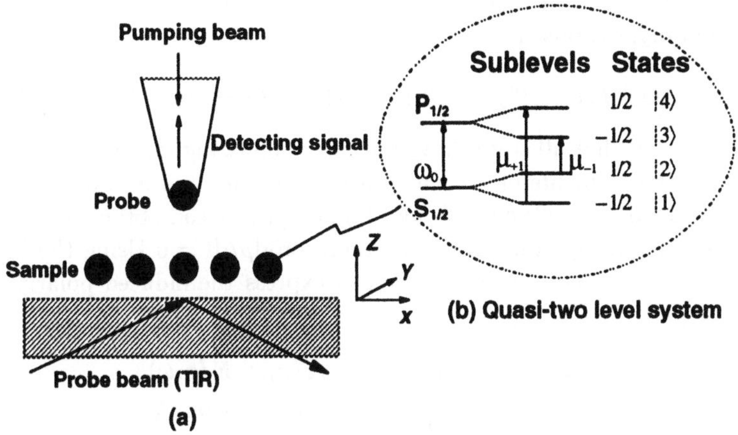

Fig. 12.8. a Schematic illustration of spin-polarization detection. b Sample quasi-two-level system

field optical spectroscopy – local polarization spectroscopy – based on the microscopic approach [78].

Let us consider a strong-coupling system of probe apex and sample, schematically illustrated in Fig. 12.8a. The sample is locally illuminated by a light field with circular polarization created in the near zone of the probe apex, and is resonantly excited (pumped) from the ground state to the excited state. To sense the state of the sample, we use a linearly polarized evanescent field on the substrate, generated by total internal reflection, which corresponds to a probe beam. An effective field is then induced at the probe apex as a result of interaction among the sample, substrate, and probe apex, triggered by the evanescent field, as described in Sect. 12.2.1:

$$\boldsymbol{E}(\boldsymbol{r},\omega)=\boldsymbol{E}_0(\boldsymbol{r},\omega)+\sum_{i=\text{probe}}^{\text{sample}}\int_{r'\in i}\mathbb{T}(\boldsymbol{r},\boldsymbol{r}',\omega)\,\chi_i(\boldsymbol{r}',\omega)\,\boldsymbol{E}(\boldsymbol{r}',\omega)\,d^3r' \quad (12.65)$$

If we rewrite the equation formally for the effective field $\boldsymbol{E}(\boldsymbol{r},\omega)$ as

$$\mathrm{A}(\boldsymbol{r},\omega)\,\boldsymbol{E}(\boldsymbol{r},\omega)=\boldsymbol{E}_0(\boldsymbol{r},\omega) \quad (12.66)$$

it follows that $\det(\mathrm{A}(\boldsymbol{r},\omega))=0$ gives information about the state of the sample when the sample–probe system is resonantly coupled in terms of frequency and sample–probe separation. We are now interested in the internal degrees of freedom of the sample, such as the spin/angular momentum, and therefore we derive the susceptibility tensor of the sample, $\chi_s(\boldsymbol{r},\omega)$, in a similar way to that described in Sect. 12.2.4. For simplicity, we choose the quasi-two-level system shown in Fig. 12.8b. We assume that there is initially no population in the excited state, and that the ground state coherence (m_x, m_y) and spin polarization m_z, expressed as

$$m_x = \rho_{12} + \rho_{21}, \quad m_y = i(\rho_{21} - \rho_{12}), \quad m_z = \rho_{22} - \rho_{11} \tag{12.67}$$

are not zero, and that all states have the same dephasing rate Γ. Noting that non-zero matrix elements of H_1 and $\boldsymbol{\mu}$ are given only by the states between $|1\rangle$ and $|3\rangle$, $|1\rangle$ and $|4\rangle$, $|2\rangle$ and $|3\rangle$, and $|2\rangle$ and $|4\rangle$, we can obtain the density operator in the steady state from the condition $d\rho/dt = 0$ Using the dipole operator and the density operator, we can express the induced polarization $\boldsymbol{P}(\boldsymbol{r},\omega)$ as [90]

$$\begin{aligned} \boldsymbol{P}(\boldsymbol{r},\omega) &= N\mathrm{Tr}(\boldsymbol{\mu}\rho) = \chi_s(\boldsymbol{r},\omega)\boldsymbol{E}(\boldsymbol{r},\omega) \\ \chi_s(\boldsymbol{r},\omega) &= \chi_0 \begin{pmatrix} 1 & im_z & -im_y \\ -im_z & 1 & im_x \\ im_y & -im_x & 1 \end{pmatrix} \\ \chi_0 &\equiv \frac{2\mu_E^2 g(\delta,\Gamma)}{\hbar\Gamma} N \end{aligned} \tag{12.68}$$

where μ_E is the radial overlap integral of the dipole moment, N is the density, and $g(\delta,\Gamma)$ denotes the frequency-dependent factor given by the detuning δ and the dephasing rate Γ as

$$g(\delta,\Gamma) = -\frac{1}{\frac{\delta}{\Gamma} + i} = \left(\frac{-\frac{\delta}{\Gamma} + i}{1 + \left(\frac{\delta}{\Gamma}\right)^2}\right) \tag{12.69}$$

It is worth recalling that the off-diagonal parts of the susceptibility (except m_z) disappear if the transition probability between sublevels is not taken into account. The explicit expression for $\chi_s(\boldsymbol{r},\omega)$ indicates that the coherence m_x and spin polarization m_z are mainly selected by the s-polarized evanescent field, while the coherences m_x and m_y are predominantly selected by the p-polarization evanescent field. It is also possible to describe the susceptibility of the probe apex $\chi_P(\boldsymbol{r},\omega)$ in a similar way, but for simplicity we approximate it by a classical macroscopic value as follows:

$$\begin{aligned} \chi_P(\boldsymbol{r},\omega) &= \alpha_P(\omega) \begin{pmatrix} 1 & 0 & 0 \\ 0 & 1 & 0 \\ 0 & 0 & 1 \end{pmatrix} \\ \alpha_P(\omega) &= \left(\frac{\epsilon_P(\omega) - 1}{\epsilon_P(\omega) + 2}\right) a^3 \delta(\boldsymbol{r} - \boldsymbol{R}_P) \end{aligned} \tag{12.70}$$

In the following, we employ as a probe apex a sphere of radius $a = 10$ nm located at a varying position $\boldsymbol{R}_P$. A sample is assumed to be a collection of

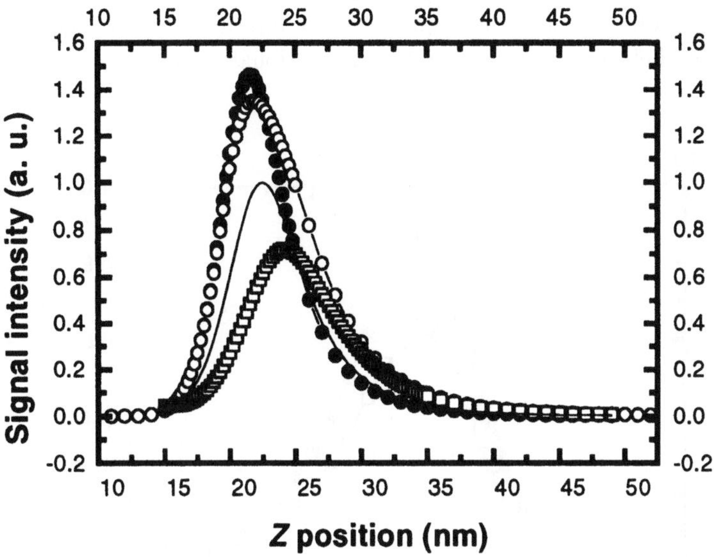

Fig. 12.9. Approach curves for s-polarization. *Solid lines* with *filled circles*, *open circles*, *rectangles*, and no marks show the results for the states (0,0,0), (1,0,1), (1,0,0), and (0,0,1), respectively

points that have only internal degrees of freedom (the quasi-two-level system shown in Fig. 12.8b), and that are located at 10 nm from the substrate surface. Regarding the coupling strength, we use μ_E = 3–10 debye (D), $N = 1/(10\ \text{nm})^3$, $\Gamma^{-1} = 10$ ns, and $\delta = 0$ unless explicitly stated otherwise. To simplify the calculation, we will choose a single point from the sample, and use an apex sphere for the sample–probe system. As an initial evanescent field normalized at the substrate surface, we use

$$\boldsymbol{E}_0(\boldsymbol{r},\omega) = \begin{cases} (0,1,0)\exp\left(\frac{-z}{r_{\text{eff}}}\right)\exp(-i\omega t)\,, & \text{for s-polarization} \\ (0,0,1)\exp\left(\frac{-z}{r_{\text{eff}}}\right)\exp(-i\omega t)\,, & \text{for p-polarization} \end{cases} \tag{12.71}$$

where r_{eff} represents the decay length of the evanescent field, which is set as 10 nm unless otherwise stated. The signal intensity as a function of the probe apex position is shown in Figs. 12.9 and 12.10 when the probe apex is scanned in the z-direction (μ_E = 10 D). As typical characteristics, the peak intensities of the (1,0,0) and (0,0,1) states are reduced and shifted for s-polarization (see Fig. 12.9).

For p-polarization, the peak intensities of the (1,0,0) state and the (0,1,0) state are reduced, but not shifted (Fig. 12.10). Here it should be noted that the signal-to-background (noise) ratio in a recent experiment [81] was reported to be about 3:1, where the fluorescence detection intensity as a function of the sample–probe separation was measured by photon counting. These results suggest that it is possible optically to distinguish the sample states in the sample–probe configuration, and might be interpreted as a consequence of a coupling between frequency and configurational resonances.

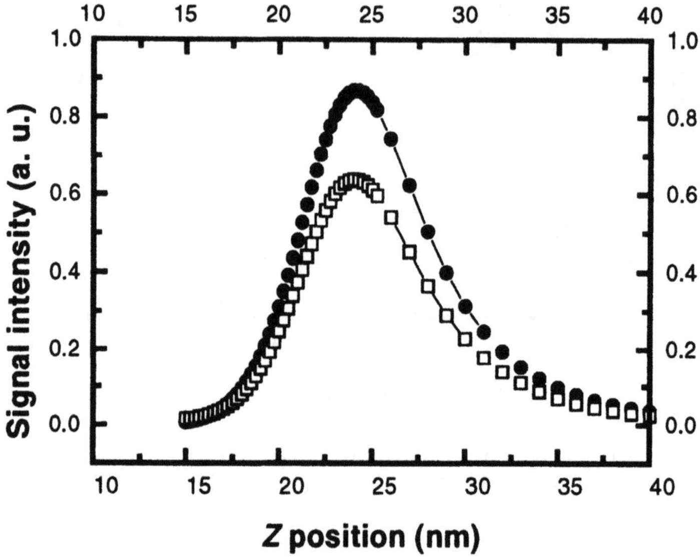

Fig. 12.10. Approach curves for p-polarization. The results for the states (0,0,0) and (1,0,0) are shown by *solid lines* with *filled circles* and *open rectangles*, respectively

12.4 Effective Field and Massive Virtual Photon Model

We have so far outlined a field propagator method that can be applied to optical near-field problems, and given some numerical results followed by qualitative discussion. We have also shown that potential study of near-field optical microscopy (NOM) can be discussed by relying on the microscopic or semi-microscopic approach within some approximations. At this stage, we should not forget an intuitive model – the massive virtual photon model [2,6, 80] – based on profound physical insight, which will provide a very instructive and critical way of thinking for future developments of optical near-field-related theories. Let us therefore consider this model (with nonzero effective mass) and discuss the physics behind it.

The model is based on the analogy of a "virtual particle" or a "normal mode" in quantum field theory (see, for example, [88]), which violates the usual energy conservation principle to survive for a very short time determined by the Heisenberg uncertainty principle. In vacuum, electromagnetic interaction governed by the Helmholtz equation (cf. Eq. 12.7) is due to the exchange of virtual photons (with mass = 0). If we can assume the Klein–Gordon equation to be

$$\left(\nabla^2 - \lambda_c^{-2}\right)\phi(\boldsymbol{r}) = 0, \quad \lambda_c = \left(\frac{m_{\text{eff}}c}{\hbar}\right)^{-1} \sim a << \lambda \tag{12.72}$$

after renormalizing the induced source term (light-matter interaction) empirically, we can obtain the well-known Yukawa function:

$$\phi(r) = \frac{\exp(-r/\lambda_c)}{r} \tag{12.73}$$

This gives us a screened and finite-range interaction, owing to virtual photon exchange with a nonzero mass (m_{eff}). Since an evanescent field has a wave number parallel to a material surface (larger than ω/c) and one perpendicular to the surface that is purely imaginary, the effective field due to the interaction is found to be highly localized, in close proximity to a material, and highly dependent on the size of sample or probe. From the model, we can easily obtain the near-field distribution, for example, in a circular aperture with a radius smaller than the wavelength of the light used. Moreover, the resolution and contrast of the NOM, as well as the size-dependent signal behavior, can be discussed as clearly as in Chap. 2. This coincidence might be due to the resemblance between $1/r^3$ (quasi-static dipole interaction) and the Yukawa function.

We now make a brief comment on the physics underlying the model. It is well known that the bath system represented by the Hamiltonian

$$\begin{aligned} H_{\text{bath}} &= \sum_{\boldsymbol{k}\lambda} \hbar\omega_0 b_{\boldsymbol{k}\lambda}^{\dagger} b_{\boldsymbol{k}\lambda} + \sum_{\boldsymbol{k}\lambda} \hbar\omega_{\boldsymbol{k}} a_{\boldsymbol{k}\lambda}^{\dagger} a_{\boldsymbol{k}\lambda} \\ &- \sum_{\boldsymbol{k}\lambda} i\hbar C \left(b_{-\boldsymbol{k}\lambda} a_{\boldsymbol{k}\lambda} - b_{\boldsymbol{k}\lambda} a_{\boldsymbol{k}\lambda}^{\dagger} + b_{\boldsymbol{k}\lambda}^{\dagger} a_{\boldsymbol{k}\lambda} - b_{-\boldsymbol{k}\lambda}^{\dagger} a_{\boldsymbol{k}\lambda}^{\dagger}\right) \end{aligned} \tag{12.74}$$

can in principle be diagonalized by a unitary transformation [85–87], or an exciton–polariton operator ($\xi_{\boldsymbol{k}\nu}^{\dagger}$, $\xi_{\boldsymbol{k}\nu}$), as

$$H_{\text{bath}} = \sum_{\boldsymbol{k}\nu} \hbar\omega_{\boldsymbol{k}\nu} \xi_{\boldsymbol{k}\nu}^{\dagger} \xi_{\boldsymbol{k}\nu}, \quad [\xi_{\boldsymbol{k}\nu}, \xi_{\boldsymbol{k}'\nu'}^{\dagger}] = \delta_{\boldsymbol{k}\boldsymbol{k}'}\delta_{\nu\nu'} \tag{12.75}$$

Here, each term on the right-hand side of Eq. 12.74 describes a two-level atomic or molecular system formed by excitons ($b_{\boldsymbol{k}\lambda}^{\dagger}$, $b_{\boldsymbol{k}\lambda}$), free photons ($a_{\boldsymbol{k}\lambda}^{\dagger}$, $a_{\boldsymbol{k}\lambda}$), and their interactions, respectively. The coupling strength is denoted as $\hbar C$.

Consider a sample–probe system coupled with a light source and detector via interacting photons and a two-level atomic (molecular) host system. Since the interaction between the free photons and the host system can be diagonalized, as explained above, the sample–probe system can be expressed in terms of exciton–polariton, or a normal mode. This view is very similar to the massive virtual photon model, but the propagator of the virtual particle has a different functional form from the Yukawa function. Thus, establishment of a theoretical foundation for the model remains an open problem.

12.5 Future Direction

Before closing this chapter, we review the current status of theoretical approaches to optical near-field problems, and comment on what needs to be done in the near future. We have reached the stage of interpreting and predicting qualitatively high-resolution imaging of a variety of samples, as seen in previous sections. On the other hand, only very limited spectroscopic information has been obtained so far, although there have been interesting experimental results, as mentioned in Chaps. 5, 9, 10 and 11. There will be scope for theoretical studies of the near-field spectroscopy of a nanometric/mesoscopic sample, or a quantum system, in order to obtain more than the dipole distribution in such a sample.

From a more theoretical viewpoint, a comprehensive quantum theory for optical near-field problems – consistent treatment from excitation to detection – is anticipated, although it will require a lot of careful examination and involve many challenges. Such an approach should be size-consistent and applicable to both weakly and strongly correlated systems, and it should also allow us to perform ground-state as well as excited-state calculations. One candidate would be the method of projection operators [83, 84], which is a simple but unified and systematic approach ranging from elementary/nuclear physics and atomic/molecular physics and chemistry to solid-state physics.

We may expect that such trials will provide a foundation for the virtual photon model described in Sect. 12.4, and will reveal the essence of effective sample–probe interaction via the optical near field.

References

1. D. Van Labeke, D. Barchiesi, in *Near Field Optics*, D. W. Pohl, D. Courjon (eds.), p. 157 (Kluwer, Dordrecht, 1993)
2. M. Ohtsu, J. Lightwave Technol. **13**, 1200 (1995)
3. C. Girard, A. Dereux, Rep. Prog. Phys. **59**, 657 (1996)
4. M. A. Paesler, P. J. Moyer, *Near-Field Optics: Theory, Instrumentation, and Applications* (Wiley, New York, 1996)
5. J. P. Fillard, *Near Field Optics and Nanoscopy* (World Scientific Publishing, Singapole, 1996)
6. M. Ohtsu, H. Hori, *NEAR-FIELD NANO-OPTICS—From Basic Principles to Nano-Fabrication and Nano-Photonics—* (Plenum, New York, 1998)

7. Lord Rayleigh, Phil. Mag. **43**, 259 (1897)
8. A. Sommerfeld, *Lectures on Theoretical Physics* (Academic, New York, 1964)
9. M. Born, E. Wolf, *Principles of Optics*, 5th edn. (Pergamon, London, 1975)
10. J. D. Jackson, *Classical Electrodynamics*, 2nd edn. (Wiley, New York, 1975)
11. H. A. Bethe, Phys. Rev. **66**, 163 (1944)
12. J. Bouwkamp, Rep. Prog. Phys. **17**, 35 (1954)
13. Y. Leviatan, J. Appl. Phys. **60**, 1577 (1986)
14. A. Roberts, J. Appl. Phys. **70**, 4045 (1991)
15. L. Novotny, C. Hafner, Phys. Rev. E **50**, 4094 (1994)
16. D. W. Pohl, L. Novotny, J. Vac. Sci. Technol. B **12**, 1441 (1994)
17. L. Novotny, D. W. Pohl, P. Regli, Ultramicroscopy **57**, 180 (1995)
18. L. Novotny, D. W. Pohl, B. Hecht, Opt. Lett. **20**, 970 (1995)
19. L. Novotny, R. X. Bian, X. S. Xie, Phys. Rev. Lett. **79**, 645 (1997)
20. G. Mie, Ann. Phys. **25**, 377 (1908)
21. H. C. van de Hulst, *Light scattering by small particles* (Dover, New York, 1981)
22. R. G. Newton, *Scattering theory of waves and particles*, Chap. 2 (McGraw-Hill, New York, 1966)
23. D. Barchiesi, D. Van Labeke, J. Mod. Opt. **40**, 1239 (1993)
24. S. Prasad, G. Wei, Opt. Commun. **136**, 447 (1997)
25. R. Petit, *Electromagnetic Theory of Gratings* (Springer-Verlag, Berlin, 1980)
26. D. Van Labeke, D. Barchiesi, J. Opt. Soc. Am. A **9**, 732 (1992)
27. A. Sentenac, J. J. Greffet, Ultramicroscopy **57**, 246 (1995)
28. J. P. Goudonnet, E. Bourillot, P. M. Adam, F. de Fornel, L. Salomon, P. Vincent, M. Neviere, T. L. Ferrell, J. Opt. Soc. Am. A **12**, 1749 (1995)
29. D. Barchiesi, C. Girard, O. J. F. Martin, D. Van Labeke, D. Courjon, Phys. Rev. E **54**, 4285 (1996)
30. D. Van Labeke, D. Barchiesi, J. Opt. Soc. Am. A **10**, 2193 (1993)
31. S. Wang, M. Xiao, Opt. Rev. **4**, 228 (1997)
32. K. Tanaka, M. Tanaka, J. Opt. Soc. Am. A **13**, 1362 (1996)
33. S. Kawata, T. Tani, Opt. Lett. **21**, 1768 (1996)
34. A. Dereux, D. W. Pohl, in *Near Field Optics*, D. W. Pohl, D. Courjon (eds.), p. 189 (Kluwer, Dordrecht, 1993)
35. F. Depasse, M. A. Paesler, D. Courjon, J. M. Vigoureux, Opt. Lett. **20**, 234 (1995)
36. K. Jang, W. Jhe, Opt. Lett. **21**, 236 (1996)
37. A. Zvyagin, M. Ohtsu, Opt. Commun. **133**, 328 (1997)
38. A. V. Zvyagin, J. D. White, M. Ohtsu, Opt. Lett. **22**, 955 (1997)
39. D. A. Christensen, Ultramicroscopy **57**, 189 (1995)

40. J. L. Kann, T. D. Milster, F. F. Froehlich, R. W. Ziolkowski, J. B. Judkins, J. Opt. Soc. Am. A **12**, 501 (1995)
41. J. L. Kann, T. D. Milster, F. F. Froehlich, R. W. Ziolkowski, J. B. Judkins, J. Opt. Soc. Am. A **12**, 1677 (1995)
42. H. Furukawa, S. Kawata, Opt. Commun. **132**, 170 (1996)
43. R. X. Bian, R. C. Dunn, X. S. Xie, P. T. Leung, Phys. Rev. Lett. **75**, 4772 (1995)
44. W. Denk, D. W. Pohl, J. Vac. Sci. Technol. B **9**, 510 (1991)
45. I. Banno, H. Hori, T. Inoue, Opt. Rev. **3**, 454 (1996)
46. H. Weyl, Ann. Phys. **60**, 481 (1919)
47. E. Wolf, M. Nieto-Vesperinas, J. Opt. Soc. Am. A **2**, 886 (1985)
48. C. J. R. Sheppard, H. Fatemi, M. Gu, Scanning **17**, 28 (1995)
49. A. Madrazo, M. Nieto-Vesperinas, J. Opt. Soc. Am. A **12**, 1298 (1995)
50. S. Bozhevolnyi, S. Berntsen, E. Bozhevolnaya, J. Opt. Soc. Am. A **11**, 609 (1994)
51. T. Inoue, H. Hori, Opt. Rev. **3**, 458 (1996)
52. R. Uma Maheswari, H. Kadono, M. Ohtsu, Opt. Commun. **131**, 133 (1996)
53. W. Lukosz, R. E. Kunz, J. Opt. Soc. Am. **67**, 1607 (1977); **67**, 1615 (1977)
54. M. A. Taubenblatt, T. K. Tran, J. Opt. Soc. Am. A **10**, 912 (1993)
55. L. Novotny, B. Hecht, D. W. Pohl, J. Appl. Phys. **81**, 1798 (1997)
56. L. Novotny, J. Opt. Soc. Am. A **14**, 105 (1997)
57. W. Jhe, K. Jang, Ultramicroscopy **61**, 81 (1995)
58. R. Ruppin, Surf. Sci. **127**, 108 (1983)
59. K. Kobayashi, O. Watanuki, J. Vac. Sci. Technol. B **14**, 804 (1996)
60. K. Kobayashi, O. Watanuki, Opt. Rev. **3**, 447 (1996)
61. K. Kobayashi, O. Watanuki, J. Vac. Sci. Technol. B **15**, 1966 (1997)
62. G. S. Agarwal, Phys. Rev. A **11**, 230 (1975)
63. C. Girard, D. Courjon, Phys. Rev. B **42**, 9340 (1990)
64. B. Labani, C. Girard, D. Courjon, D. Van Labeke, J. Opt. Soc. Am. B **7**, 936 (1990)
65. C. Girard, X. Bouju, J. Opt. Soc. Am. B **9**, 298 (1992)
66. O. Keller, M. Xiao, S. Bozhevolnyi, Surf. Sci. **280**, 217 (1993)
67. C. Girard, A. Dereux, Phys. Rev. B **49**, 11344 (1994)
68. C. Girard, A. Dereux, O. J. F. Martin, Phys. Rev. B **49**, 13872 (1994)
69. Z. Li, B. Gu, G. Yang, Phys. Rev. B **55**, 10883 (1997)
70. O. J. F. Martin, A. Dereux, C. Girard, Opt. Soc. Am. A **11**, 1073 (1994)
71. O. J. F. Martin, C. Girard, A. Dereux, Phys. Rev. Lett. **74**, 526 (1995)
72. O. J. F. Martin, C. Girard, A. Dereux, J. Opt. Soc. Am. A **13**, 1801 (1996)
73. O. J. F. Martin, C. Girard, Appl. Phys. Lett. **70**, 705 (1997)
74. K. Cho, Prog. Theor. Phys. Suppl. **106**, 225 (1991)
75. K. Cho, Y. Ohfuti, K. Arima, Jpn J. Appl. Phys. Suppl. **34-1**, 267 (1995)

76. C. Girard, O. J. F. Martin, A. Dereux, Phys. Rev Lett. **75**, 3098 (1995)
77. K. Cho, Y. Ohfuti, K. Arima, Surf. Sci. **363**, 378 (1996)
78. K. Kobayashi, Appl. Phys. A Suppl. **66**, S391 (1998)
79. J. M. Vigoureux, C. Girard, D. Courjon, Opt. Lett. **14**, 1039 (1989)
80. H. Hori, in *Near Field Optics*, D. W. Pohl, D. Courjon (eds.), p. 105 (Kluwer, Dordrecht, 1993)
81. T. Saiki, M. Ohtsu, K. Jang, W. Jhe, Opt. Lett. **21**, 674 (1996)
82. A. D. Stone, A. Szafer, IBM J. Res. Develop. **32**, 384 (1988)
83. K. Kobayashi, T. Kohmura, Prog. Theor. Phys. **71**, 327 (1984) and references therein
84. P. Fulde, Springer Series in Solid-State Sciences **100**, *Electron Correlations in Molecules and Solids*, 3rd edn. (Springer-Verlag, Berlin, 1995)
85. J. J. Hopfield, Phys. Rev. **112**, 1555 (1958)
86. J. Knoester, S. Mukamel, Phys. Rev. A **40**, 7065 (1989)
87. G. Juzeliūnas, D. L. Andrews, Phys. Rev. B **49**, 8751 (1994)
88. J. J. Sakurai, *Advanced Quantum Mechanics* (Addison-Wesley, Reading, 1967)
89. A. L. Fetter, J. D. Walecka, *Quantum Theory of Many-Particle Systems* (McGraw-Hill, New York, 1971)
90. M. D. Levenson, S. S. Kano, *Introduction to Nonlinear Laser Spectroscopy*, 2nd edn. (Academic, Florida, 1988)
91. B. Hanewinkel, A. Knorr, P. Thomas, S. W. Koch, Phys. Rev. B **55**, 13715 (1997)
92. R. Kubo, J. Phys. Soc. Jpn **12**, 570 (1957)

Index

MIX
Papier aus verantwortungsvollen Quellen
Paper from responsible sources
FSC® C105338

If you have any concerns about our products,
you can contact us on
ProductSafety@springernature.com

In case Publisher is established outside the EU,
the EU authorized representative is:
Springer Nature Customer Service Center GmbH
Europaplatz 3, 69115 Heidelberg, Germany

Printed by Libri Plureos GmbH
in Hamburg, Germany